American Passage

American Passage

The History of Ellis Island

VINCENT J. CANNATO

HARPER

An Imprint of HarperCollins*Publishers*
www.harpercollins.com

HarperCollins books may be purchased for educational, business, or sales promotional use. For information, please write: Special Markets Department, HarperCollins Publishers, 10 East 53rd Street, New York, NY 10022.

Grateful acknowledgment for permission to reproduce illustrations in the insert is made to the following: EMMET COLLECTION, MIRIAM AND IRA D. WALLACH DIVISION OF ART, PRINTS AND PHOTOGRAPHS, THE NEW YORK PUBLIC LIBRARY, ASTOR, LENOX AND TILDEN FOUNDATIONS: insert page 1, top. PICTURE COLLECTION, THE BRANCH LIBRARIES, THE NEW YORK PUBLIC LIBRARY, ASTOR, LENOX AND TILDEN FOUNDATIONS: page 1, bottom; page 5, top; page 7, bottom; page 8, top. PHOTOGRAPHY COLLECTION, MIRIAM AND IRA D. WALLACH DIVISION OF ART, PRINTS AND PHOTOGRAPHS, THE NEW YORK PUBLIC LIBRARY, ASTOR, LENOX AND TILDEN FOUNDATIONS: page 4, top; page 8, bottom. MILSTEIN DIVISION OF UNITED STATES HISTORY, LOCAL HISTORY & GENEALOGY, THE NEW YORK PUBLIC LIBRARY, ASTOR, LENOX AND TILDEN FOUNDATIONS: page 5, bottom; page 9, top. LIBRARY OF CONGRESS: page 2, top and bottom; page 3, top and bottom; page 13, top and bottom; page 14, top and bottom. WILLIAM WILLIAMS PAPERS, MANUSCRIPTS AND ARCHIVES DIVISION, THE NEW YORK PUBLIC LIBRARY, ASTOR, LENOX AND TILDEN FOUNDATIONS: page 3, middle; page 4, bottom; page 6, top; page 7, top; page 10, bottom; page 11, top and bottom. THE LA GUARDIA AND WAGNER ARCHIVES, LA GUARDIA COMMUNITY COLLEGE/THE CITY UNIVERSITY OF NEW YORK: page 6, bottom. GEORGE EASTMAN HOUSE: page 9, bottom. THE AMERICAN CATHOLIC HISTORY RESEARCH CENTER AND UNIVERSITY ARCHIVES, THE CATHOLIC UNIVERSITY OF AMERICA: page 12, top. THE PHILIP LIEF GROUP, INC.: page 12, bottom. CORBIS: page 15, top; page 16, top and bottom. MEYER LEIBOWITZ/THE NEW YORK TIMES/ REDUX: page 15, bottom.

FIRST EDITION

Designed by Emily Cavett Taff

Library of Congress Cataloging-in-Publication Data

Cannato, Vincent J.
American passage: the history of Ellis Island /Vincent J. Cannato. — 1st ed.
p. cm.
Includes bibliographical references and index.
ISBN: 978-0-06-074273-7
1. Ellis Island Immigration Station (N.Y. and N.J.)—History. 2. Ellis Island (N.J. and N.Y.)—History. 3. Immigrants—New York (State)—New York—History. 4. Immigrants—United States—History. 5. United States—Emigration and immigration—History. 6. New York (N.Y.)—Emigration and immigration—History. I. Title.
JV6484.C366 2009
325.73 dc22   2008052245

09 10 11 12 13   OV/RRD   10 9 8 7 6 5 4 3

In Memory of
My father
Vincent John Cannato
(1930–2008)

and

My grandfather
Vincent Joseph Cannato
(1893–1983)

Contents

PART IV
DISILLUSION AND RESTRICTION

PART V
MEMORY

Introduction

Ellis Island is one of the greatest human nature offices in the world; no week passes without its comedies as well as tragedies.
 —William Williams, Ellis Island Commissioner, 1912

Ellis Island was the great outpost of the new and vigorous republic. Ellis Island stood guard over the wide-flung portal. Ellis Island resounded for years to the tramp of an endless invading army.
  —Harry E. Hull, Commissioner-General of Immigration, 1928

BY 1912, THIRTY-THREE-YEAR-OLD FINNISH CARPENTER Johann Tyni had had enough of America. "I wish to go back to Finland. I didn't get along well in this country," he admitted less than three years after he and his family had arrived. The married immigrant with four children was depressed and unemployed. "I worked too hard and I am all played out," he said. "I am downhearted all the time and the thoughts make me cry."

The Reverend Kalle McKinen, pastor of Brooklyn's Finnish Seamen's Mission, had had enough of Johann Tyni. For the previous year and a half, Finnish charities had been taking care of the Tyni family. "This man has been crazy since he landed here," McKinen wrote immigration officials. "It is to be regretted that his family were [sic] ever admitted to this country." He also complained that Tyni's wife was not very bright and could no longer care for her children. Out of a mixture of desperation, pity, and anger, Reverend McKinen brought the Tyni family to Ellis Island.

After observing Johann on the island's psychiatric ward, immigration officials decided that they too had had enough of the Tyni family. Doctors at Ellis Island diagnosed Johann with "insanity characterized by depression, sluggish movements, subjective complaints of pain in the head and a feeling of inefficiency." They also declared that Johann's nine-year-old son, John, was a "low grade imbecile" who showed "the characteristic stigmata of a mental defective."

The family had originally arrived at Ellis Island under much happier circumstances. With three children in tow, Johann and his wife arrived with $100 and presented themselves to authorities in good physical and mental health. Less than three years after coming to America, Johann, his wife, two Finnish-born sons, and two American-born children were deported back to Finland from Ellis Island, anxious to get back to Johann's mother-in-law to rebuild a life that did not make sense in America.

Something had clearly happened since they arrived. Though two more children were born after their arrival, the Tynis lost their two-year-old Finnish-born son, Eugen, while living in Brooklyn. Perhaps the shock of his son's death, combined with a new, harsh, and unfamiliar environment, was enough to push Johann Tyni into a deep psychological abyss.

Immigration officials were not interested in the reasons for Tyni's mental illness. They were only concerned that he could no longer work and support his family. In the official terminology, the entire Tyni family was deemed "likely to become public charges," a designation that allowed officials to deport them back to their native Finland. Two-year-old David and infant Mary, both citizens by reason of their birth on American soil, were not technically deported and could have remained in the country, but obviously joined their parents and siblings on the return trip to Finland.

By this time, the government could not only exclude immigrants at the border but also deport them after their arrival if they came under an excludable class. The specter of Ellis Island haunted not just those newly arrived immigrants awaiting inspection but also those who managed to land initially who could be threatened with deportation for three years after.

Unlike the Tyni family, some immigrants never got the chance to set foot on the American mainland before being sent back home. Eighteen-year-old Hungarian Anna Segla arrived a few months after the Tyni family in 1910. After the inspection at Ellis Island, doctors certified her as possessing "curvature of spine, deformity of chest," as well as being a dwarf. They believed that those physical defects would prevent Anna from gaining meaningful employment in America. Anna Segla was ordered excluded.

Anna had been headed to live with her aunt and uncle in Connecticut. The childless couple had promised to take care of Anna and offered to post a bond for her release. For nearly two weeks, Anna was detained at Ellis Island while her case was appealed to officials in Washington. In a letter most likely written by her aunt and which Anna signed with an X, Anna eloquently made her case for admittance. "I beg to say that the hunchback on me never interfered with my ability to earn my living as I always worked the hardest housework and I am able to work the same in the future," the letter stated. "I pray Your Honor permit me to land in the United States." Despite her pleas, Anna was sent back to Europe.

Other immigrants were detained for even longer periods of time at Ellis Island, although many were eventually allowed to enter the country. When Louis K. Pittman came through Ellis Island in 1907 as a young boy, doctors discovered that he suffered from trachoma, a mildly contagious eye disease against which medical officials were especially vigilant. Rather than being deported, Pittman was allowed to stay in the island's hospitals while doctors treated his condition. Decades later, Pittman remembered his stay at Ellis Island as "very pleasant," with toys, good food, playmates, and very lax supervision by adults. After seventeen months in custody at Ellis Island's hospital, Pittman was allowed to rejoin his family on the mainland.

Others, luckier than Pittman, were detained for shorter periods. Frank Woodhull's experience at Ellis Island began in 1908 when he returned from a vacation to England. The Canadian-born Woodhull, who was not a naturalized American citizen, was heading back to New Orleans where he lived. As he walked single file with his fellow passengers past Ellis Island doctors, he was pulled aside for further inspection. The fifty-year-old was of slight build with a sallow complexion. He wore a black suit and vest, with a black hat pulled down low over his eyes and covering his short-cropped hair. His appearance convinced the doctors to test Woodhull for tuberculosis.

Woodhull was taken to a detention ward for further examination. When a doctor asked him to take his clothes off, Woodhull begged off and asked not to be examined. "I might as well tell you all," he said. "I am a woman and have traveled in male attire for fifteen years." Her real name was Mary Johnson. She told her life story to officials, about how a young woman alone in the world tried to make a living, but her manly appearance, deep voice, and slight mustache over her thinly pursed lips made life difficult for her. It had been a hard life, so at age thirty-five Johnson bought men's clothing and started a new life as Frank Woodhull, working various jobs throughout the country, earning a decent living, and living an independent life. Mary Johnson's true sexual identity was a secret for fifteen years until Frank Woodhull arrived at Ellis Island.

Johnson requested to be examined by a female matron, who soon found nothing physically wrong with the patient. She had enough money to avoid being classified as likely to become a public charge, was intelligent and in good health, and was considered by officials, in the words of one newspaper, "a thoroughly moral person." Ellis Island seemed impressed with Johnson, despite her unusual life story. Nevertheless, the case was odd enough to warrant keeping Johnson overnight while officials decided what to do. Not knowing whether to put Johnson with male detainees or female detainees, officials eventually placed her in a private room in one of the island's hospital buildings.

"Mustached, She Plays Man," said the headline in the *New York Sun*. Despite her situation, officials deemed Johnson a desirable immigrant and allowed her to enter the country and, in the words of the *Times*, "go out in the world and earn her living in trousers." There was nothing in the immigration law that excluded a female immigrant for wearing men's clothing, although one can imagine that if the situation had been reversed and a man entered wearing women's clothing, the outcome might have been different.

Before she left for New Orleans, Johnson spoke to reporters. "Women have a hard time in this world," she said, complaining that women cared too much about clothes and were merely "walking advertisements for the milliner, the dry goods shops, the jewelers, and other shops." Women, Johnson said, were "slaves to whim and fashion." Rather than being hemmed in by these constraints, she preferred "to live a life of independence and freedom." And with that Frank Wood-hull left Ellis Island to resume life as a man.

But the vast majority of the 12 million immigrants who passed through Ellis Island between 1892 and 1924 did not experience any of these hassles. Roughly 80 percent of those coming to Ellis Island would pass through in a matter of hours.

For these individuals, Arthur Carlson's experience is probably closer to their own. A Swedish immigrant who arrived in 1902, Carlson spent about two hours at Ellis Island before being allowed to land. "I was treated very well," Carlson reminisced later in his life. "Nothing shocked me. I was so thrilled over being in a new country." Destined for New Haven, Connecticut, Carlson originally planned to travel there by boat, but officials suggested that the train would be faster. Soon thereafter, Carlson had his train ticket and was on his way to be reunited with his brother.

Each of these people experienced Ellis Island in a different way. Their experiences ran the gamut of stories: admitted (Carlson), detained then admitted (Woodhull/Johnson), hospitalized then admitted (Pittman), admitted then deported (the Tyni family), and excluded (Segla).

No one story encapsulates the Ellis Island experience; there are literally millions. For most immigrants, Ellis Island was a gateway to a new life in America. It was an integral part of their American passage. It would become a special place for some immigrants and their families, while others retained only faint memories of the place or saw it as a site of unimaginable emotional stress filled with stern government officials who possessed the power to decide their fate. For a small percentage of people, Ellis Island was all they would see of America before being sent back home.

For immigrants like the Tyni family, Frank Woodhull, Arthur Carlson, Louis Pittman, and Anna Segla, why did the passage to America have to run through this inspection station on a speck of an island in New York Harbor, and why did their experiences differ so dramatically?

IN 1896, THE MAGAZINE *Our Day* published a cartoon entitled "The Stranger at Our Gate." It featured an immigrant seeking entrance into

America. The man makes a pathetic impression: short, hunched over, sickly, toes sticking out of his ragged shoes. Literally and figuratively, he is carrying a lot of baggage. In one hand is a bag labeled "Poverty" and in the other a bag labeled "Disease." Around his neck hangs a bone with the inscription "Superstition," signifying his backward religion and culture. On his back are a beer keg with the words "Sabbath Desecration" and a crude bomb labeled "Anarchy."

The man has come upon a gate that provides entry past a high stone wall. A pillar at the gate reads: "United States of America: Admittance Free: Walk In: Welcome." Standing in the middle of the gate is Uncle Sam. Much taller than the immigrant, the unhappy Uncle Sam is decked out in full patriotic regalia. He is holding his nose, while looking down contemptuously at the man standing before him. Holding one's nose implies the existence of a foul odor, but it also means that one is forced to do something that one does not want to do. And that's just the fix that Uncle Sam is in.

"Can I come in?" the immigrant asks Uncle Sam.

"I s'pose you can; there's no law to keep you out," a disgusted Uncle Sam replies.

According to this cartoonist, the gates to America were wide open to the dregs of Europe, and the government could do nothing to stop them. Although a powerful idea to many Americans, by 1896 this notion had become outdated. Congress was now creating a list of reasons that immigrants could be excluded at the nation's gates, and that list would grow longer as the years passed.

To enforce those new laws, the federal government built a new inspection station. Almost 80 percent of immigrants to America passed through the Port of New York, and this new facility was located on an island in New York Harbor called Ellis Island.

The symbolism of the gate is important. Each day, inspectors, doctors, and other government officials stood at the gate and examined those who sought to enter the country. They deliberated over which immigrants could pass through and which would find the gate closed.

At the gate, Ellis Island acted as a sieve. Government officials sought to sift through immigrants, separating out the desirable and the undesirable. America wanted to keep the nation's traditional welcome to immigrants, but only to those it deemed desirable. For undesirables, the gates of America would be shut forever. Federal law defined such categories, but the enforcement and interpretation of those laws were left up to officials at places like Ellis Island.

The process at Ellis Island was not a happy event, wrote Edward Steiner, but rather "a hard, harsh fact, surrounded by the grinding machinery of the law, which sifts, picks, and chooses; admitting the fit and excluding the weak and helpless." To another observer of the process, this sifting process resembled "the screening of coal in a great breaker tower."

The central sifting at Ellis Island occurred at the inspection line. All immigrants would march in a single-file line toward a medical officer. Sometimes having to process thousands of immigrants a day, these officials had only a few seconds to make an initial judgment. They would pay careful attention to the scalp, face, neck, hands, walk, and overall mental and physical condition. The immigrant would then make a right turn in front of the doctor that allowed a rear and side view. Often, doctors would touch the immigrants, feeling for muscular development or fever, or inspect hands that might betray more serious health concerns. They might also ask brief questions. Doctors developed their own methods of observation. As one noted, "Every movement of the body has its own peculiar meaning and that by careful practice we can learn quickly to interpret the significance of the thousand-and-one variations from the normal."

After 1905, all immigrants would then pass before another doctor whose sole job was to perform a quick eye exam. If any of these medical officers found any sign of possible deficiency, they would use chalk to mark the immigrant with a letter. *L* stood for lameness and *E* stood for eye problems, for instance. Those chalk-marked immigrants, some 15 to 20 percent of all arrivals, would then be set aside for further physical or mental testing.

Immigration officials largely based their decisions of the desirability of immigrants on their mental, physical, and moral capacities. To modern ears, the notion of classifying any human being as "undesirable" is an uncomfortable one that smacks of discrimination and insensitivity, but we should be careful not to judge the past by modern-day standards. Instead, it is important to understand why Americans went about classifying people in this manner, however unpleasant that process might seem to us.

First, they were concerned that immigrants would become "public charges," meaning they would not be able to take care of themselves. In the days before a federal welfare system and social safety net, this meant being wards of private charity or local institutions like poorhouses, hospitals, or asylums. If immigrants were to be allowed into the country, they needed to prove they were healthy and self-sufficient.

Second, immigrants were meant to work. Specifically, they were to be the manual labor that fueled the factories and mines of industrial America. Such tough work demanded strong physical specimens. Sickly, weak, or mentally deficient immigrants were deemed unlikely to survive the rigors of the factory.

Lastly, scientific ideas that would reshape the modern world were beginning to seep into the public consciousness in the late nineteenth century and affect the way Americans saw immigrants. Darwin's theory of evolution and primitive genetic theory offered Americans dark lessons about the dangers of the wrong kinds of immigrant. Many Americans considered poverty, disease, and illiteracy to be hereditary traits that would be passed on to future generations, thereby weakening the nation's gene pool and lowering the vitality of the average American, not just in the present, but for generations to come.

All of these ideas assume that it is acceptable for a nation to exclude immigrants it deems undesirable. Then, as now, Americans have grappled with the question: Is everyone in the world entitled to enter America? This question lies at the heart of the history of Ellis Island. At the time, most native-born Americans believed that they had the right to decide this as a matter of national sovereignty. Senator Henry Cabot Lodge of Massachusetts summarized this view in a 1908 speech:

> Every independent nation has, and must have, an absolute right to determine who shall come into the country, and secondly, who shall become a part of its citizenship, and on what terms. . . . The power of the American people to determine who shall come into the country, and on what terms, is absolute, and by the American people, I mean its citizens at any given moment, whether native born or naturalized, whose votes control the Government. . . . No one has a right to come into the United States, or become part of its citizenship, except by permission of the people of the United States.

Even though Lodge was an unabashed believer in the superiority of white, Anglo-Saxon Protestants, his ideas about national sovereignty strike at the heart of how any nation deals with those who knock at its gates.

The nation's immigration law was predicated on the idea that a self-governing people could decide who may or may not enter the country. But that idea came into conflict with other ideals, such as America's traditional history of welcoming newcomers. More importantly, it conflicted with the idea that the rights guaranteed in the Constitution were universal rights. How could the Declaration of Independence's basic creed that all individuals were created equal mesh with the idea that some immigrants were desirable and others undesirable? That conflict between American ideals is central to an understanding of why Ellis Island was created in the first place.

TRADITIONAL HISTORIES OF THE Ellis Island period, like John Higham's classic *Strangers in the Land*, focus on the rise and fall of nativism, which the historian defined as the "intense opposition to an internal minority on the ground of its foreign . . . connections." Yet Higham would soon come to see the shortcomings of his own analysis. Shortly after the publication of his book, he asked: "Shall I confess that nativism now looks less adequate as a vehicle for studying the struggles of nationalities in America, than my earlier report of it, and other reports, might indicate?" He later admitted: "Repelled as I was not only by the xenophobias of the past but also by the nationalist delusions of the Cold War that were all around me, I had highlighted the most inflammatory aspects of ethnic conflict."

The "nativist theme, as defined and developed to date, is imaginatively exhausted," Higham concluded. By overemphasizing the psychological interpretations of American attitudes toward immigrants, he diminished the rationality of individuals and reduced their reactions to complex social changes down to primitive and primordial emotional reflexes. That does not mean downplaying the often ugly anti-immigrant sentiment that has characterized certain periods of American history. Higham is mostly correct that such feelings were rising in the late nineteenth century as the demographics of immigration shifted from northern Europeans to southern and eastern Europeans. He is also correct that World War I brought a significant opposition to foreigners.

However, both of these periods also saw a larger shift in American society. The former occurred during the dawning of the era of Progressive reforms, with the beginnings of the federal administrative state designed to enact those reforms. The latter occurred at a time of great disillusionment with reform and government in the wake of the Great War. As the Progressive impulse to regulate society ebbed, Americans instead tried to restore a lost world that had been overtaken by the rise of modern, industrial America.

By looking past the mere expressions of anti-immigrant sentiment and focusing on the implementation of immigration policy, we find that much of the debate surrounding Ellis Island was not as polarized as we might imagine. Despite the heated rhetoric, this debate took place within the proverbial forty-yard lines of American political life. There was considerable consensus about immigration. Most Americans found themselves in the political middle on the issue. That debate took place most famously at Ellis Island for more than three decades.

Few Americans argued for a completely open door to all immigrants and few argued for their complete exclusion. Allan McLaughlin, a doctor with the U.S. Public Health Service, put forth the parameters of the debate:

> There are extremists who advocate the impossible—the complete exclusion of all immigrants or the complete exclusion of certain races. There are other extremists who pose as humanitarians and philanthropists and who advocate an act of lunacy—removing all restrictions and admitting all the unfortunate—the lame, the halt, the blind and the morally and physically diseased—without let or hindrance. Neither of these extreme positions is tenable. The debarring of all immigrants, or the unjust discrimination against any particular race, is illogical, bigoted and un-American. On the other hand, the indiscriminate admission of a horde of diseased, defective and destitute immigrants would be a crime against the body politic which could not be justified by false pretense of humanity or a mistaken spirit of philanthropy.

Americans rarely challenged the government's right to exclude or deport immigrants, but rather fought over the legitimate criteria for exclusion and how strictly government should enforce those laws at immigration stations like Ellis Island.

Take the opinions of two men active in the debate during this time. Max Kohler was a lawyer for the American Jewish Committee who doggedly defended the rights of Jewish immigrants and criticized the strict enforcement of the law at Ellis Island. Nevertheless, he admitted that the immigration law at the time was appropriate in barring those deemed undesirable. "We do not want aliens to be admitted of any race or creed," Kohler said, "suffering from loathsome or contagious diseases, mentally or morally defective, contract laborers or paupers or persona likely to become public charges in fact." What he opposed was both the stricter enforcement of the law and the passage of any more restrictive measures by Congress to exclude immigrants.

On the other side was Commissioner-General of Immigration Frank Sargent. The former labor leader favored closer inspection and tighter restriction of immigrants, but conceded that he "would not advocate a 'closed-door' policy . . . as we still have need for a high class of aliens who are healthy and will become self-supporting." For him and other like-minded individuals, the present law was fine, but needed to be more strictly enforced. The debate, then, was not one over the restriction of immigrants, but instead over the regulation of who may be allowed to enter the country.

"We desire to emphasize at this point that the immigration laws of the United States," noted the American Jewish Committee in recommendations it made to the U.S. Immigration Commission, "have always been enacted to regulate immigration." Both sides of the immigration debate agreed on the need for the United States to continue to accept immigrants and for the need to sort through those who arrived and reject those deemed undesirable. They differed, however, in how strictly to regulate immigration. In practice, this allowed almost three decades of continuous immigration, mostly from Europe, at levels that remain historic highs in American history. For all the talk about exclusion and restriction, less than 2 percent of individuals who knocked at its gate were ultimately excluded at Ellis Island.

The laws that dealt with European immigrants, as well as smaller numbers of Middle Eastern and Caribbean immigrants, were in marked contrast to the law directed toward Chinese immigrants. For the Chinese and other Asians, American immigration policy *was* one of restriction. This proved the exception to the larger rule of immigration regulation, and Americans at the time were quite conscious of this differential treatment and at pains not to replicate it with other immigrant groups. For Asians, their near-complete exclusion from the country was based on race; for all others seeking entry, officials would try to weed out supposedly undesirable immigrants based not on race, but rather on individual characteristics. Prejudice against southern and eastern Europeans certainly existed, but it was not written into the law until the quotas of the 1920s.

CONTRARY TO MUCH THAT is written about American immigration, this book does not see this history strictly through the jaundiced interpretive lens of nativist sentiments or the sentimental notions of Ellis Island as a chronicle of American bounty and frothy idealism. Instead, this book looks at how actual people created, interpreted, and executed immigration laws at Ellis Island.

This is a story about the growing pains of a modern nation that was struggling with vast and seemingly disturbing changes. In response, America engaged in a debate about who could become an American. It was heated, loud, and often nasty. Raw emotions and blunt opinions were expressed in language that is often discomforting to modern readers.

In response to this debate, Congress translated these concerns into laws that were carried out at Ellis Island and other, smaller immigrant inspection stations around the country, where officials were confronted with the very real mass of humans who washed upon America's shores daily.

Guarding the borders became the key to defining the character of the nation itself. Ellis Island represents the dawning of a new age: the rise of the United States as a modern nation-state. After the Civil War, it would become an industrial powerhouse, achieve a unified nation from coast to coast, and expand its power on the world stage by extending its sphere of influence into Asia, the Caribbean, and Latin America. To manage this economic, military, and political behemoth, a new federal government had to be created almost from scratch. Immigration control should be placed in the context of the rise of this modern state.

The immigration service that ran inspection stations like Ellis Island was one of the country's first large government programs. The strong federal government that we know today was in its infancy in the late 1800s. As the federal government devoted more time, energy, money, and manpower to inspecting immigrants, it created a larger and larger administrative system. Such a system created its own set of rules.

Instead of seeing the work of Ellis Island in terms of immigration restriction, it is better to see it as a form of regulation. The relatively unobtrusive federal government of the nineteenth century evolved into a system of greater regulation by the twentieth century, one that did not end capitalism, but sought to control its excesses. Over that same period, the laissez-faire attitude of the federal government gave way to a system that did not end immigration, but regulated it in the public interest.

The impulse behind immigration control was the same impulse that banned child labor, regulated railroads and monopolies, opened settlement houses, created national parks, battled the corruption of urban political machines, and advocated for temperance. It was these reforms of the Progressive Era that drove the expansion of the federal government to ensure that it would regulate private business in the public interest.

In this sense, immigration control fits well as a Progressive reform. To many reformers, big business, together with selfish steamship companies and aided by corrupt political bosses, sought to keep the faucet of immigration open full blast as a source of cheap labor to power the new industrial economy and provide voters for urban political machines. Reformers wanted to temper this by regulating immigration, not ending it. They believed that a large industrial and urban society needed to be actively molded and shaped, and that the older laissez-faire philosophy of the nineteenth century was inadequate to deal with the problems of the modern era.

Much of the political history of twentieth-century America was a battle over the extent of government regulation. Historians generally agree that the spirit of Progressive reform temporarily died out after World War I, and it is no surprise that this period also sees the end of the kind of immigration regulation practiced at Ellis Island for three decades. This regulatory approach to immigration would be replaced by the blunt instrument of immigration quotas by the 1920s. This new mechanism would not try to sift desirable from undesirable immigrants, but instead severely limit immigrants based on where they came from. America did not completely shut down immigration from Europe, as it had done earlier to immigration from China, but the era of mass immigration was effectively ended. Ellis Island had lost its raison d'être.

When a new spirit of reform came with the New Deal and the federal government again began to intervene actively in the private sector, immigration was left out of the equation. The nation's conflicting views toward government power would find itself mirrored in its immigration laws.

Ellis Island would become little more than a prison for enemy aliens during World War II and for noncitizen aliens with radical beliefs during the Cold War. In the flush of postwar prosperity, the government abandoned Ellis Island in 1954 and left it to rot. Not until the 1980s, when the nation began to witness the rise of a new era of mass migration, did the country again pay attention to Ellis Island. By then, the former inspection station had evolved into an emotional symbol to millions of Americans, a new Plymouth Rock. Parts of the old facility were rehabilitated and reopened as a museum of immigration history. Ellis Island had now entered the realm of historical memory.

THIS BOOK IS A biography, not of a person, but of a place, of one small island in New York Harbor that crystallized the nation's complex and contradictory ideas about how to welcome people to the New World. It traces the history of Ellis Island from its days of hosting pirate hangings in the nineteenth century to its heyday as America's main immigration station where some 12 million immigrants were inspected from 1892 to 1924. The story continues through the detention of aliens at Ellis Island during World War II and the Cold War and concludes with its rebirth as an immigration museum and a national icon. Long after Ellis Island has ceased to be an inspection station, the debates that once swirled around it continue to be heard.

Today, Ellis Island has become a tired cliché for some, a story about the pluck and perseverance of those "poor huddled masses yearning to be free" who found freedom at the end of the inspection line. It is a nostalgic ode to our hardy ancestors who achieved success in spite of their experiences at the infamous Isle of Tears, where bigoted officials made their lives miserable and changed the family's name from something with six syllables and no vowels to Smith.

In reality, Ellis Island was the place where the United States worked out its extraordinary national debate over immigration for more than three decades. Inspectors, doctors, and political appointees wrestled every day with the problems of interpreting the nation's immigration laws while being personally confronted with hundreds of thousands of living, breathing individuals. The dry enterprise of executing the law came into direct conflict with the mass of humanity seeking to make new lives in America.

Ellis Island embodies the story of Americans grappling with how best to manage the vast and disruptive changes brought by rapid industrialization and large-scale immigration from Europe. It is the story of a nation struggling with the idea of what it meant to be an American at a time when millions of newcomers from vastly different backgrounds were streaming into the country.

Americans need a history that does not glorify the place in some kind of gauzy, self-congratulatory nostalgia, nor mindlessly condemn what occurred there as the vicious bigotry of ugly nativists. Instead, this book seeks to understand what happened at Ellis Island and why it happened.

This island, so small in size, has imprinted itself on the minds of so many Americans. It is a gritty and tumultuous history, but one that helps to explain why millions of immigrants had to make their American Passage through Ellis Island and how that passage in turn helped shape this nation.

Part I

BEFORE
THE DELUGE

Chapter 1

Island

———— ◆ ————

FIFTY THOUSAND NEW YORKERS CLOGGED THE INTER-
section of Second Avenue and 13th Street on the afternoon of April 2,
1824. Nearly one-third of the city's population was there to witness the
public hanging of a convicted murderer named John Johnson.

City officials were not happy with the scene. They were less concerned about the question of whether a civilized city should play host
to such a gruesome event than they were about the gridlock created
by the public spectacle. The city would later order future executions
moved to nearby Blackwell's Island (now Roosevelt Island). But the
public could not get enough. At the next execution, they arrived in
boats so numerous they shut down river traffic and caused a number
of boating accidents. The city council then ordered that all future executions take place in the city prison, out of public view.

The city did not have jurisdiction over all executions. The crime of
piracy on the high seas was a federal offense and common enough to
occupy the minds of federal authorities. While the city banned public
executions, the federal government continued to offer such grotesque
displays to New Yorkers for a few more years on a small island it controlled in the harbor. Nineteenth-century New Yorkers knew the place
as Gibbet Island, but under another name it would later become one of
the most famous islands in the nation: Ellis Island. However, its early
history can best be described as ignominious.

Pirates bring to mind images of eye-patched swashbucklers, skull-
and-crossbones flags, and loads of treasure, but real-life piracy was a

more mundane, if still violent, pastime. When caught for their crimes, pirates often faced a death sentence. Pirate hangings were not merely about punishment; they were also about deterrence. After death, the damned would be hung in iron chains for an unspecified time, a warning to those who would dare wreak havoc and chaos on the commerce of the seas. The post on which the dead bodies were hung was called a gibbet, hence the island's chilling name.

When Washington Irving published his great satire of New York history under the pen name Diedrich Knickerbocker in 1809, he included a number of references to Gibbet Island. Mixing real history with myth, he wrote of a settler named Michael Paw who, according to Irving, "lorded it over the fair regions of ancient Pavonia and the lands away south, even unto the Navesink mountains, and was moreover patroon of Gibbet Island." While Paw probably did own the area, the three-acre rock and sand island granted him little by way of power or prestige and was not a possession of which to boast.

Gibbet Island and the legend of pirate hangings also eerily appear in another Irving tale, "Guests from Gibbet Island." In this ghost story, two pirates row out to Gibbet Island and find three of their fellow conspirators "dangling in the moonlight, their rags fluttering, and their chains creaking, as they were slowly swung backward and forward by the rising breeze." When one of the pirates returns home, waiting for him are "the three guests from Gibbet Island, with halters round their necks, and bobbing their cups together." The other living pirate would soon die, his body found "stranded among the rocks of Gibbet Island, near the foot of the pirates' gallows."

Pirate hangings on Gibbet Island were more than the stuff of ghost stories. Just after noon on June 11, 1824, a black sailor named Thomas Jones was hanged at Gibbet Island for his part in the murder of his ship's captain and first mate. "There appears to be no doubt on the mind of those who attended him, that he has gone to the realms above," according to a pamphlet written just after Jones's execution. "He closed his life leaving to the world a past example of a great sinner, and also a proof of the richness of divine grace, and the willingness of Jesus Christ to save sinners."

By the time of Jones's hanging, the guilty were no longer left on gibbets, but the public still needed to draw lessons from these executions. Rather than being a lesson of vengeance, these widely distributed pamphlets emphasized the notion of Christian redemption, as the accused always repents of his sins and accepts the salvation of Jesus Christ. The pamphlets not only provided the public with gruesome accounts of murder and piracy, but also a soothing tale in which even the most wicked criminals confessed their sins before death in order to save their souls from eternal damnation.

A similar tale was told when William Hill was hanged at Gibbet Island two years later. But the Hill case was decidedly different from that of Jones. Both men were black, but while Jones was a freeman and a sailor, Hill was a twenty-four-year-old Maryland slave arrested after an unsuccessful escape attempt. Frederick Douglass, once a Baltimore slave, described what happened to Maryland slaves who misbehaved: "If a slave was convicted of any high misdemeanor, became unmanageable, or evinced a determination to run away, he was brought immediately here, severely whipped, put on board the sloop, carried to Baltimore, and sold to Austin Woolfolk, or some other slave-trader, as a warning to the slaves remaining." That is what happened to William Hill.

On the night of April 20, 1826, Austin Woolfolk placed Hill and thirty other slaves bound in chains on the *Decatur*. From Baltimore, the ship would sail for New Orleans, where the slaves would be sold off to work on the large plantations of the Deep South. Rather than accept their fate, Jones and a number of other slaves managed to free themselves, take control of the ship, and throw the ship's captain and first mate overboard. It is a tale familiar to readers of Herman Melville's story "Benito Cereno" or viewers of the movie *Amistad*.

The slave mutineers were captured, but only Hill was convicted for the crime. He felt no malice toward the murdered captain, but said he and his fellow mutineers were only seeking their freedom. In fact, he felt so bad about his role in the captain's death that he wished that he had jumped overboard himself rather than kill another man.

On December 15, 1826, Hill was sent to Gibbet Island to face death. According to one account, "All the way in the Steam Boat, to his place of Execution, he appeared to be perfectly resigned to God; and continually praying and singing—On his arriving at the island, he was showed his Coffin; he said that was only for my body not for my Soul; that has gone to GLORY, with my beloved Saviour."

Present at the execution was Austin Woolfolk. While on the gallows, Hill spied the slave trader and in his final words on Earth forgave Woolfolk and said he hoped they would meet again in heaven. In response, Woolfolk cursed the doomed man saying he was going to get what he deserved. Members of the crowd, shocked at Woolfolk's outburst, quieted him down. Then, the slave-turned-pirate was "launched into eternity."

More executions followed. The most famous were the dual hangings of pirates Charles Gibbs and Thomas Walmsley in 1831. On a spring day in April, the harbor was again filled with boats whose passengers badly wanted to witness the executions. Gibbet Island was "crowded with men and women and children—and on the waters around, were innumerable boats, laden with passengers, from the steamboat and schooner, down to the yawl and canoe." In the chaos of the crowded harbor, a few boats were overturned.

Confusion reigned. The *Commercial Advertiser* noted that it had received a call from a man who had given one of his clerks the day off to watch the execution and that clerk had not been heard from since. The *Workingman's Advocate* also ran a notice about the mysterious disappearance of a thirty-six-year-old man who left his house the day of the hangings and never returned. His friends assumed that he went to the harbor to witness the executions and drowned. It is unclear whether either man actually drowned or whether they were just playing hooky from work, but an unidentified dead body was found the following day floating up to the Coffee House Slip at the foot of Wall Street.

Gibbs was a white man in his midthirties, reputedly from a respectable Rhode Island family. By one exaggerated account, Gibbs and his men were responsible for capturing more than twenty ships and murdering almost four hundred people. Gibbs, Walmsley, a twenty-three-year-old stout mulatto, and their accomplices took control of the ship *Vineyard* in November 1830, killing the captain and first mate. Making off with the money on board, they grounded the ship off the coast of Long Island and headed ashore. Three of the conspirators drowned before making it to land. Gibbs and Walmsley were soon arrested and fingered as the ringleaders by one of their colleagues who seemed unhappy with his share of the stolen loot.

At the trial, Walmsley, who had been the ship's steward, seemed to make the case for his innocence, pointing to racial prejudice. "I have often understood that there is a great deal of difference in respect of color, and I have seen it in this Court," he testified. Nevertheless, on April 22, 1831, Gibbs and Walmsley, according to one account, "paid the forfeit which the laws demand from those who perpetrate such crimes as they have been convicted of." Speaking to the gathered crowd at Gibbet Island, Gibbs addressed the crowd from the gallows for nearly a half hour. Both men acknowledged the justice of their death sentences. Rather than being dropped from a scaffold, the two men were killed by being slung up on a rope, on whose other end was tied heavy weights. While Walmsley died almost immediately, Gibbs suffered a much slower and more painful death because the knot on his neck had not been properly placed.

Their dead bodies swung on the gallows for nearly an hour, after which they were handed over to surgeons for autopsies. Before the surgeons took the bodies, a sculptor took a cast of Gibbs's head so that phrenologists could "examine minutely the skull of one of the greatest murderers ever known." Phrenologists believed that measuring the size and shape of skulls would reveal the character and mental capacity of the individual.

The island's last execution occurred on June 21, 1839, when New Yorkers watched a pirate named Cornelius Wilhelms die. It would be their last chance to witness such a horrific spectacle at Gibbet Island, although two decades later some ten thousand New Yorkers, most in boats, would come to nearby Bedloe's Island to watch the hanging of pirate Albert Hicks.

By the end of the nineteenth century, pirate hangings were a thing of the past and both Bedloe's Island and Gibbet Island would be transformed from their earlier dubious history into America's mythic historical pantheon. By then, on the site of the gallows from which Albert Hicks was hanged, would stand the base of the Statue of Liberty. Gibbet Island would shed its notorious name and history and revert back to a previous name: Ellis Island. By the late 1800s, it would attract many more people than had ever come to witness a pirate execution.

NEW YORK CITY IS an archipelago, a Philippines on the Hudson River, the handiwork of a glacier thousands of years ago. It is an island empire consisting of nearly six hundred miles of shoreline. Only one borough—the Bronx—is actually attached to the mainland. There are some forty islands in addition to Manhattan, Staten Island, and Long Island. These minor islands are nestled in the bays, rivers, harbor, and other waterways that encase the city. One of the largest, Roosevelt Island, is a city within a city, 2 miles long and 800 feet wide, with a population of over eight thousand. Just south of its tip is one of the city's smallest islands, measuring just 100 feet by 200 feet and named for former secretary of the United Nations U Thant.

Many of the city's islands once served important social functions and some still do. As the city grew northward up the island of Manhattan, along with it came the pesky social problems that afflict any budding metropolis. Under such circumstances, these islands became *cordons sanitaires*, in the words of writer Phillip Lopate, "where the criminal, the insane, the syphilitic, the tubercular, the orphaned, the destitute . . . were quarantined." It is no surprise that they were also handy places for pirate hangings.

Among these exile islands were Hart Island, which became the city's largest potter's field, the last resting spot for the anonymous poor; Blackwell's Island, which once housed a mental hospital for prisoners, as well as a city hospital; North Brother Island, where a hospital for the treatment of infectious diseases was "Typhoid Mary" Mallon's home for nearly three decades; Ward's Island, the site of more mental institutions; and Rikers Island, which is still a city jail, with nearly fifteen thousand inmates housed in ten buildings, one of the largest such facilities in the country.

In upper New York Harbor, just a few hundred yards from the shore of New Jersey, sits Ellis Island. During the last Ice Age, a thick blanket of ice covered most of New York. When the glaciers beat a retreat some twelve thousand years ago, they left behind a big marshland dotted with pockets of high ground. The coastline was some hundred miles farther out in the Atlantic. Much of what is harbor and sea today was once dry land. A person could have strolled from today's Ellis Island to neighboring Liberty Island to the high ground of Staten Island and not have gotten his feet wet.

As the waters continued to rise, the harbor was formed and much of the high ground became New York's islands. Today Ellis Island consists of around twenty-seven acres, but for much of its modern history it was little less than a three-acre bank of sand and mud—"by estimation to high water mark, two acres, three roods, and thirty-five perches"—that barely kept its head above high tide.

Seals, whales, and porpoises once swam in the waters near the island. And then there were the oysters. New York Harbor and the lower Hudson River were once home to 350 square miles of fertile oyster beds, supplying more than half of the world's oysters. They were prized as delicacies, while cheap and abundant enough to be a staple of the workingman's diet. A 1730 map of New York harbor shows the entire Jersey shore section of the harbor to be "one gigantic oyster reef."

In deference to the edible treasures that could be found in the waters surrounding the sandy outcrop, European colonists named the tiny island in the harbor Little Oyster Island, while its larger neighbor was dubbed Great Oyster Island.

Little Oyster Island would figure into a small piece of early New Amsterdam history. In 1653, Peter Stuyvesant, the director general of the West India Company and de facto ruler of New Amsterdam, was ordered by his bosses to create a municipal government. In February 1653, the new city government met in Fort Amsterdam.

One of the first orders of business that day was a complaint from Joost Goderis, the twenty-something son of a minor Dutch painter. In late January, Goderis had gone in a canoe with a boy "for oysters and pleasure" at Oyster Island. Goderis was interrupted and accosted by Isaack Bedloo and Jacob Buys, who taunted Goderis by shouting: "You cuckold and horned beast, Allard Antony has had your wife down on her back." Another man, Guliam d'Wys, taunted Goderis that he should let d'Wys have a "sexual connection" with Goderis's wife, since Antony already had done so. When Goderis, whom one historian had deemed "excitable" and "ill-balanced," confronted Bedloo at his house, he slapped him. In turn, Bedloo drew a knife and cut Goderis on the neck.

Goderis decided to take his case before the new local government to restore the good name of his wife and the pride of his family. He also hauled in a number of other men, friends of the defendants, who reportedly had witnessed the incident. The witnesses refused to cooperate against their friends and the case dragged on for weeks. One of the men hearing the case was none other than Allard Antony, the alleged cuckolder himself.

Goderis and the others have vanished into history, but Isaack Bedloo lives on. He became a wealthy merchant and later joined other prominent leaders of New Amsterdam in 1664 to convince Stuyvesant to turn over control of New Amsterdam to England. It was a purely business decision. In return, Bedloo received political patronage in the new British colony and was able to purchase Great Oyster Island. Bedloo, like other Dutch settlers under British rule, Anglicized his name to "Bedlow," which later generations corrupted to "Bedloe," the name that would eventually attach itself to the island that in 1886 became home to the Statue of Liberty.

Little Oyster Island would also become known as Dyre Island and then Bucking Island in the eighteenth century. Ownership of the island from the late 1690s until 1785 was unclear. In that latter year, an advertisement appeared in a local newspaper offering for sale "that pleasant situated Island, called Oyster Island, lying in York Bay, near Powles' Hook, together with all its improvements, which are considerable." In addition to the island, the seller offered two lots in Manhattan, a "few barrels of excellent shad and herrings," "a quantity of twine," and "a large Pleasure Sleigh, almost new."

The seller was Samuel Ellis, a farmer and merchant who resided at 1 Greenwich Street. It is not known when Ellis bought the island, though a notice was found in a 1778 newspaper publicizing the fact that a boat had been found adrift at "Mr. Ellis's Island."

Ellis died in 1794, still in possession of his island. His daughter, Catherine Westervelt, was pregnant at the time and Samuel's will made clear that if she had a boy, it was his wish "that the boy may be baptized by the name of Samuel Ellis." Ellis was clearly interested in his posterity. With three daughters, he most likely feared his name would not live on past his death, and having a grandson named Samuel Ellis Westervelt was the next best thing. His plans were tragically thwarted. Though Catherine's child was a boy and christened as his grandfather had ordered, Samuel Ellis Westervelt died young. Yet through the agency of history and luck, the name Ellis would still attach itself to one of the nation's most famous islands.

Even during Samuel Ellis's life, the island's ownership became a matter of some controversy and confusion, as the new government of the United States became interested in the island. In the 1790s, tensions with England continued and the War Department began to devise a strategy for defending its shores. In New York, the military began to fortify the islands of New York Harbor to ward off a possible British naval attack.

Before Samuel Ellis passed away, the city granted to New York State the right to the soil around the island from the high-water mark to the low-water mark. The city felt it had the right to that land, even though the island proper was in private hands.

Over the next few years, the state built earthen fortifications on the island, some of them intruding upon private property. In 1798, Colonel Ebenezer Stevens advised the War Department that a troop barrack there had been completed, along with twelve large guns. However, he reminded his superiors that the island was still in private hands. "I think something ought to be done with respect to purchasing it and the State will cede the jurisdiction to the Federal Government," Stevens wrote. In 1800, New York State transferred jurisdiction over all the fortified islands in New York Harbor to the federal government, even though it still did not have legal rights over Ellis Island.

In 1807, Lieutenant Colonel Jonathan Williams, chief engineer of the United States Army, declared that the fortification at Ellis Island was "totally out of repair." He drew up new plans for a fortified New York Harbor that included a new fort at Ellis Island. But first the title of the island needed to be settled. The New York governor, Daniel Tompkins, wrote to Williams that although Samuel Ellis had agreed to sell the island, he had died before the deed could be executed. The military works constructed there, wrote Tompkins, "are occupied merely by the permission of the owner whose ancestor assented to it and whose first permission has never been withdrawn by his descendants."

In response, on April 27, 1808, the sheriff of New York County and a group of selected New Yorkers visited Ellis Island to appraise its value, eventually settling on the figure of $10,000, which astounded Colonel Williams. What the appraisers found on Ellis Island gives us some idea why it may have interested Samuel Ellis as an investor.

> It is found to be one of the most lucrative situations for shad fishing by set netts [sic] within some distance of this place, yielding annually from 450 to 500 dollars to the occupant from this single circumstance. The Oyster banks being in its vicinity affords an income in the loan of boats, rakes, etc. . . . besides this a considerable advantage results to the occupant

from a tavern in the only possible place of communication for people engaged there, between the oyster banks and this city.

Despite Colonel Williams's reluctance, the government agreed to pay the money to clear up the confusion, and the state then transferred the deed to the federal government. The nation would soon be at war with England, yet when the War of 1812 ended, not a shot had been fired in anger from any of the forts of New York Harbor.

NATURE BLESSED NEW YORK'S island empire in many ways, especially with its four-mile-wide harbor sheltered from the rough Atlantic waters. The sand banks that line the Lower Bay south of Coney Island to Sandy Hook act as a natural breakwater, while the Narrows, a two-mile-long bottleneck passageway between Staten Island and Brooklyn, protects the placid harbor from stormy seas and ocean waves. Standing at the Battery, staring at the expansive harbor, one cannot help but be soothed by its calm waters.

Having such a natural port was only part of the equation. Although New York had been a major port for the young Republic, the opening of the Erie Canal in 1825 secured the city's position as the country's dominant commercial outpost. A chain was now formed from the Atlantic Ocean, through the harbor, up the Hudson River, west across the new canal, into the Great Lakes, to the American heartland.

New York City was to become the commercial fulcrum of the new nation, connecting the booming Midwest with the markets of Europe and beyond. In the thirty-five years after the opening of the canal, Manhattan's population went from 123,000 to 813,000. During that same period, 60 percent of all imports and one-third of all exports passed through the Port of New York.

New York imported woolen and cotton clothing from the factories of England, and expensive silk, lace, ribbons, gloves, and hats for upscale female shoppers. Sugar, coffee, and tea also came through the port. Much as New York monopolized the import of these goods, it also led the way in another kind of European import: immigrants.

Between 1820 and 1860, 3.7 million immigrants entered through the portal of New York Harbor—some 70 percent of all immigrants to the

United States during this time. Those ships streaming up the Narrows into New York Harbor, packed with immigrants, would keep coming throughout the nineteenth century, but to those newcomers Ellis Island meant nothing.

For the next few decades, Ellis Island would exist in relative obscurity, used by the army and the navy mostly as a munitions depot. Destined to be little more than a footnote in the city's history, the island did have a front row seat for the unfolding drama that took place across the harbor on the island of Manhattan. It stood watch as a small city began evolving into an urban colossus.

For immigrants coming to New York in the second half of the nineteenth century, the words on their lips were not Ellis Island, but Castle Garden.

Chapter 2

Castle Garden

———•———

*The present management of this very important department
[Castle Garden] is a scandal and reproach to civilization.*
        —Governor Grover Cleveland, 1883

*Castle Garden is one of the most beneficent institutions in
the world.*
        —*Harper's New Monthly Magazine*, June 1884

ON A HOT AUGUST NIGHT IN 1855, A LINE OF OIL LAMPS
lit the early evening sky on lower Broadway in Manhattan. Torch-bear-
ing New Yorkers proceeded down the short hill, past Bowling Green,
the tiny oval patch of grass surrounded by a wrought-iron fence, and
into the Battery. It was a joyous and raucous affair, part political protest and part social outing, with loud shouting, fireworks, and even the
firing of cannons as the crowd marched around the Battery carrying
banners in German and English. By the time they had arrived, their
numbers had grown to some three thousand people.

These men, women, and children were responding to an advertisement that had been posted around the city:

INDIGNATION MEETING!
citizens of the first ward
Assemble in your Might, and vindicate your Rights!
citizens
Do you wish to have

plague and cholera in your midst!
Do you wish to have your Children laid low with Small Pox
and Ship Fever?
New-yorkers
Will we have our most honored and sacred spot desecrated
by the sickly and loathsome Paupers and Refugees of European
Workhouses and Prisons?

Populist mobs were a regular feature in American cities dating back to revolutionary-era protests like those over the Stamp Act. Indignation meetings allowed citizens to blow off steam and flex their collective muscles to authorities.

The object of the crowd's indignation on this night was the recent opening of a brand-new immigration depot on a rocky outcropping just off the Battery and connected to it by a footbridge. Castle Garden stood on the site of a fort built in 1811 as part of the defensive fortifications of New York Harbor. When the Marquis de Lafayette visited America in 1824, he first arrived at the fort, where more than five thousand guests welcomed him.

The old fort was later converted into a music hall where Jenny Lind, the "Swedish Nightingale," made her American debut in 1850 as part of her cross-country tour financed and publicized by the irrepressible P. T. Barnum. The same seats where the city's elite once sat to hear Lind were now occupied by immigrants from Ireland and Germany awaiting their chance to enter the country.

The new immigration station riled the crowd. Organizers billed the protest as an "anti-cholera meeting," playing on the fears of New Yorkers who had endured a number of cholera outbreaks in years past and blamed immigrants for the disease. "Knaves and speculators," the notice warned, were "introducing paupers and emigrants infected with cholera, small-pox, ship fever, and all the vices of foreign prisons and workhouses." The advertisement also appealed to the crowd's patriotism, calling on New Yorkers to protest the desecration of the hallowed ground of Castle Garden, where Presidents George Washington and Andrew Jackson once stood.

The indignation meeting succeeded in drawing a large and lusty crowd. When the assembly had settled down at the Battery, someone read a resolution against Castle Garden, and a number of speakers came forth to voice their opposition. One of them was Captain Isaiah Rynders, who began his speech to raucous cheers and the explosions of roman candles and rockets. As the crowd quieted, Rynders told them he had not originally been invited to speak and was sorry that the crowd "did not call upon somebody else, better able than I am to address you."

This was an exercise in false modesty, for Rynders was no ordinary speaker and he most clearly belonged at that rally. In fact, Rynders himself was likely the brains behind the protest. Theodore Roosevelt, in his history of New York City, would later describe Rynders as one of "the brutal and turbulent ruffians who led the mob and controlled the politics of the lower wards" who "ruled by force and fraud, and were hand in glove with the disorderly and semi-criminal classes."

Born in upstate New York to a German-American father and an Irish Protestant mother, Rynders gained the title "Captain" not for his war exploits, but from his time running a ship along the Hudson River. A classic "sporting man" of the 1830s and 1840s, Rynders held no steady job, but devoted himself to the leisurely and manly pursuits of gambling, horses, and politics. At one point, he earned a living as a riverboat gambler on the Mississippi River.

He established a political club called the Empire Club, whose crew of "shoulder hitters" was the political muscle for New York City Democrats. He and his men became a force not only in the seedy underworld of gambling, taverns, and brothels but also in local and national politics. They intimidated voters, broke up opponents' rallies, and forcibly brought voters to the polls to vote for Democratic candidates. The money brought in from gambling houses and brothels helped support a political organization that could bring out the vote on election day, intimidate opponents, and have enough money left over at the end of the day to make men like Rynders wealthy.

Many credit Rynders with helping James K. Polk win the presidency in 1844. The Tennessee Democrat would have lost the election had he not won New York by a slim margin. The Captain sealed his fame when he helped instigate the bloody 1849 Astor Place Riot. The following year, he tried to break up a meeting of the American Anti-Slavery Society led by William Lloyd Garrison, when he stormed the stage to challenge Frederick Douglass, who was in the middle of a speech.

Why Rynders would oppose the opening of an immigration station speaks to another of his roles. Despite its rhetoric, the mob was not really concerned about the tainting of the patriotic memory of Castle Garden or the health dangers posed by the immigrant station. The anti-immigrant tone was made all the more puzzling considering that much of the crowd was first- and second-generation New Yorkers and that many of the banners were in German. In reality, the protest was about money and control. As it turns out, Rynders was more than just a political operative; he was also the chief of the city's so-called immigrant runners.

Midnineteenth-century New York was a rough and tumble city where the civilizing effects of modernity had not yet smoothed the rough edges of many of its citizens. The struggle for survival predominated, and much of that struggle revolved around business. In the booming commercial emporium of nineteenth-century New York, some people found their business not in trading goods but in another import: greenhorns.

Though it would only later come specifically to define new immigrants, the term "greenhorn" signified anyone new and unfamiliar to the ways of the big city. One's clothes, one's accent, and that faraway—part dazzled and part confused—look in the eyes were a signal to savvy New Yorkers that a greenhorn had arrived.

There were certainly a lot of greenhorns on the streets of New York. Between 1820 and 1839, New York received about 25,000 immigrants a year. The numbers kept growing every year. During the 1840s, some 1.2 million people came through New York, which handled three-quarters of the nation's immigrant arrivals. These numbers may not seem that large, until one considers that the population of Manhattan in 1850 was only slightly more than half a million.

Many New Yorkers looked on these greenhorns with a mix of pity, bemusement, and contempt, but for others these newcomers meant money. The wharves and docks where these immigrants first set foot on American soil were crowded and chaotic. Men like Rynders found opportunity in the chaos. There was profit to be had by exploiting the immigrants' lack of knowledge and naïveté.

Rynders was at the top of a corrupt totem pole of politicos, gangsters, gamblers, railroad companies, forwarding agents, tavern owners, boardinghouse keepers, and prostitutes. Their base of operations was the taverns and boardinghouses that lined Greenwich, Washington, and Cedar Streets in lower Manhattan. This area, according to one eyewitness, was home to "one hundred and thirty-nine immigrant runners, drinking at boarding houses for immigrants, prostitutes, rummies, watch stuffers, thimble riggers and pocketbook droppers." There was money to be made in selling railroad tickets at inflated prices, charging exorbitant rates for rooms at boardinghouses, overcharging immigrants for their baggage by playing with the scales, or even outright thievery and extortion. Confusion was the ally of the runner and the enemy of the immigrant.

As soon as a ship docked, runners would board it. If the immigrants were from Germany, the runners would speak German; Irish immigrants would encounter runners who hailed from the old sod. If immigrants were not immediately taken in by these entreaties, runners would forcibly take their luggage to a nearby boardinghouse for "safekeeping." When immigrants tried to claim their baggage, they were often induced to stay at the boardinghouse with the promise of cheap lodging and meals. When their stay had ended and it was time to move on, these greenhorns would be handed an excessive bill for their room and food and the storage of their luggage. If they could not pay the inflated bill, lodging house owners would keep the baggage as collateral. It was a prosperous racket, and much of the money made in fleecing immigrants went up the chain to Rynders, who was able to run his operations with little interference from city officials. They were all making a good living from immigration, and now Castle Garden was in danger of putting them out of business.

A committee of the New York State Assembly investigated the situation in the mid-1840s. It had heard the rumors and read the newspaper reports about how runners preyed on immigrants, but the committee confessed that it could not "have believed the extent to which these frauds and outrages have been practiced" until it began to investigate them.

The federal government was largely uninterested in immigration. Occasionally, Congress would be prodded into action to address the overcrowding that afflicted immigrants traveling across the Atlantic in steerage, but it did little in the way of regulating the flow of immigrants. Despite an undertone of anti-immigrant and anti-Catholic sentiment, the growing nation welcomed European immigrants to help settle the country. In the 1840s, President John Tyler lauded "emigrants from all parts of the civilized world, who come among us to partake of the blessings of our free institutions and to aid by their labor to swell the current of our wealth and power." However, the slaveholding Tyler made clear that his message was for white Europeans only.

The job of regulating immigration was left to states like Massachusetts and New York, which passed laws continuing colonial policies restricting the immigration of criminals, paupers, or those with contagious diseases. States charged ship owners a head tax for each immigrant to pay for the care of poor and sick immigrants and required the posting of a bond for those immigrants deemed likely to become public charges. Although state laws would foreshadow the future of federal immigration regulation, they were weakly enforced, and few immigrants were excluded.

It would be up to private individuals and organizations to protect immigrants from abuse. Ethnic solidarity prompted the creation of immigrant aid societies. New York's Irish already had some success in this endeavor, forming the Irish Emigrant Society in 1841 to "afford advice, information, aid and protection, to emigrants from Ireland, and generally to promote their welfare." In 1847, it teamed up with the German Society and lobbied New York State to create the Board of Commissioners of Emigration, which consisted of the mayors of New York and Brooklyn, the heads of the German and Irish Emigrant Societies, and six others appointed by the governor.

A head tax of $1 would be assessed on each immigrant, to be collected by the board. With the money, the board opened the Emigrant Hospital and Refuge on Ward's Island to care for sick immigrants. By 1854, the board was caring for over 2,500 immigrant patients.

The timing of the idea could not have been better. In 1847, the potato famine in Ireland had begun to drive out large numbers of Irish. For the next few years, poor Irish refugees, fleeing starvation and death, flooded American ports. Nearly 3 million immigrants landed in the United States from 1845 to 1854. Many of them ended up in New York City. Between 1840 and 1850, Manhattan's population increased by 65 percent; by 1855 over one-half of the city's 629,904 residents were immigrants and over one-quarter of New Yorkers hailed from Ireland.

If the Board of Commissioners was going to be successful in protecting this flood of immigrants from the predations of runners, it would need its own reception center for new arrivals, a place where immigrants would be processed, their needs met, and their interests protected. For this purpose, in April 1855, the board chose Castle Garden as its immigration depot.

The Board of Commissioners laid out the major benefits of Castle Garden. First and foremost, it would allow for a quicker and easier landing for immigrants and free them from the clutches of immigrant runners, allowing them to land "without having their means impaired, their morals corrupted, and probably their persons diseased." The board would also begin keeping track of the numbers of immigrants arriving and where they were heading.

The altruism of the board and its interest in the welfare of immigrants was genuine. Not surprisingly, it ran into a good deal of resistance to its idea of converting what had formerly been the city's premier music hall into an immigration-processing station. City officials were leery of the idea. This would be a state-run program—generating lots of money through the head tax—right in their backyard, and all local officials would get were two seats on the ten-person board.

Wealthy New Yorkers and businessmen in the city's First Ward also opposed the plan, fearing that an immigrant depot in their neighborhood would cause a decline in property values. They worried that immigrants would bring "pestilential and disagreeable odors" that would blow into the windows of respectable homes in the summertime. Many had hoped that the newly expanded Battery around Castle Garden would become a pleasant harbor-view promenade, but the board had thwarted those plans.

The *Times* editorialized against the plans for Castle Garden, writing that "one of the delights of the City for nearly thirty years" would "be a delight no more. Hereafter it is to be a nuisance . . . an offence to the eye, and an ugly obstacle to a view of the magnificent moving panorama of our glorious Bay." One of those prosperous New Yorkers unhappy with Castle Garden was railroad magnate Cornelius Vanderbilt, who lived across the street from the Battery and would lend his name to the August indignation meeting. Even with such opponents, Castle Garden opened as planned.

On Castle Garden's first day, as immigrants streamed through its doors, a group of runners gathered outside, shouting at and intimidating Castle Garden workers. One member of the board was forced to pull a gun on the rowdies. That first night, after midnight, a handful of runners—a "foul brood of villains who have so long fattened upon the plunder of emigrants," one newspaper called them—tried to crash through the doors of Castle Garden to wreak havoc, but were turned away.

Having failed to stop the opening of Castle Garden, Rynders and his supporters took to the streets in mid-August three days after the station opened. Rynders claimed to be merely seeking "open, fair competition among the emigrant forwarders" and opposed any attempt by the state to grant a monopoly over the business of handling immigrants. In other words, the state of New York and the Board of Commissioners were squeezing Rynders and his men out of business.

After the final speaker addressed the gathering, the crowd left the Battery to the strains of "Yankee Doodle" and took their torchlight procession back through the streets of the city's First Ward. The *Times* was not fooled by the patriotic rhetoric or the claims that Castle Garden would endanger the city's health. The organizers of the indignation meeting, the paper informed its readers, "were the emigrant runners, baggage smashers, boarding-house keepers, and other professional gentry, who have long filled their own pockets by robbing emigrants upon their first arrival." The *Daily Tribune* was even more blunt, seeing the meeting as a way "to devise means to throw the immigrants again into the hands of the thieves . . . who have grown rich by robbing strangers."

Throughout the fall, runners would try to cause trouble at Castle Garden or steer immigrants away from the depot, but they were fighting an uphill battle. By December 1855, Castle Garden had been operating for four months and one reporter who visited the depot was pleased with what he saw. The entrance was heavily guarded and those without letters of introduction were turned away. "There is no public undertaking in the city more wise and benevolent." The reporter continued, "It redeems our city to know that anything so judicious and benevolent could be executed by it."

The harassment of Castle Garden officials continued for another year, but the runners and their allies were never able to shut the station down. Between January and April 1856, Castle Garden processed over 16,000 immigrants arriving on 106 ships. In its annual report of that year, the Board of Commissioners made oblique reference to the troubles, stating that "where violence threatened with a strong hand to lay waste and destroy, the police . . . effectually checked the thoughtless and lawless in their course and preserved a valuable property from destruction or damage."

Some reports claimed that immigrant runners, faced with failure in New York, had left for California to seek their fortune or else had joined private military expeditions to Mexico and Nicaragua. Rynders managed to cling to political power and was named U.S. Marshall for New York in 1857 as a reward for his work in helping to elect James Buchanan president.

At Castle Garden, immigrants received reliable information about travel, jobs, and housing. The newcomers could exchange foreign money for American currency and buy railroad tickets without fear of fraud. An employment bureau helped immigrants find work around the country. The sick and disabled were provided with medical care. Immigrants' baggage was carefully handled and boardinghouses were screened, licensed, and supervised by the board. Decent food at decent prices was available.

With the runners seemingly vanquished, the Board of Commissioners won lavish praise. Friedrich Kapp, a member of the board, described the institution he helped manage as "one of the most benevolent establishments in the civilized world . . . it forestalls untold misery, need and suffering." One English emigrant called it "a great national refuge for the emigrant from all lands. . . . It stands alone in its noble and utilitarian character." In William Dean Howells's 1890 novel A *Hazard of New Fortunes*, the book's main character, Basil March, describes how well officials at Castle Garden treated newcomers. "No one appeared troubled or anxious; the officials had a conscientious civility," March mused. A journalist called Castle Garden "one of the most beneficent institutions in the world."

Despite the accolades, Kapp had trouble understanding the country's laissez-faire attitude toward immigration. "People look with indifference at this colossal immigration of the European masses," Kapp wrote in 1870, "whose presence alone will exercise a powerful influence on the destinies of the Western World; National and State legislators care little or nothing for the direction which is given to this foreign element."

That would soon change. By the 1880s, it seemed as if all of America had become interested in—even obsessed with—immigration. The industrial revolution was transforming the way Americans worked and lived. The United States was now a continental nation from the Atlantic to the Pacific, unified by transcontinental railroads. The nation saw its population nearly double between 1870 and 1900, while the gross national product increased sixfold. The United States was transforming itself overnight from a predominantly rural, agrarian society into an urban, industrial nation.

Between 1860 and 1910, the number of Americans living in cities rose from 6 million, or 20 percent of the population, to 44 million, or 40 percent of the population. In 1885, a Protestant minister named Josiah Strong wrote a best-selling book, *Our Country*, where he complained that cities were "a serious menace to our civilization" and possessed "a peculiar attraction for the immigrant." Census data showed that these cities had become foreign territories. Immigrants and their children would soon account for nearly 80 percent of the population of cities like New York and Chicago.

A few years after Strong published his jeremiad, historian Frederick Jackson Turner looked at the 1890 Census and declared that the American frontier was officially closed. Open land, at least in theory, was disappearing. To Turner, open land had made earlier immigration possible, as the frontier became the crucible in which "immigrants were Americanized, liberated, and fused into a mixed race."

If the frontier was now closed, where would new immigrants go? Critics feared that the city would be the new frontier, but without the same ability to assimilate newcomers. Overcrowded cities populated by those who spoke in foreign tongues marked the end of the Republic, as the United States was in danger of becoming just like Europe: corrupt, overindulgent, class-ridden, contemptuous of republican government, and doomed to revolution. Political corruption, alcoholism, and socialism would reign.

A writer in the *Atlantic Monthly* worried in 1882 that "our era . . . of happy immunity from those social diseases which are the danger and the humiliation of Europe is passing away." The new immigrants evoked not just fears of overcrowded and corrupt Old Europe, but also ancient Rome, which had been threatened by an urban rabble and an increasingly non-Roman citizenry: "In spite of the magnificent dimensions of our continent, we are beginning to feel crowded," said a writer in 1887. "Our cities are filling up with a turbulent foreign proletariat, clamoring for *panem et circenses*, as in the days of ancient Rome, and threatening the existence of the republic if their demands remain un-heeded."

Daily newspapers, middle-class magazines, and highbrow intellectual journals devoted increasing space to immigration. The *North American Review*, the voice of conservative, Northern, native-born Protestants, published two lengthy articles in the early 1880s that detailed how many immigrants were coming and where they were coming from. The articles displayed a marked ambivalence. Immigrants brought great ma-terial benefits to the United States, but it was "inevitable, however, that much moral and physical evil will be brought hither by the multitudes who come."

Newspapers throughout the country chimed in. The *Ohio State Journal* asked, "What statesman will be wise enough to sift the hun-dreds of thousands of emigrants crowding over from Europe and say which . . . should be admitted and which, in the exercise of the sacred right of self-protection, should be excluded?" Such a statesman would be needed since, according to the *Philadelphia Telegraph*, " a large per-centage of the foreigners to whom we have given welcome are unworthy of it," because they were "often idle, vicious socialists and anarchists, social pests and incendiaries."

There were fears that America was becoming a dumping ground for Europe's unwanted peasants. The *Chicago Times* warned that the "pres-ence among us of a large body of socialists, anarchists, nihilists, luna-tics, pordioseros [beggars], and other social dregs from the old world is a danger that threatens the destruction of our national edifice by the erosion of its moral foundations." The *New York Times* wondered how long would "the people of this country submit cheerfully to this burden shifted to their shoulders from the Old World?"

A more benign view of immigration still continued to be heard. The *Boston Pilot* reminded its largely Irish Catholic readers that calls for restriction were an "unsavory reminder of the dark days" of the nativist Know-Nothings. The *Milwaukee Journal*, with its large German readership, saw immigration as "natural in its movements as the flow of the tides. It is a movement to restore the human equilibrium of the globe." The paper's laissez-faire prescriptions rejected government interference, calling it "un-American" and arguing that natural forces

DORBOLO, HALEH

Unclaim : 8/11/2018

Held date   : 8/4/2018
Pickup location : Cedar Mill Community Library

Title   : American passage : the history
 of Ellis Island
Call number   : 325.73 CANNATO
Item barcode   : 33614040270869
Assigned branch : Cedar Mill Community Library

Notes:

would slow or increase immigration based on market forces and social conditions.

Immigration was "giving us the best blood in the world," according to the *Milwaukee Journal*, alluding to the benefits of an expanded gene pool. "American humanity in the end promises to be an advance on all other humanity that has yet appeared on the planet." Americans were a "composite people," according to the *St. Louis Republican*. "Our Americanism is continually changing. It is not today what it was a generation ago, and it will not be a generation hence what it is to-day."

Others used genetic theory for darker purposes. Senator Justin Morrill argued that the effects of new immigrants were "more dangerous to the individuality and deep-seated stamina of the American people. . . . I refer to those whose inherent deficiencies and iniquities are thoroughbred, and who are as incapable of evolution, whether in this generation or the next." Morrill, a well-respected Vermont Republican, argued that Americans "must not be coerced to support the weak, vile, and hungry outcasts from hospitals, prisons, and poor-houses, landed here not only to stay themselves but to transmit hereditary taints to the third and fourth generation."

It took Episcopal bishop and poet A. Cleveland Coxe to bring this theory to its logical conclusion. Coxe called the new immigrants "invaders" who "come with weapons of fatal import to our civilization and to our race." America was under attack from "hordes of barbarians," for which, Coxe warned, historically minded Americans, there was ample precedent. Past invasions made Spain "a mongrel race, and have fastened upon her a chronic state of decay and imbecility." Of course, the great historical example was the Roman Empire, with "Goths and Vandals pouring into the sunny south." Coxe argued that new immigrants were hereditarily indisposed to democracy and could endanger the nation's experiment in self-government.

Despite Coxe's florid rhetoric, most people were trying to find a way to reconcile a vision of America as a refuge for immigrants with a desire to accept only desirable newcomers. The *Troy Times* welcomed "the intelligent, industrious, and honest foreigners who come here to establish homes," but wanted "the highest and strongest barriers . . . raised" for the "worthless human rubbish." "What we need," argued the *Minneapolis Tribune*, "is the inspection and sifting of intending immigrants." Such a system, the *New York Commercial Advertiser* believed, would allow the nation to "defend itself against undesirable additions to its population without crowding out immigrants who are qualified to become good citizens."

By the late 1880s, changes in the way America handled immigration were inevitable. Castle Garden had become an anachronism, a quaint relic of a disappearing world. It was a state-run institution trying to deal with a nationwide problem. It was an institution run by machine politicians and private citizens looking to protect the interests of immigrants, not disinterested professionals looking out for the greater good of society. It was a nineteenth-century solution to a (soon to be) twentieth-century problem.

Castle Garden officials would find themselves under nearly continuous assault. Where Isaiah Rynders and his immigrant runners failed years before, the Supreme Court, the Congress, the Treasury Department, the governor of New York, and crusading journalists succeeded.

The Supreme Court struck the first blow in 1875 with *Henderson v. Mayor of New York*, which declared that state laws requiring the immigrant head taxes were unconstitutional because they usurped Congress's constitutional powers to regulate immigration. The Constitution is fairly oblique in its references to immigration, and Congress had shown little desire to exercise that right previously. A decade after the victory of the Union Army at Appomattox, the Supreme Court was unsympathetic to the idea of states' rights. "The laws which govern the right to land passengers in the United States from other countries ought to be the same in New York, Boston, New Orleans, and San Francisco," the Court declared.

Shortly after the decision, Congress responded by passing the first federal law restricting immigration. The Immigration Act of 1875 banned prostitutes, criminals, and Chinese laborers. However, it was an odd law. Though Congress declared its authority to exclude immigrants, the federal government showed little interest in enforcing the new law and left the task to the states. Back in New York, the Board of Commissioners at Castle Garden, without the revenue from the immigrant head tax, was in debt and could no longer take care of immigrants. For the next six years, Congress ignored pleas from New York State for financial help to enforce federal law. Frustrated, the Board of Commissioners threatened to close down Castle Garden.

It was not until 1882 that Congress again acted on immigration when it passed two important pieces of legislation. The first placed the power to regulate immigrants more firmly in the hands of the U.S. Treasury Department. As some 476,000 immigrants passed through Castle Garden that year, the Immigration Act of 1882 imposed a head tax of 50 cents on all incoming immigrants. More importantly, it expanded the exclusionary categories to include any "convict, lunatic, idiot, or person unable to take care of himself or herself without becoming a public charge." Congress was expanding the classifications of undesirable immigrants, and the list would soon grow even longer in years to come.

That same year, Congress passed another law with a different intent. The Chinese Exclusion Act barred the entry of nearly all Chinese immigrants. The number of Chinese immigrants was small—some 250,000 arrived between 1851 and 1880, and they represented less than 3 percent of all immigrants arriving each year—yet Congress succumbed to racial fears, as well as concerns that cheap Chinese labor would lower the standard of living for native-born workers.

In 1885, Congress again heeded the wishes of labor by passing the Foran Act, also known as the Alien Contract Labor Law, which made it illegal "to assist or encourage the importation or migration of aliens . . . under contract or agreement," thereby outlawing the recruitment of immigrants whose passage was prepaid by a third party, usually a business agent. Skilled workers, artists, actors, singers, and domestic servants were exempt from the ban on contract labor.

Even with these new laws, more tolerant attitudes toward immigrants still ran deep in the American psyche. One congressional supporter of the contract-labor law emphasized that the law "in no measure seeks to restrict free immigration; such a proposition would be odious, and justly so, to the American people." With these laws, Congress made clear that the method for dealing with European and Asian immigrants would be very different. When it came to European immigrants, Americans tried to balance concern for the impact of these new immigrants with a national mythology that welcomed newcomers. The Chinese, however, faced near exclusion solely by their race.

How to enforce these laws was still an open question. Neither the Treasury Department nor anyone else in Washington had the capacity to monitor, investigate, and examine hundreds of thousands of immigrants. To solve this problem, the secretary of the Treasury simply contracted with state governments and groups like the Board of Commissioners to continue what they had already been doing.

The Board of Commissioners was being asked to shoulder a greater burden at the same time that it was under increasing criticism. In 1883, New York's newly elected Democratic governor, Grover Cleveland, attacked Castle Garden as "a scandal and a reproach to civilization," a place where "barefaced jobbery has been permitted, and the poor emigrant, who looks to the institutions for protection, finds that his helplessness and forlorn condition afford the readily seized opportunity for imposition and swindling."

Although Castle Garden had been created in an altruistic spirit, it soon became enmeshed in a battle between Republican state officials and Democratic city officials. Cleveland was echoing partisan criticisms that Castle Garden had been a Republican patronage pot in the middle of Democratic New York City. Many people were angry that the Board of Commissioners was constantly demanding more money from the state for the operation of Castle Garden, while private companies reaped profits inside it thanks to the monopoly granted them by their Republican patrons.

Of the estimated one hundred workers at Castle Garden, 90 percent were Republicans. Profits were being made there, and New York Democrats had little to say over who got the spoils. Privileges at the immigrant depot, such as railroad tickets and money changing, were given away to politically connected firms. It was estimated that railroads did over $2.5 million worth of business at Castle Garden in 1886.

To many, this cried out for intervention from the federal government. The *Times* editorialized that if only "foreign immigration were taken in hand by the national government . . . it is certain that great waste would be prevented, many scandals be avoided, and an important public interest would be placed where it properly belongs."

Yet there was more to the criticism and the demands for a federal takeover than just blind partisanship. No matter the good intentions of those administering Castle Garden, the situation had certainly deteriorated, so much so that by the 1880s Jewish immigrants coined a new Yiddish phrase—*kesel garten*—that became synonymous with chaos. The old vigilance against runners and others sharks had weakened, and immigrants could not be guaranteed complete security from scams and thieves.

In 1880, a twenty-two-year-old English miner named Robert Watchorn arrived at Castle Garden. With gnawing hunger, he spied a pie stand. After dropping 50 cents at the counter, Watchorn devoured his 10-cent piece of apple pie. When he asked for his change back, the salesman refused. Watchorn tried to jump the counter to retrieve his money, but a policeman intervened and threatened to charge him with assault. In one telling of the story, he got his money back and went on his way, but in another he did not get his money back, but gained "a great deal of sad experience." Either way, it was an incident that would remain with Watchorn even a quarter-century later when he would become the man in charge of processing immigrants in New York.

Another sign that conditions at Castle Garden were deteriorating was the creation of the Catholic Church's Mission of Our Lady of the Rosary's Home for Irish Immigrant Girls in 1883. The Home was founded by Father John Riordan and located directly across from the Battery at 7 State Street. Writing in 1899 of the early days of the Home, Father M. J. Henry made clear that even with the protection of Castle Garden, the old predators of immigrants still survived outside its walls.

> Thieves, blackmailers, and agents of bawdy-houses made their harvest on many a hapless immigrant. As long as the immigrants remained in Castle Garden they had protection and also the privilege of a labor bureau established by the Irish Emigrant Society. Once, however, they left the landing depot to seek relatives or friends or to secure boarding houses, they had to run the gantlet of these scheming wretches.

Run by Catholic priests, the Home gave these Irish girls a safe place to stay. The priests watched over the girls from the time of their arrival at Castle Garden. The main concern was the protection of the sexual virtue of these young, single, Catholic girls, and the fear that they might be unwittingly ensnared into the life of prostitution by the leeches who roamed the Battery. In its first sixteen years of operation, an estimated seventy thousand Irish girls were guests at the Home after having first passed through Castle Garden.

Public concern about the affairs at Castle Garden continued to grow in the 1880s when Joseph Pulitzer, editor of the *New York World*, launched a blistering crusade against Castle Garden. A Hungarian immigrant who had come through Castle Garden decades earlier, Pulitzer turned his newspaper into a forum for populist pursuits. In 1884, he had led his "people's paper" in a campaign to raise money for the completion of the pedestal for the Statue of Liberty. While shaming the wealthy for not giving more, Pulitzer promised to list the name of every person who made a contribution, no matter how small the donation. In response, over $100,000 was raised, the circulation of the *World* increased, and Pulitzer's reputation as a crusader grew.

In 1887, Pulitzer trained the cannons of his broadsheet at Castle Garden and never let up. In the first of many articles over a year's time, the *World* scored Castle Garden as a monopoly, arguing that the immigrant depot had become a "cumbrous and unwieldy institution." Railroads, the *World* charged, were fleecing immigrants with the consent of the board. The paper headlined another editorial on Castle Garden: "Purification Needed." The commissioners did not take the accusations lying down. One of them called Pulitzer "a mean, dirty, contemptible coward" who "ran away to Europe to save himself from incarceration," and sued the paper for libel.

Soon after the *World*'s exposés appeared, Washington took action. Grover Cleveland, who as New York governor had harsh words to say about Castle Garden, was now sitting in the White House. In August 1887, his secretary of the Treasury ordered an investigation. Not only was Castle Garden accused of granting monopolies to companies that cheated immigrants, but it was also accused of not strongly enforcing the 1882 law barring certain classes of undesirable immigrants. J. C. Savory of the American Emigrant Society called Castle Garden "a delusion to the public and a snare to the immigrant."

The next to pile on Castle Garden was Congress. During the 1880s, it proved unwilling to sit on the sidelines of this increasingly national issue. In an era predating Theodore Roosevelt's bully pulpit and the imperial presidency, Congress was the true power in Washington. Responding to the ever-growing debate about the meaning of immigration, Congress began to assert its authority.

In 1888, Rep. Melbourne Ford of Michigan chaired a congressional committee to investigate immigration. When Ford brought his committee to New York, Pulitzer's *World* was there to greet it and splash the testimony of witnesses on its front page. The committee released its report early the following year. It foreshadowed a coming change in how the nation dealt with immigration. The report described how immigrants were processed at Castle Garden in 1888.

> When the vessel containing them has been moored to her dock, the immigrants are transferred to barges, which are towed to Castle Garden. There they disembark, and are required to pass in single file through narrow passage-ways, separated from each other by wooden railings. In about the center of each of these passage-ways there is a desk at which sits a registry clerk who interrogates the immigrant as to his nationality, occupation, destination, etc.—questions calculated to elicit whether or not he is disqualified by law from landing. . . . These questions must be asked rapidly, and the inspection is necessarily done in a very hurried manner, in order that there may be no undue delay in landing them.

The process was simply not thorough enough to comply with existing immigration law. According to the Ford Report, "large numbers of persons not lawfully entitled to land in the United States are annually received at this port." The committee reported that one of the Castle Garden commissioners had even called its operations "a perfect farce."

The report did not stop there. It concluded with some general observations. After paying homage to the benefits of past immigrants settling the West, succeeding with their "industry, frugality, and thrift," the report asked whether the same could be "said of a large portion of the immigrants we are now receiving." The congressmen answered their own question: "The committee believe not."

The committee believed that the "class of immigrants who have lately been imported and employed in the coal regions of this country are not such . . . as would make desirable inhabitants of the United States." It described these Slavs and Italians as having low intelligence. Their purpose in the United States was to "accumulate by parsimonious, rigid, and unhealthy economy" enough money to return home. They lived "like beasts" and ate food that "would nauseate and disgust an American workman. . . . Their habits are vicious, their customs are disgusting."

The Ford Report echoed much of the contemporary concern about immigration. First, it differentiated between desirable and undesirable immigrants. Government policy, it argued, should sift through these immigrants and separate the wheat from the chaff.

Second, the language of immigration regulation closely mirrored the parallel discussion of economic regulation of trusts, monopolies, and railroads. The vast social changes that Americans experienced could be pinned upon the greed of businessmen who put profit before public interest. Reformers sought to use government power to exert the public interest and reign in selfish private interests. According to the Ford Report:

> For the purpose of greed these men have exaggerated the
> advantages and benefits to be derived by persons immigrating
> to this country, and have been guilty of erroneous statements
> in order to secure their commission upon the price of
> a passage ticket to such an extent that some localities in
> Europe have been nearly depopulated, and the poor deluded
> immigrant has come to the United States, arriving here
> absolutely penniless, to find out that the statements made
> by the steam-ship agents were absolutely false, and, in many
> instances, after a short time, he has become a public charge.

In a time of growing disillusionment with laissez-faire economic theory, immigration restrictionists found their enemies in greedy steamship companies and American businesses that contracted with low-wage immigrants to take jobs from native-born workers.

Third, the Ford Report did not call for the debarment of immigrants from specific countries or races, nor did it call for the suspension and ending of all immigration. In terms of Chinese immigration, the report included only one line, saying that it made no effort to investigate it. The regulation of European immigration would be categorically different from the rigid and near-complete banning of the Chinese.

There was one dissenter. Rep. Francis Spinola, a Democratic congressmen from Brooklyn and one of two Italian-American generals in the Union Army during the Civil War, made it clear that he opposed any attempt to restrict "honest immigration." However, even Spinola agreed with efforts "to shut out paupers, lunatics, idiots, cripples, and thieves, as well as all other evil-doers, who come here to practice their wickedness and fill our poor-houses and prisons."

Congress never acted upon the "Bill to Regulate Immigration" that the Ford Committee recommended. However, both the House and the Senate established permanent standing committees on immigration for the first time, thereby assuring continued congressional interest.

As conditions at Castle Garden continued to worsen, its critics became more vocal, driving one member of the Board of Commissioners to the point of despair. "So far as Castle Garden is concerned, the country would be better off if it were wiped out of existence," Edmund Stephenson told the *New York Sun* in 1889. He felt understandably beleaguered, caught between those who wanted tougher restriction of immigration, defenders of immigration who wanted lax enforcement, and the usual predators looking to take advantage of any immigrant who made it outside the walls of Castle Garden.

At the end of 1889, Secretary of the Treasury William Windom ordered another report on Castle Garden. The Treasury report also found the inspection of immigrants at Castle Garden inadequate and the arrangement between state and federal officials in the regulation of immigration unsatisfactory. It recommended that the federal government take complete control over the regulation of immigrants. Windom accepted that advice, and in February 1890, he notified the Board of Commissioners at Castle Garden that he was terminating their contract in sixty days.

The decision was inevitable. A Republican named Colonel John B. Weber, who was soon to oversee the federal control of immigration, visited Castle Garden in its waning days. He found that boardinghouse runners were having their way with the confused and bewildered immigrants, with seemingly little interference from officials. A new direction was in order.

On Friday, April 18, 1890, the steamers *Bohemia* and *State of Indiana* were the last two ships to drop passengers at Castle Garden, landing 465 people that day. In a spiteful mood, members of the Board of Commissioners had refused to allow the Treasury Department to use Castle Garden until new facilities could be found. A makeshift immigrant depot was set up at the Barge Office on the other side of the Battery. Castle Garden was now closed for business.

TREASURY SECRETARY WINDOM WAS determined to begin anew and erase the memory of Castle Garden by building a new facility for processing immigrants completely under the control of the federal government. Some two weeks after announcing the termination of the Castle Garden contract, Windom made public his desire to place the new immigrant station at Bedloe's Island in New York Harbor.

Bedloe's Island was also the home of the newly erected Statue of Liberty. Once again, Pulitzer used the pages of the *New York World* to defend Lady Liberty. For weeks, the *World* hammered away, warning that the island would "be converted instead into a Babel." The paper even tracked down Auguste Bartholdi, the statue's sculptor, who called the decision a "downright desecration."

In response, a joint House and Senate committee selected another island in New York Harbor for the home of the new federal immigration station. Congress appropriated $75,000 to improve Ellis Island for the purposes of creating a new immigrant depot. The island was a perfect choice in many ways. It was already in the possession of the federal government as an underused munitions depot. Its island location meant that the immigrant runners and other predators could be kept at a distance, but it was only a quick ferry ride to Manhattan or the railroads on the Jersey side of the harbor.

Before the low-lying island could be made usable, a good deal of work was needed. While immigrants were being processed at the Barge Office—mostly by ex–Castle Garden inspectors—work had begun on dredging a deeper channel to Ellis Island. Docks were constructed on the island, as well as a two-story wooden building, which would be the main reception area. It would take nearly two years to complete the project. Meanwhile, the national debate over the meaning of immigration only intensified.

"Give us a rest," thundered Francis A. Walker. He worried that "no one can be surely enough of an optimist to contemplate without dread the fast rising flood of immigration now setting in upon our shores." Walker was no average citizen. He was the nation's most esteemed economist, a late-nineteenth-century combination of Milton Friedman and John Kenneth Galbraith.

A well-bred Bostonian descended from generations of Anglo-Saxon stock, Walker possessed an envious résumé that mirrored the great transformations of nineteenth-century America. A Union general in the Civil War by his midtwenties, Walker was in charge of the 1870 and 1880 Censuses and then taught economics at Yale. At the time of his musings on immigration, Walker was president of both MIT and the American Economic Association.

Walker found the new immigrants "ignorant, unskilled, inert, accustomed to the beastliest conditions with little of social aspiration, with none of the expensive tastes for light and air and room, for decent dress and homely comforts." They were lowering the country's wages and standard of living.

Walker also saw the birth rates of native-born Americans shrinking, while immigrant families produced more and more children. While most social scientists now see birth rates as a function of class, with birth rates shrinking as incomes rise, Walker had a different explanation: Immigrants brought down the nation's standard of living, and native-born Americans revolted against this situation by refusing to bring more children into such a degraded world. Walker's thesis neglected the fact that native-born American birth rates had been declining since the early nineteenth century, with seemingly little correlation to immigration rates.

Walker's views were echoed by a younger man with an even more distinguished pedigree. Forty-one-year-old Henry Cabot Lodge had already established himself in academia, gaining the first PhD in political science from Harvard. Though he would continue to write, especially about the glory of Anglo-Saxon culture, it was politics, not academia, that beckoned. In the 1890s, first as a congressman and then as a senator, Lodge began sounding the alarm about immigration. He hoped to prove that current immigration showed "a marked tendency to deteriorate in character."

Lodge used the occasion of the March 1891 lynching of eleven Italian immigrants in New Orleans to argue that changes were needed in the nation's immigration law. The cause of the attack was not anti-immigrant sentiment, Lodge argued, but rather "the utter carelessness with which we treat immigration to this country." For Lodge, the lynchings were one more piece of evidence showing that America could no longer "permit this stream to pour in without discrimination or selection or the exclusion of dangerous and undesirable elements." He called for moderate restriction that did not "exclude a desirable immigrant who seeks in good faith to become a citizen of the United States."

Whatever benefits immigration might bring, there were other values that took precedence. "More important to a country than wealth and population is the quality of its people," wrote Lodge. He was articulating the attitude of upper-class Americans dismayed by both the extravagances of the Gilded Age as well as the squalor and poverty brought about by urbanization. Like his close friend Theodore Roosevelt, Lodge was critical of crass materialism. Though this attitude was a luxury confined to those living on inherited wealth, it also reminded Americans that the public interest could not always be calculated by figures in a ledger book.

Walker and Lodge had tapped into a larger national concern. Newspaper headlines in 1891 screamed: "Lunatics and Idiots Shipped from Europe" and "The World's Dumping Ground." Alabama Congressman William C. Oates, who had led the Confederate charge up Little Round Top at the Battle of Gettysburg, summed up the growing belief in the undesirability of new immigrants.

> A house to house visit to Mulberry Street, in New York
> [the city's Little Italy], will satisfy any one that there are
> thousands of people in this country who should never have
> been allowed to land here. . . . Many of the Russian Jews who
> inhabit other streets in New York, and other cities are of no
> better class than the Italians just referred to. Many of the
> mining towns and camps of Pennsylvania and other states
> are overrun with the most beastly, ignorant foreign laborers
> who herd together almost as animals and are disgraceful to
> civilization.

The atmosphere was ripe for a major change in immigration policy. In 1891, while workers were busy constructing the physical edifice of Ellis Island's facilities, Congress was building the legal structures that would govern what would occur there.

The 1891 Immigration Act expanded the types of undesirable immigrants listed in the 1882 law to include "idiots, insane persons, paupers or persons likely to become public charges, persons suffering from a loathsome or a dangerous disease, persons who have been convicted of a felony or other infamous crime or misdemeanor involving moral turpitude, polygamists."

Excluded immigrants would be shipped back home at the expense of the steamship company that brought them. The burden of inspecting immigrants would lie not just with American officials, but with steamship companies who now had a financial incentive not to bring over immigrants who would not pass muster at American ports. For the first time since the Alien and Sedition Acts one hundred years earlier, the federal government laid out a method for deporting immigrants.

Immigration was now completely under the control of the federal government. Embedded deep within the law, Congress granted vast powers to this new federal agency. "All decisions made by the inspection officers or their assistants touching on the right of any alien to land . . . shall be final unless appeal be taken to the superintendent of immigration, whose action shall be subject to review by the Secretary of the Treasury." Though the language seemed innocuous, this provision would prove to be the most controversial. Congress had effectively declared that immigrants could not appeal their exclusions in court. Instead, all appeals had to be made through the executive branch, with a final decision made by the secretary of the Treasury.

The new immigration system represented a big step for Washington. The federal government of the nineteenth century had been a rather sleepy enterprise. The locus of power was in the political parties that controlled patronage for the few jobs that did exist, as well as the judiciary system. The federal government was a weak shell whose main responsibilities were to deliver the mail and pay the pensions of retired Civil War veterans and their widows. More than half of the federal government's workforce was employed by the Postal Service.

The growing complexity of the American economy would change all that. Within three years, Congress passed two landmark laws—the Interstate Commerce Act (1887) and the Sherman Anti-Trust Act (1890)—which set the stage for the regulation of private business by the federal government. In many ways, the 1891 Immigration Act deserves to be mentioned with the other two landmark laws. Like those other two laws, the immigration bill was enacted to address what many considered to be a failure of the free market. Almost immediately, it created an immigration service larger in size and manpower than the Interstate Commerce Commission or the anti-trust division of the Justice Department and bestowed upon it greater powers.

Between 1875 and 1891, Congress had seized control of the immigration issue by passing sweeping laws banning the entry of most Chinese immigrants, defining classifications of undesirable immigrants, prohibiting the recruitment and contracting of immigrant laborers, and creating a system that would enforce these measures, with a centralized office in Washington overseen by congressional committees, and federal immigrant inspection stations at ports throughout the nation. The most important of these stations was at Ellis Island. The era of big government was dawning.

THE CONTRAST BETWEEN CASTLE GARDEN and Ellis Island is instructive. Castle Garden was a state operation, created largely at the behest of immigrant aid societies, designed to protect and aid new arrivals to America. Ellis Island was a federal operation, created in response to the national uproar at perceived changes in the type and nature of immigration at the end of the nineteenth century. Its raison d'être was neither the protection of immigrants nor their complete exclusion, but rather their regulation so that only the fittest, ablest, and safest would be permitted to land.

Castle Garden has long since receded from the national memory. As the years passed, the old immigrant station evolved first into the city's aquarium, then into neglect, and then into a historical reconstruction of the original fort. From here, modern tourists buy tickets for the ferry ride to Ellis Island and the Statue of Liberty.

Over 8 million immigrants passed through Castle Garden between 1855 and 1890. Many of their descendants know little of its history, thinking their forebears entered at Ellis Island.

Despite the corruption that plagued Castle Garden, one historian has called it "not only a monumental work, but also a great human expression, which can be placed among the shining achievements of American history during the nineteenth century."

Yet it is Ellis Island, not Castle Garden and its albeit imperfect history of benevolence and service, which takes center stage in the nation's tale of immigration.

Part II

THE SIFTING
BEGINS

Chapter 3

A Proper Sieve

That there is such a problem [immigration] no one doubts. It is in the air. It is in the conversation and the speculation of all parts of the country.
—New York World, 1892

A great system has been perfected on Ellis Island for sifting the grain from the chaff . . . not as a dam to keep out good and bad alike, but as a sieve fine enough in the mesh to keep out the diseased, the pauper, and the criminal while admitting the immigrant with two strong arms, a sound body and a stout heart.
—Dr. A. J. McLaughlin, 1903

AS SHE EXITED THE BARGE *JOHN E. MOORE*, YOUNG ANNIE Moore tripped across the gangplank landing her ashore. One could forgive her nervous clumsiness. It was the first morning of 1892 and the fifteen-year-old from County Cork, Ireland, was being rushed from the gaily decorated barge to the new immigration station on Ellis Island.

Even though Washington officials had nixed the idea of a large celebration for the grand opening, the sound of bells and shrieking whistles could still be heard over the noise coming from the crowd of newsmen and government officials. Having spent twelve days at sea, Moore would have been startled by the reception and was likely a little anxious as officials ushered her into the building. Amid the confusion and commotion, she made sure not to lose sight of her two younger brothers: eleven-year-old Anthony and seven-year-old Phillip.

Annie Moore had no idea she would be entering history books as the first immigrant to arrive at Ellis Island. After a brief inspection, she was signed into the entry books by an official from the Treasury Department and given a ten-dollar gold piece by Colonel John Weber, the commissioner of Ellis Island. "Is this for me to keep, sir," a blushing Annie asked Weber, embarrassed by all of the attention. She then thanked him, saying that it was the largest amount of money she had ever seen and she would keep it forever as a cherished memento of the occasion.

She was soon reunited with her father, Matt, who had sent for his children. Their destination was Monroe Street on the Lower East Side of Manhattan, a neighborhood teeming with immigrants like the Moores and where, just a few blocks from the family's apartment, a nineteen-year-old budding politician named Al Smith was beginning to make his way in the world.

How Annie became the first official immigrant at Ellis Island is unclear. One story claims that officials had rushed her ahead of a male Austrian immigrant. Another claimed that a fellow passenger named Mike Tierney, in a "spark of Celtic gallantry," pulled the Austrian away from the gangplank by his collar, shouting "Ladies first," and let young Annie pass.

Annie Moore's story is an oft-told tale and ultimately it is impossible to know whether her selection as the first arrival at Ellis Island was pure luck or a conscious decision by officials. It would not be surprising if officials had picked Moore out early for special treatment. After all, one of the main purposes of Ellis Island was to reassure increasingly anxious Americans that public officials were on guard to sift out undesirable immigrants. It is hard to argue that Annie Moore and her rosy cheeks, with younger brothers in tow, wasn't a reassuring image—although fifty years earlier, the arrival of County Cork's surplus population would not have been looked upon with much approval.

But a look at the rest of her shipmates tells a more complicated story. Annie and her two brothers and four other Irish immigrants all embarked on the steamship *Nevada* at the port of Queenstown. When the seven Irish travelers got on the *Nevada* in late December 1891, the ship already had 117 passengers who had boarded in Liverpool. With 124 passengers, the *Nevada* made its way toward New York Harbor. Twenty of the passengers, including ten American citizens, traveled in cabin accommodations, while the rest settled for steerage.

Roughly one-third of the *Nevada*'s passengers hailed from northern Europe: twelve English, seven Irish, two German, two French, and fourteen Swedish. The vast majority of the passengers accompanying Annie Moore were Russian Jews, fleeing the increasingly oppressive measures of the czar. Seventy-seven men, women, and children—over 60 percent of the passengers—had made their way from Russia to England and were now taking their final journey to America. If Annie Moore's prominence on that day was a bow to the old immigrants from a half-century earlier, the reality of most of the *Nevada*'s passengers was decidedly new immigrant.

Most of the travelers were in their twenties and thirties. The oldest was a fifty-year-old tailor from Russia, while the youngest was four-month-old Sara Abramowitz. Most would end up staying in New York, but some headed for Pennsylvania, Maryland, Minnesota, and even Wyoming. Most listed their professions as farmers and laborers, while others were skilled laborers like tinsmiths, bookmakers, machinists, and tailors.

When the *Nevada* entered New York Harbor, it did not head directly for the immigration station. The waters around the island were too shallow and its pier could not accommodate even the smallest transatlantic ship, meaning that ships would have to dock in Manhattan and unload their passengers. From there, immigrants would board smaller ferries, like the *John E. Moore*, which would take them to Ellis Island, where they would undergo the formal inspection process.

Not all of the passengers who sailed into New York Harbor alongside Annie Moore would end up at Ellis Island. The ship's twenty cabin passengers could head directly to their destinations on the mainland, whether U.S. citizens or not. An inkling of the rationale for such differential treatment can be found on the ship's manifest. Whereas steerage passengers listed their occupations with proper plebian titles like laborer or farmer, the twenty cabin passengers on the *Nevada* were marked down simply as "Gentleman" or "Lady," signifying their more rarified social status.

As Annie Moore was entering the facility at Ellis Island, two more ships—the *City of Paris* and the *Victoria*—were waiting in the harbor. By noon, some 700 steerage passengers from all three ships were at Ellis Island. It is doubtful that immigration officials would have chosen any of the passengers from the *Victoria* as their celebrated first immigrant. Having just arrived from the ports of Palermo and Naples, 311 of the *Victoria*'s 313 steerage passengers hailed from southern Italy.

Colonel Weber noted that while officials processed 700 immigrants the first day, the new facilities could handle thousands more in a day. Most people assumed that such a capacity would never be reached. For now, even the *New York World* was pleased with the new facilities, noting that during the first day's business, "everything worked like a charm . . . under the new conditions the comfort and safety of the immigrants will be all that can be desired."

The immigrants who followed Annie Moore entered the immigrant depot—which was located closer to the ferry slip than its later brick replacement—and then headed up a double staircase to the second floor. Vigilant medical inspectors would watch them as they climbed the stairs, on the lookout for cripples and other invalids.

Once on the second floor, immigrants were herded into ten lines, each of which ended at the desk of a clerk whose job was to cross-examine the immigrants, verifying information from the ship's manifest and making sure the immigrant did not fall into one of the categories for exclusion. On the second floor were places to buy railroad tickets, information bureaus, telegraph counters, money exchanges, and a lunch counter.

A reporter from *Harper's Weekly* visiting Ellis Island in 1893 found it "suggestive of a prison in many of its aspects," with uniformed guards keeping order. When inspectors questioned immigrants, the reporter found some to be "nervously defiant" while others looked frightened. Still others were "angry, and some stolid with indifference or stupidity." The long-awaited opening of the federal inspection station did not end the debate over immigration. This five-acre scratch of land sticking out of New York Harbor would now become the focal point for everyone concerned with immigration.

Politicians, journalists, union leaders, and private citizens would now make their way to Ellis Island with their own agendas. The Women's Christian Temperance Union soon complained about the alleged existence of saloons at Ellis Island. Samuel Gompers, head of the American Federation of Labor, wanted more inspectors to enforce the contract-labor law and visited Ellis Island to make his case to Colonel Weber. Immigration restrictionists went to Ellis Island to make sure the laws were properly enforced. Immigration defenders visited to make sure that newcomers were fairly treated.

"The existing immigration law was framed to sift incomers—to draw a dividing line between desirable and undesirable immigrants," the superintendent of immigration said in his first annual report, adding that "it is not the serious intention of the Government to prohibit immigration, but from time to time to prohibit the people whom experience has demonstrated fail in some important direction in entering beneficially into American citizenship."

From the moment Annie Moore stumbled from the gangplank onto Ellis Island, this kind of sifting of desirable and undesirable immigrants proved to be a matter of trial and error. The opening of Ellis Island raised more questions than it answered. How would a nation manage the inspection, regulation, and sometimes exclusion of immigrants on a daily basis? Were immigration laws too strict or too lenient? Should the government create more classes of excludable immigrants or should immigrant inspectors interpret the law in a more generous manner?

In 1875, the Supreme Court placed the control of immigration with the federal government. Now it would decide just how far that power could go. In 1889, three years before the opening of Ellis Island, the Court heard the case of a Chinese immigrant named Chae Chan Ping, who was denied reentry into the country. It dismissed Ping's challenge to his exclusion, arguing that Congress had the power to make the rules that governed immigrant admissions and the courts should be deferential to that expression of democratic will. "The power of exclusion of foreigners being an incident of sovereignty belonging to the government of the United States as a part of those sovereign powers delegated by the Constitution," wrote Justice Stephen Field, "the right to its exercise at any time . . . cannot be granted away or restrained on behalf of anyone."

Three years later, just a little over two weeks after Annie Moore's arrival at Ellis Island, the Supreme Court went further. Nishimura Ekiu was a twenty-five-year-old woman from Japan who landed in San Francisco in 1891. With $22 in her pocket, she claimed to be headed to meet her husband, who was already in the country. Immigration officials believed that she had not been honest in her answers and refused her permission to land. Under the 1891 Immigration Act, the Japanese immigrant was declared "likely to become a public charge." Ekiu argued that the immigration officials in San Francisco had deprived her of due process in a court of law, but the Supreme Court ruled against her, arguing that Congress had entrusted upon officials the "sole and exclusive" right to exclude or admit aliens and immigrants had no recourse to the courts.

Although these two decisions related to events on the West Coast, they would deeply influence what went on at Ellis Island for the next sixty years. If an immigrant felt that officials had unfairly excluded her, the only recourse was up the administrative chain of command in the executive branch, not to the courts. This was called the plenary power doctrine that would dominate American immigration law for more than a century. Congress and the executive branch would have exclusive authority over immigration, and immigrants would be limited in their ability to challenge that authority in federal courts.

The implication was that immigrants who had not yet been approved to land had fewer constitutional rights. Admission to the United States was a privilege, not a right. A sovereign nation had a right to define its borders and decide who may or may not enter the country. "It is an accepted maxim of international law that every sovereign nation has the power," the Court argued in the *Ekiu* decision, "to forbid the entrance of foreigners within its dominions, or to admit them only in such cases and upon such conditions as it may see fit to prescribe." Congress could decide what kinds of immigrants could enter America, and the executive branch—represented by immigration stations like Ellis Island—had the right to execute those laws. Courts would have little say in the matter.

COLONEL JOHN B. WEBER never imagined that his life's path would lead him to Ellis Island. A former Republican congressman from the Buffalo area, Weber had been appointed commissioner of immigration by President Benjamin Harrison, taking over duties the day after the official closing of Castle Garden in April 1890. As Weber admitted, it was a classic patronage appointment: "I took the business of immigration commissioner with as little knowledge of it as a man who never had seen an immigrant."

Weber's parents were immigrants from Alsace, loyal subjects of

France at the time. Hearty peasants who were able to read and write, they separately arrived in upstate New York in the 1830s and married there. John Baptiste Weber's name symbolized the controversial history of the Alsace. Though his last name was solidly German, his middle name attested to the French influence of the region.

Forged in the fire of the Civil War, Weber's life resembled that of many Northerners at the time. At fourteen, he volunteered as the color bearer for the local militia. When war broke out a few years later, the eighteen-year-old volunteered for service. He survived the war without a scratch, despite having seen his share of combat, and made his way quickly through the ranks, going from private to colonel before his twenty-first birthday. In 1864, he helped organize and lead the 89th United States Colored Infantry. At twenty-one, Weber left the army, returned home to New York, and prepared to take an active role in civic affairs.

He started a family and became a large landowner and a grocer. Like many of the Union soldiers who survived the killing fields of the Civil War, Weber's postwar life was defined by membership in the local post of the Grand Army of the Republic and the Republican Party. Weber ran for Erie County sheriff in 1870, but narrowly lost to a Democrat named Grover Cleveland. Weber would later win the post in his second attempt, then go on to serve two terms in the House of Representatives.

In return for helping a fellow Civil War officer named Benjamin Harrison win the Republican presidential nomination in 1888, Weber was given the job of commissioner of immigration at the Port of New York, overseeing the construction of the new facilities at Ellis Island as well as the processing of immigrants at the Barge Office until Ellis Island was open. He was also given an additional task. In 1891, Treasury Secretary Charles Foster asked him to chair a five-man commission to travel to Europe and report on immigration. This was the first time the federal government sought to investigate the reasons why Europeans were emigrating to America.

The government wanted answers to very specific questions. Why were Europeans coming to the United States? Was immigration being "promoted or stimulated by steamship or other carrying companies or their agents for the resulting passenger business"? To what extent were "criminals, insane persons, idiots, and other defectives, paupers or persons likely to become a public charge, and persons afflicted with loathsome or dangerous contagious disease" encouraged to emigrate?

There was an additional reason for the trip. Before Weber set sail for Europe, President Harrison summoned him to a seaside cottage in Cape May, New Jersey, where the president was vacationing. Harrison wanted Weber to investigate the condition of Russian Jews. Upon arriving in London, where he would meet the other four members of the commission, Weber would choose one of the four to accompany him to Russia. Weber had to leave for Europe only three days after he received his instructions. The other commissioners, whom Weber had never met, were already waiting in London.

Once in London, Weber chose Dr. Walter Kempster as his traveling companion, leaving the other three members of the commission free to conduct their own investigations. Kempster was born in London and arrived in upstate New York as a young boy. Like Weber, he was a former Civil War officer, having served at Gettysburg. After the war, Kempster became one of the nation's leading alienists, making the study of the human brain his specialty. He worked at a number of mental hospitals in New York before moving to Wisconsin. In 1881, Kempster served as one of the witnesses for the prosecution in the trial of Charles Guiteau, the assassin of President James A. Garfield.

Beginning in late July 1891, Weber and Kempster city-hopped from Liverpool to Paris, Antwerp, Amsterdam, Berlin, St. Petersburg, Moscow, Minsk, Wilna, Bialystok, Grodno, Warsaw, Cracow, Budapest, and Vienna. They ended their trip in early October in Bremen, Germany. Along the way, Weber and Kempster met with consular officials, visited local neighborhoods, and spoke with officials from steamship companies. With all of his official duties, Weber had little time for sightseeing, complaining that all he managed was one hour at the Tower of London.

Weber and Kempster released their report in January 1892 and concluded that individuals left Europe largely because of "superior conditions of living in the United States . . . and the general belief that the United States present [sic] better opportunities for rising to a higher level than are furnished at home." Europeans mostly received these ideas not from the agents of steamship companies looking to drum up business, but from "the relatives or friends who have preceded and are established in the United States, and who, through letters and newspapers sent from this country, furnish such information."

Many argued that new immigrants were assisted or involuntary immigrants brought here on contract by American businesses or enticed by agents of steamship companies. By their estimates, Weber and Kempster concluded that some 60 percent of immigrants did come to America with prepaid tickets. However, these were tickets largely bought by friends and relatives in America and sent to the potential immigrant in Europe. Weber and Kempster were describing chain migration, the process by which recent immigrants, through letters, newspaper clippings, and money, entice family and friends to join them in the New World.

Weber noted that the country needed these immigrants because Americans traditionally shun hard manual work. "When the foreigner came in, the native engineered the jobs, the former did the shoveling," he argued. "The foreigner plows and sows, the native reaps; the one builds railroads, the other runs them and waters the stock; one digs canals, the other manages the boats; one burrows in the mines, the other sells the product." Relying on the connection between immigration and free labor for the health of the economy, Weber asked: "Stop the stream, and where will the new material come from which with a little training and experience develops into useful domestic help?"

Weber concluded that "the evils of immigration are purely imaginary in some features, greatly exaggerated in others, and susceptible of nearly complete remedy by the amendment of existing laws." He saw only the need for "rigid inspection at our ports," to enforce the 1891 law. Of course, just what constituted rigid inspection would become a matter for debate for every immigration official at Ellis Island throughout its history.

Following his instructions from Harrison, Weber paid special attention to the plight of Jews. The situation in Russia was beginning to have repercussions for the United States. Mary Antin had already emigrated and wrote a memoir of her family's journey from Russia to America. During those bleak times, she wrote, "America was in everybody's mouth. Businessmen talked of it over their accounts . . . people who had relatives in the famous land went around reading their letters for the enlightenment of less fortunate folk . . . children played at emigrating . . . all talked of it, but scarcely anyone knew one true fact about this magic land." The number of immigrants coming from Russia, the vast majority being Jewish, was increasing dramatically. From 1890 to 1891, the number increased from 41,000 to 73,000.

The emigration of Russian Jews was rooted in the turmoil of late-nineteenth-century Russia. Many of the problems can be traced to 1881, when Czar Alexander II, who had inaugurated an era of relative liberalism in Russia, was assassinated by a group of revolutionaries. Jews bore the brunt of the anger of the Russian people and of the new czar, Alexander III, who pursued anti-Jewish policies with a vengeance. Life in the Pale of Settlement, where much of the Jewish population was forced to live, became harder. Jews who had left the Pale in decades past to earn their livings in cities were now being forced out. Petty harassments increased along with the restrictive edicts.

His time among Russia's oppressed Jews had a deep impact on Weber. He and Kempster witnessed the expulsion of Jews from Moscow to the Pale. They met an elderly man with paralysis and partial blindness who was suffering in his own bed because he was refused entry into a Moscow hospital. Many who had lived in Moscow for decades now found their businesses failing. "Among those ordered out while I was there were cashiers, clerks, correspondence chiefs, and bookkeepers of banks; heads of business departments; manufacturers," Weber remembered.

The two Americans traveled extensively through towns such as Minsk, Wilna, Bialystok, and Grodno. The stories from Russian Jews were "sad and pitiful in the extreme. . . . Everywhere there was gloom and dejection." Weber encountered pronounced and deep-seated misery. "The emaciated forms, the wan faces, the deep sunken cheeks," he later remembered about the experience, "the pitiful expression of those great staring eyes reminding one of a hunted animal, are ever present and will never leave me." Weber was haunted by nightmares of the tragic Jewish figures he encountered and sometimes wondered if he was not suffering from hallucinations.

Weber and Kempster's report was full of sympathetic observations of Jewish life. They argued that Jewish immigration was forced largely by the religious and ethnic persecution found in Russia. They described in detail life in Jewish ghettos and the history of laws that made life difficult for those of the Jewish faith. After a visit in London with Baron de Hirsch, who used part of his vast fortune to assist Jews fleeing

Russia, Weber and Kempster admitted that the case of Russian Jewish immigrants was decidedly different from that of other immigrants.

By the 1890s, Russian Jewish paupers had increased by nearly 30 percent and some estimates counted as many as 40 percent of the Jewish population as luftmenschen, people without jobs, skills, or prospects, floating through the Pale and surviving as best as they could. Here seemed proof of what immigration restrictionists claimed: that assisted immigrants came to America with tickets paid for them by third-party philanthropic groups. They needed to be helped because so many had become paupers.

While Weber and Kempster admitted that some cases of paupers emigrating to America did exist, "that the movement assumes any sort of proportions [as believed by restrictionists] is not warranted by our investigations nor is it believed." The case of Russian Jews could not be seen simply through the prism of paupers dumped onto U.S. shores. Instead, Weber and Kempster asked that Americans look beyond the temporary condition of immigrants.

> A person who by reason of unexpected misfortunes or persecutions is deprived of his accumulations, who has been subjected to pillage and plunder while fleeing from the burdens which have become unbearable, if capable of supporting himself and family, if he has one, with a reasonable certainty after obtaining a foothold, and that foothold is guaranteed by friends or relatives upon landing or strong probable surrounding circumstances, is not, according to our definition, a pauper.

Weber and Kempster's report was a sharp rebuke to immigration restrictionists.

However, instead of a unified report on European conditions, the committee released four separate ones. Weber's three other colleagues had conducted their own tours of Europe. The report of Judson Cross most closely resembled the conclusions of Weber and Kempster. Writing about Italian immigration, Cross also described the process of chain migration. Italian immigrants "are constantly bestirring others to go. Each Italian in the United States can easily secure a place for a friend and the process is ever being repeated." Contrary to some of the reports about these new immigrants, Cross found southern Italians "sober, industrious, and economical and fond of their children." These Italians left their homelands because of a lack of land, not because they were encouraged to leave by the government.

Joseph Powderly, the brother of famed union leader Terrence V. Powderly, was labor's representative on the committee, and his report mirrored the concerns of many native-born workingmen. He was concerned that workers from eastern Europe were coming to western Pennsylvania and competing with native-born workers in the mines and factories, driving down wages and the quality of life. Unless immigration was restricted, Powderly argued, the native-born American would be driven from the coal mines or else he "will have to come down from his extravagant standard, and be contented with one room for himself, wife, and children in which to live, eat, and sleep."

The commission's final member, Herman J. Schulteis, took issue with the nuanced notion of pauperism found in Weber and Kempster's report. Schulteis complained that recent immigrants were coming to America with the help of immigrant aid societies and other associations that encouraged paupers and criminals to emigrate. He also reported on the widespread involvement of Italian banks and labor agents in the distribution of prepaid tickets for Italian immigrants. As for whether steamship companies could be trusted to screen out immigrants who might be disqualified under the 1891 Immigration Act, Schulteis answered with an emphatic no, claiming to have witnessed the "sham inspection" of immigrants at the port of Naples.

While Weber was sympathetic to the plight of Russian Jews, Schulteis wrote of the "alleged" persecutions in Russia, which only existed in the minds of "Russophobists and of persons who have never looked into the economic situation in Russia." Schulteis approved of Russia's anti-Jewish edicts, writing that they were "in the interest of the general welfare of the Russian people." After all, Schulteis noted, while Jews were only 5 percent of the population, they owned half of the wealth of Russia. "This is a matter of general notoriety in Russia and has an important bearing on the social status of the Hebrew," he concluded.

It is no surprise that someone who would recycle the anti-Semitism of Russian officials would conclude that throughout Europe, "there are many persons engaged in the business of transferring from the moribund systems of European misgovernment vast members of their 'dangerous' pauperized, diseased, decrepit, and criminal population, not only a safety valve to their own overstrained machinery, but to serve as an element of weakness in this Republic, the greatness of which they view with growing alarm."

Despite his insensitivity, Schulteis never called for a ban on immigration or the selection of immigrants only from desirable races. Instead, his recommendations included having American inspectors at European ports inspect and approve potential immigrants; a bigger head tax on immigrants; the end of prepaid tickets; and the granting of emergency quarantine powers to the president.

These dueling reports lay out a spectrum of attitudes toward immigration. To Weber and Kempster, newcomers fled poverty and prejudice in search of opportunities in the New World, where they were certain to be molded into independent and productive citizens. By contrast, Powderly voiced the concerns of workingmen eager to protect their wages from the competition of cheap foreign labor brought to America by greedy businesses. Lastly, Schulteis articulated the darker vision of immigration, seeing newcomers as Europe's refuse dumped on America's shores—a losing equation that would only weaken the Republic, while strengthening Europe.

Rather than a final answer on the root causes and nature of immigration, the Treasury secretary got more of the same contentious debate of Americans grappling with the changes that were wrenching the nation into the modern world and showed no signs of abating. Ellis Island, created as the "proper sieve" to weed out undesirable immigrants, would soon become a lightning rod in this debate.

Chapter 4

Peril at the Portals

———◦•◦———

There lies the peril at the portals of our land. . . . In careless
strength, with generous hand, we have kept our gates wide
open to all the world. . . . The gates which admit men to the
United States and to citizenship in the great Republic should
no longer be left unguarded.
—Senator Henry Cabot Lodge, 1896

COLONEL JOHN WEBER WAS BURSTING WITH RIGHTEOUS
anger. The commissioner of Ellis Island had braved the cold January winds coming off New York Harbor to witness the procession of
some seven hundred immigrants who had arrived on the steamship
Massilia. What disturbed Weber was that many of those passing by
him were clearly in bad health. These were no rosy-cheeked young Irish
girls like Annie Moore. Instead, many of those now crossing the same
gangplank to Ellis Island as Moore had four weeks earlier were sick
and emaciated—not the hearty lot most Americans hoped for in their
future neighbors.

Weber was not resentful toward these predominantly Russian
Jewish immigrants. A few months earlier, he had witnessed the tragic
plight of oppressed Jews throughout the Pale of Settlement and seeing
a similar neglect and obvious lack of concern here at an American port
fueled his anger. He was upset that steamship officials had forced these
sickly passengers to cross the harbor to Ellis Island in an open barge
in frigid weather. The normally even-tempered Weber was so outraged
at the lack of care given to these newcomers that he fired off an angry

letter to the steamship company, accusing them of "inhuman, if not criminal" behavior and promising to fight for legislation to punish steamships for any future incidents of "brutality and inhumanity."

Among those worn refugees parading past Weber was the Mermer family: Fayer, her husband Isaac, and their five young children. The Mermers had managed to survive both the trip to Ellis Island and the inspection process and would soon begin their lives in America at a temporary lodging house at 5 Essex Street on Manhattan's Lower East Side, provided for them by the United Hebrew Charities.

Twelve days after their arrival, the Mermers' world was thrown into even greater turmoil. City health officials forcibly entered their Essex Street tenement, dragging Fayer, already sick with fever, out of the building kicking and screaming. Along with her son Pincus and daughter Clara, Fayer was forced into quarantine as city officials moved quickly and brusquely to deal with a highly contagious typhus fever outbreak. One week later, Fayer would be dead, though her children would recover. This outbreak was believed to have originated with the *Massilia* immigrants and was spreading throughout the lodging houses of lower Manhattan.

How the Mermer family and 265 other Russian Jews ended up in America in the first place—and how their case ignited a national panic—is a story of its time.

The *Massilia* had departed from the port of Marseilles on January 1, 1892, with 270 Russian Jewish passengers. Since the spring of 1891, many had been wandering the continent, landless and countryless, unwanted throughout Europe. Some of them were originally Turkish subjects who, years earlier, had migrated to Russia for better opportunities. With the increasing repression of the Jews in Russia, they found themselves expelled. After their expulsion, they landed in Constantinople with hopes of heading to Palestine. Instead, Turkish authorities refused them passage. For three months, they were trapped in the city's Jewish ghetto. In December 1891, Turkish officials expelled them, and they headed for Smyrna. Their travails had exhausted what few funds they originally had, leaving them paupers. With the assistance of the Baron de Hirsch Fund, founded that year to assist east European Jews, they made their way from Smyrna to Marseilles and from there boarded the *Massilia* for New York.

Along with its Jewish passengers, the *Massilia* also carried fine

French wine headed for sale in New York. Instead of leaving directly for America, the ship next made its way southeast toward Naples, where it loaded up on more cargo: macaroni, fruit, and 457 Italians traveling in steerage to America.

The ship would be at sea for over three weeks. By all accounts, it was a stormy trip in the brutal depths of winter. To keep out the cold air, the ship's hatches were battened down for the entire journey and passengers were kept in close quarters, rarely able to go above deck to stretch their legs in the fresh air.

Considering the traumas of their nearly yearlong trek, it is no surprise that many of the Jewish migrants succumbed to illness. Twenty-three-year-old Julia Hoch, for example, suffered from uterine hemorrhaging on the trip, leading ship doctors to prescribe a treatment of "purgative clysters [enemas] two times a day for obstinate constipation, hot sedative vaginal injections. Internally, a solution of extract of ergot in cognac and peppermint water, strengthening nutrition."

Despite the treatment, Hoch somehow managed to recover. Young Isaac Holinsky was not so lucky. Seven days out from Marseilles, the nine-year-old Russian boy became afflicted with chronic nephritis, a kidney condition. Doctors subjected him to "a milk diet, to constant applications for wet hot flaxseed poultices on the renal region and on the chest." The treatment did not work, and four days later Isaac passed away, his body thrown overboard "with all due formality of a sea burial," according to the ship's log.

When the ship finally arrived in New York Harbor, Weber made sure that the sick immigrants were immediately taken to the Ellis Island hospital. Besides the sick passengers, officials put aside nearly seventy other *Massilia* immigrants for further inspection, fearing that their poverty would likely lead to their becoming public charges.

Despite these concerns, nearly all of the 270 Russian immigrants were eventually allowed to land, thanks to a sympathetic ruling by Colonel Weber and the intervention of another Jewish aid society. In the cases of those suspected of not meeting inspection standards, Weber accepted the posting of bonds by the United Hebrew Charities, which then placed the immigrants in boardinghouses on the Lower East Side. Although a few of *Massilia*'s Jewish passengers scattered across the country after leaving Ellis Island, most landed in this growing Jewish ghetto.

The story of the *Massilia* should have ended there. Colonel Weber's charity would have gone unnoticed. The poor treatment of the sickly Jewish travelers by ship officials would have been largely ignored, except for a mild reprimand. On the following day, more ships would have entered New York Harbor, bringing with them more personal stories and more decisions for immigration officials. But the *Massilia* would not fade so quickly into the city's past.

On the morning of February 11, 1892, Dr. Cyrus Edson, the chief sanitary inspector of the New York City Health Department, arrived at his office to find four postcards waiting for him. All four were sent by Dr. Leo Dann of the United Hebrew Charities, and dealt with four cases of typhus fever that Dann had discovered among *Massilia* passengers at a boardinghouse at 42 East 12th Street on the Lower East Side.

Often confused with typhoid fever, typhus had similar symptoms, including high fever, dizziness, muscle ache, nausea, and the outbreak of a reddish-purple rash. Typhus was a fast-spreading disease that had threatened the city in previous years. In 1851, almost a thousand New Yorkers had died from the disease, but since 1887, only five people had succumbed to typhus. City officials were anxious to prevent any new outbreak, so Edson and his staff made the trek that afternoon to the East 12th Street tenement where they found not four but fifteen cases of what Edson later called "that most dreaded of all contagious diseases."

The thirty-five-year-old Edson, a direct descendant of Rhode Island founder Roger Williams, was also the politically savvy son of a former New York City mayor with strong ties to Tammany Hall. Now Edson was in charge of a potential public health crisis, in a city and a nation already uneasy about immigration.

It quickly became apparent that the disease could be traced to the *Massilia*. Edson and his team of inspectors, with the help of officials from the United Hebrew Charities, set out to track down every passenger who had arrived on the *Massilia* and test them for typhus. The task was made easier by the fact that nearly all the *Massilia* Jews were being housed in eight lodging houses on the Lower East Side.

By nightfall, Edson's team had inspected residents at all eight tenements and diagnosed nearly seventy with typhus fever, including Fayer Mermer and two of her children. These men, women, and children were then escorted to the foot of East 16th Street at the East River where, in six separate trips, they were forcibly removed to the city's quarantine hospital on North Brother Island, off the Bronx coast. Unwanted in Russia, Turkey, and France, these poor individuals were hastily and roughly herded into quarantine and must have wondered whether they were even welcome in America.

Within two days, every Russian Jewish immigrant in the city from the *Massilia* was located. While those with symptoms were sent to North Brother Island, *Massilia* passengers without symptoms, and anyone else who had lodged in the same tenements as those afflicted with the disease, were rounded up and placed in temporary quarantine at two boardinghouses at 5 Essex Street and 42 East 12th Street, with police stationed outside to prevent anyone from entering or leaving. Health officials fumigated the empty lodging houses by burning sulfur in iron receptacles suspended in water, with the steam aiding the distribution of the sulfur. The rooms were then aired out and scrubbed with a disinfectant of bichloride of mercury.

Meanwhile, the *Massilia* was at sea heading back to Marseilles. On the return trip, the ship's fireman, baker, and several sailors came down with serious fevers and delirium, all symptoms of typhus. They survived the trip back to France, but it is unclear what happened to these crew members after they landed.

Edson's next task was to track down the 457 Italians who had also entered the country on the *Massilia*. Unlike the Jewish immigrants, who nearly all stayed in Manhattan, as many as a hundred Italians were already scattered across the nation, some as far away as Chicago, Fort Wayne, Indiana, and Bryan, Texas.

Health officials in Trenton, New Jersey, were able to track down two of the *Massilia*'s Italian passengers and bring them by cattle car to Edson's office in New York for inspection. Edson was unhappy with the Jersey officials, although it is not clear whether he was more concerned about the forcible taking of the Italians or the fact that potential typhus carriers were brought into the city.

However, only three of the *Massilia* Italians were eventually found to have the disease. The Italians were lucky to have been almost entirely segregated from Jewish passengers throughout the entire twenty-three-day journey, mingling only when they were transferred by ferry to Ellis Island.

Edson and his staff, with the help of the United Hebrew Charities, had acted quickly and aggressively. "I do not believe there is the slightest danger of an epidemic. We have the situation entirely in our hands," the young doctor predicted. In addition to the vigorous actions of Edson's office, Dr. William Jenkins, the health officer of the Port of New York, ordered that all ships with Russian Jewish passengers be held in quarantine, despite the fact that the typhus among the *Massilia* passengers originated in Turkey, not Russia. Even so, seven cases of typhus would be detected among incoming immigrants at quarantine in the coming months.

Despite his actions, Edson could not completely prevent the spread of disease to other city residents. Less than three weeks after the arrival of the *Massilia*, a carpenter named Max Busch took sick at his Bowery lodging house. He was diagnosed with typhus and taken to North Brother Island. Each day seemed to bring more stories about more typhus cases, with victims being discovered as far away as Providence, Rhode Island; Newburgh, New York; Baltimore, Maryland; and even St. Louis.

Meanwhile still more cases of typhus fever were being discovered among *Massilia* passengers in the two Lower East Side quarantine houses. On March 6, Edson took an even more drastic step and ordered everyone in those buildings to be removed to North Brother Island, whether they showed symptoms of the disease or not. During the height of the outbreak, New York officials transferred some thousand people to the quarantine hospital. Those without symptoms were quarantined for twenty-one days—the outer limit of the disease's incubation period—before being released.

Then the crisis ebbed. By the end of March, the outbreak had largely been contained—with the exception of a few additional cases over the summer and a small, unrelated outbreak in December. In Manhattan, with a population of nearly 1.5 million people, 241 cases of typhus were ultimately diagnosed in 1892 and the final death toll was 45. To Edson, this was a great success. He proudly compared it to the last major outbreak in 1881, when 153 people died. Not only did fewer people die in 1892, but nearly all of the deaths occurred in the first month of the epidemic. In contrast, the 1881 epidemic continued to wreak havoc for over five months.

Newspapers lauded the bold leadership of Edson and his team, though modern critics have complained about the rough and unequal treatment of these Jewish immigrants. Although the handling of the

Massilia's Jewish passengers by city health officials was often brusque and insensitive, Edson and his staff never resorted to overt anti-Semitic finger-pointing. They worked closely with the United Hebrew Charities and focused their attention as closely as possible on the *Massilia* passengers and those who may have come into contact with them. Still, it was hard to divorce the fear of immigrants from genuine concerns about protecting the public from the ravages of disease.

The actions of Edson and his staff, although excessive by modern medical and social standards, managed to slow the spread of typhus. While many were quarantined under less than ideal conditions, not only did the outbreak slow down considerably by late March, but the death rate among *Massilia* passengers was relatively low. Although more than half of the *Massilia*'s Jewish passengers had come down with typhus, only 13 of the 138 victims died from the disease and the rest recovered after receiving medical treatment. In contrast to the less than 10 percent mortality rate among *Massilia* passengers, the death rate among city residents who came down with typhus was 33 percent (27 deaths) and the mortality rate for nurses, helpers, and policemen with the disease was 38 percent (5 deaths).

The sometimes callous treatment extended beyond Jewish immigrants. An article in the *New York Times* described how a group from the city's health office, armed with "strong cigars," set off for the Bowery. Their goal was to vaccinate the single men—many of whom were native-born—who resided in Bowery flophouses. Although it is not entirely clear what kind of vaccinations these officials were administering, the article was clear that health officials sometimes forcibly entered private rooms and injected these men against their will. Many put up a fight and some managed to escape their pitiful dorms and elude the health inspectors.

In theory, Ellis Island was designed to prevent immigrants from starting such an epidemic in the first place. The 1891 immigration law specified that immigrants with "loathsome or contagious diseases" be excluded. Yet the *Massilia* immigrants were able to pass through Ellis Island with relative ease, thanks to the kindness of Colonel Weber and the work of the United Hebrew Charities. The *Massilia* incident gave immigration restrictionists an opening to push ahead with their agenda. For them, the 1891 law and the Ellis Island facility were merely opening bids in the continuing battle for the greater restriction of immigrants.

Only two months after the arrival of Annie Moore and just three weeks after typhus was first discovered on the Lower East Side, Congress began its first investigation of Ellis Island.

New Hampshire senator William Eaton Chandler led the joint House and Senate investigation. As chairman of the newly formed Senate Immigration Committee, Chandler was a leading voice for immigration restriction in the early 1890s. An old-stock New England Yankee, Chandler had a family tree loaded with Puritans who had settled New England. He was ten generations removed from the William Chandler who helped settle Roxbury, Massachusetts, in 1637, and his great-great-great-grandfather was one of the founders of Concord, New Hampshire. Despite his father's illegitimacy—he was the offspring of a married Nathan Chandler and his servant—Senator Chandler was a pure New England Yankee.

Chandler may have been descended from Puritan stock, but unlike Henry Cabot Lodge, Francis Walker, and other anti-immigrant Yankees, the New Hampshire Republican was not descended from wealth. His Yankee inheritance was one of values and pride, not dollars and cents. William's father had owned a stable and inn in New Hampshire, and although they were not poor, his sons were not raised as country gentlemen.

As easy as it may be to caricature immigration opponents like Chandler, the senator's political career shows a more complicated man. Though skeptical of the new immigration streaming into the country, Chandler was in many ways a progressive Republican of the time, with all of the contradictions that term implies.

He supported Reconstruction and black voting rights. He was suspicious of big corporations and supported legislation to regulate them. The frugal Yankee was angered by the power, wealth, and arrogance of the new class of capitalists, especially the railroad companies. In 1892, he sought unsuccessfully to incorporate a plank in the Republican platform stating that big business was not the master of the state, but would be obedient to the law. His constant battles against the railroads eventually caused him to lose his Senate seat in 1900.

Back in the spring of 1892, Chandler's mind was firmly fixed on the dangers of immigration. On the morning of March 5, with the fear of typhus still on the minds of New Yorkers, Chandler led his fifteen-man committee to Ellis Island. Its initial impression of the facility was less than enthusiastic. The congressmen found the buildings to be "very cheap-looking affairs, unsubstantial though comfortable. The wood put in them was evidently soft and poorly seasoned for in the partitions and elsewhere great seams had opened up, through some of which the hand could be placed."

Chandler's investigation, which would drag on for more than three months, was a two-pronged attack on Ellis Island, representing both political partisanship and the concerns of immigration restrictionists. At the time, the House and Senate were split between Republican and Democratic control. The year 1892 was a presidential election year and Democrats were eager to use the hearings to excoriate Treasury Department officials in the Republican administration of President Benjamin Harrison.

The Chandler hearings allowed Democrats to home in on what they argued was fraud and waste in the construction of Ellis Island. While Congress had approved $250,000 for the construction of Ellis Island, the project went way over budget and $362,000 more was needed to complete it. The committee concluded that "there has been great waste of public money in the construction of the improvements on Ellis Island" and that the buildings are "badly constructed, of inferior material, and poor workmanship."

Republican members of the committee dissented from that opinion, noting that everyone had agreed that the old system at Castle Garden was bad and needed to be fixed. They reminded their colleagues of what had been accomplished: Ellis Island was doubled in size, the land was raised, a dock was constructed, and a navigable channel and deep-water basin for ferries were constructed. Republicans admitted, however, that the facilities were "of a somewhat rough and shed-like character."

Chandler had not brought the committee all the way to New York just to examine accounting records and look at the quality of Georgia pine used in the buildings. He was concerned about the lax enforcement of the nation's new immigration laws. He believed that the 1891 law was hardly more than a reenactment of the 1882 law and barely made a dent in the problem. The *Massilia* incident gave him the opportunity to prove that point.

The hearings highlighted what would become a recurring theme throughout Ellis Island's history: the chasm between immigration law as written and immigration law as enforced. While denying "any willful negligence or dereliction of duty on the part of the commissioner of immigration at the Port of New York and his assistants in the execution of the laws relating to immigration," Chandler argued, "there have been many undesirable immigrants permitted to land, who, under a reasonable and proper construction of the laws not in force, should have been refused admission."

Weber came across as a soft touch during the hearings, a characterization he did not deny. "I appreciated the responsibilities and tried to act with the utmost fairness," Weber wrote in his autobiography. "If I leaned at all it was towards the humanitarian side." Faced daily with the mass of humanity streaming through Ellis Island, many of whom were strange and foreign even to the second-generation Weber, the commissioner worried that "the frequent scenes of misery would cause callousness on my part, but on the contrary I grew more sympathetic in my regard for these helpless and pathetic creatures with whom I came into daily contact."

Weber's tolerance bothered Chandler, who argued that too much power was being placed in the hands of the commissioner, rendering him "liable, not only to be swayed by too generous impulses from within, but also exposes him to influences by outside pressure, formidable and potent." In 1891, of the 472,000 immigrants entering the Port of New York, only 1,003 were excluded, roughly 0.2 percent. Ellis Island's new facilities, together with the 1891 Immigration Act, were supposed to lead to a stricter regulation of immigrants. Chandler was not convinced this was happening. "There is a feeling now that has gotten abroad that the Immigration Bureau is nothing more than a Census Bureau," Chandler argued. Ellis Island was supposed to do more than just provide a head count of newcomers—it was supposed to act as a sieve to filter out undesirable immigrants.

To Chandler and his supporters, the typhus outbreak highlighted a serious flaw in the inspection process. Although typhus was covered under the category of loathsome or contagious disease, carriers of the disease could pass through inspection undetected, not showing symptoms until days after their admittance. Presymptomatic typhus was not something Ellis Island doctors could spot with the naked eye. It was therefore difficult to criticize Weber and his men for failing to find typhus among the *Massilia* passengers.

Should not some of the *Massilia* passengers have been excluded for other reasons? Chandler asked. The committee discovered that the initial inspection of the *Massilia* immigrants led to the temporary detention of seventy passengers on suspicion that they were likely to become public charges. Most were allowed to enter the country under the orders of Weber.

Chandler got Weber to admit that Ellis Island inspectors marked on the cards of immigrants "P.O." (paid own passage) and "P.P." (prepaid). Many of the *Massilia* passengers were marked "P.P." since the Baron de Hirsch Fund had paid for their tickets. Here seemed proof of what restrictionists had long claimed: new immigrants differed from old immigrants in that they were assisted in coming to America and therefore were of lesser character and more likely to end up as wards of the state or private charity.

To Chandler, all "P.P." immigrants should have been given a hearing to prove they were eligible to land. Weber's explanation of his actions repeated his position on paupers as described in his report on European immigration published just weeks earlier. These Jewish immigrants "were possessed of sufficient means on starting [their journey] but it was necessarily expended for sustenance in their wanderings," Weber told the committee. "Their trials naturally made them an easy prey to disease." For Weber, these individuals were paupers by circumstance and not by character.

The intervention of the United Hebrew Charities helped sway Weber's decision. The charity signed guarantees or bonds for those immigrants thought to be paupers, promising that they would not become public charges. Weber admitted he had no idea whether the guarantees could be legally enforced. For Chandler, this was another example of sentiment trumping the "correct practice under the law."

In this, Chandler was surely correct. There was no feature for the bonding of suspect immigrants under immigration law. This brought up the issue of possible reverse discrimination. Why, one congressman wanted to know, were only Jewish immigrants allowed in with guarantees or bonds? Weber, sensitized and even traumatized by his trip to the Jewish Pale of Settlement during the previous summer, answered that the plight of Jews was different. "The assisted immigrant, however, who comes from Russia comes here assisted because of extraordinary circumstances that have reduced him practically to a state of destitution," Weber explained.

Chandler's committee concluded it was impossible to tell how many of the assisted Jewish immigrants on the *Massilia* had come down with typhus. However, the committee believed "that if its views of the present law had been enforced typhus fever would not have been introduced into the city of New York."

It is ironic that Jewish charitable agencies came in for such criticism for their role in bonding Jewish immigrants. The spread of typhus would certainly have been much worse without them. Agents of the United Hebrew Charities immediately alerted Edson's office about the typhus outbreak, and most immigrants were kept together in boardinghouses run by the charity, making it easier to contain the disease. In fact, Edson praised both the Baron de Hirsch Fund and the United Hebrew Charities for their assistance.

But this did little to ease the concerns of Chandler and his allies. They found more fuel for their crusade when the behavior of Weber's assistant, James O'Beirne, became public. Whenever Weber was away from Ellis Island, O'Beirne assumed the duties of acting commissioner. In one instance, O'Beirne came up with a requirement that all immigrants have $10 in their possession or they would be excluded as paupers. Using this formulation, he detained about three hundred immigrants, but Weber set them free upon his return.

Chandler believed this highlighted the arbitrary nature of the exclusion process. "What we want to get at is how much better or worse the immigrant fares when he is in Col. O'Beirne's hands than when he is in yours," Chandler told Weber. For Chandler, this incident, coupled with the *Massilia* incident and the activities of the Jewish charities, proved that whether or not an immigrant would be excluded "depends upon no statutory provisions, and is decided very much according to the personal feeling or judgment of the person who makes the decision."

Throughout much of his testimony, Weber was on the defensive. Had he chosen to be more combative, he could have countered that Congress provided no way to determine who was likely to become a public charge. It fell to immigration officials to decide such questions. The modern bureaucratic government had been created almost from scratch. Congress made the laws creating this administrative state, but it was up to the political appointees and bureaucrats of the separate departments to work out detailed instructions and interpretations of the law to operate their agencies. Congress could write new laws or express displeasure through public hearings, but the day-to-day operation of federal agencies was out of its hands.

Given the fact that immigration officials had such wide latitude, Chandler and his allies homed in on one issue they thought showed the inconstancies of immigration policy at Ellis Island. If an immigrant chose to appeal his case, it took the decision of four people—inspector, commissioner, superintendent of immigration in Washington, and secretary of the Treasury—to exclude him. However, it took the decision of only one individual—usually an inspector at Ellis Island—to allow an immigrant to land.

Another member of the committee, Rep. Herman Stump, who would later be put in charge of the Immigration Service in Washington, continued this line of argument. He believed that immigrants had all the protections of appeal, yet no one spoke for the masses of Americans who wanted tighter controls over immigration. "Only one overcrowded and overworked inspector guards the people against the admission of an illegal immigrant," Stump argued. "Why not then give the people, as it were, some rights on their side to say any person should not be landed while the officer now has the sole say as to whether it should be done or not?"

To that end, Chandler and Stump proposed a board of special inquiry made up of three or four inspectors who would sit in judgment of all immigrants not landed beyond a reasonable doubt. A majority vote would be needed to land any questionable immigrant. As to why this would lead to greater restrictions, Chandler noted that three men would have a harder heart than just one.

Then there was the question of what caused the outbreak of typhus. The general public, and even many doctors, were mystified by its origins. Was it something in the air? Was it caused by a poor diet? Germ theory was still not widely understood, and it was not until the early decades of the twentieth century that doctors discovered that typhus was transmitted by the common body louse. The overcrowded conditions in Constantinople's Jewish ghetto most likely led to the outbreak of typhus among *Massilia*'s passengers.

Typhus, the *New York Times* claimed, was "caused by filth, overcrowding, destitution, and neglect of the fundamental laws of sanitation. The epidemics of this fever in New York have been imported . . . from Europe in the crowded steerage quarters of steamships." The immigrants were at fault because their "habits and condition invite deadly infectious diseases." Therefore, the editorial concluded, the "dreaded disease must bring forcibly to the attention of all intelligent citizens the evils of unrestricted immigration. . . . The doors should be shut against them [diseased immigrants]."

Members of Chandler's committee continually linked the disease to immigrants and filth. Congressman Stump asked Dr. Edson: "Would a filthy person cause typhus fever—a dirty person or dirty surroundings?" Edson, while admitting that it would take "a large number of filthy persons," pointed to environmental factors as a leading cause. Stump asked Edson flat out whether "a filthy person in open air could never develop typhus." Edson replied "Never."

William Jenkins, the health officer at the Port of New York, echoed Stump's concerns. He admitted that while he could not pinpoint exactly how the disease developed, he believed that the "Hebrew passengers were a poorly nourished lot of people and from subsequent affidavits I saw very unclean." Jenkins also blamed kosher food preparation for making the problem worse.

The linkage of Jewish immigrants and filth was common at that time. In an 1888 congressional hearing, the director of the Jewish Immigration Protective Society of New York, was asked about the personal habits of Jews: Were they "nice, clean, tidy people or the reverse?" A doctor stationed in 1892 at Swinburne Island, off Staten Island, which served as a quarantine island for incoming ships, commented on the condition of immigrants he saw there: "They were mostly Russian Polish Jews and filthy beyond description, frequently covered with vermin. They seemed more like animals than human beings, and appeared to possess no desire for personal cleanliness."

Cyrus Edson, on the other hand, did not buy into these stereotypes. "There is no cause for alarm, much less panic, but there is abundant cause for careful, thorough, and scientific supervision and watchfulness," he reassured readers of the *North American Review*. While unrestricted immigration was a danger, Edson saw no need to shut off immigration of Russian Jews. Yet he was not completely sanguine about the situation, warning that America "should class all Russians and Russian goods as suspects and should treat them accordingly."

It was this kind of ambivalence that marked the final report of Senator Chandler's committee. It expressed disappointment not only that so much money had been spent at Ellis Island, but that the "expense certainly justified the expectation that the work of inspection done there should be more thoroughly and effectually conducted than that done at the other ports of entry." Less than a year into Ellis Island's career as an immigrant depot, Congress had declared it a disappointing failure.

Though the committee seemed driven by restrictionist concerns, the report did not recommend ending immigration completely or even passing stricter immigration laws. What it did call for was tighter administrative controls that would better sift the incoming immigrants, hopefully raising the exclusion rate higher than 0.2 percent.

Chandler tried to position himself in the ideological middle of the immigration debate. On one side, he placed Henry George, radical author and proponent of the Single Tax, who believed in no "restriction whatever upon the immigration of people from Europe of the Caucasian race, who are not diseased and who are not chronic paupers or criminals." On the other side, Chandler placed an anonymous citizen from New Jersey who thought immigration "should be stopped entirely and immediately; that it is dangerous to admit more till those that are here are fully Americanized, which it will take years to accomplish."

In between those extremes, Chandler rejected expanding the categories of excluded classes, but argued for tighter enforcement of current laws. He believed that no member of his committee, including himself, thought that "the time has arrived for the United States to exclude from coming to this country to become citizens, any individuals or families who would make good and valuable members of Society."

Both extremes of the debate, a political scientist noted in 1892, were "regarded by the majority of thinkers as unsatisfactory; and it is believed that the action called for is neither to bar out all nor to admit all, but rather to take a middle course and restrict immigration by some discriminating measure." *Harper's Weekly* took a similar tack, arguing that although "many of the immigrants who have come to us of late years are not of a desirable kind . . . we believe also that the evils complained of and the dangers apprehended as springing from the influx of such undesirable immigration are very much exaggerated." Americans would spend the next thirty years trying to find that middle point between complete restriction and a completely open door.

For Chandler, that middle way called for putting more responsibility on steamship companies to exclude unfit immigrants before sailing for America. His committee called for the creation of boards of special inquiry staffed by four inspectors to evaluate all questionable immigrant cases. The last major recommendation was the abolishment of the bonding system for immigrants thought liable to become public charges.

Not surprisingly, the *American Hebrew* attacked Chandler's proposal. If Chandler argued that too much power was in the hands of Ellis Island officials, the paper argued that doing away with bonding "would certainly place in the hands of prejudiced commissioners the power to exclude absolutely all immigrants." The editorial worried that "the Commissioner of Immigration would thus be invested with a degree of absolute and autocratic power, not possessed by any other official and never contemplated by the constitution to be in the possession of any administrative authority without either legal direction or judicial review." Although the *American Hebrew* editors exaggerated the extent of Chandler's proposal, they correctly noted that the federal bureaucracy was accruing more and more power, with possibly unfortunate results.

Despite the congressional investigation and the publicity that ensued, for the time being the recommendations of the Chandler committee went nowhere. Another crisis loomed later in 1892, however, that threatened to bring the restrictionist agenda back on the front burner.

Traveling from Turkey to Russia to Germany to France to England, a worldwide cholera outbreak was threatening to land on American shores. Some wanted President Harrison to order a temporary halt to immigration in hopes of stopping cholera from entering the country. But it was Congress, not the president, that held such power.

What was within Harrison's authority was to declare a strict quarantine of twenty days for all ships coming from Europe, although some complained that the president did not even possess this limited power. Still, Harrison ordered the quarantine on September 1 and local authorities carried out the plan.

If typhus fever scared New Yorkers, the possible scourge of cholera provoked a near panic across the country, especially in densely packed urban areas. A hideous disease that kills its victims by massive dehydration from diarrhea, cholera had devastated New York in the past. During the city's worst outbreak, over 5,000 people perished in 1849. In the last major outbreak, over 1,100 people died in 1866.

Now, in the fall of 1892, cholera victims were on steamships in the Atlantic heading for New York. Most of these ships were immediately put under quarantine. Within days, the quarantine hospitals at Swinburne and Hoffman Islands were filled to capacity with cholera victims and suspected victims. Other passengers would remain on their ships for the duration of the quarantine.

At the same time, cholera victims began appearing in the city. On September 6, Charles McAvoy, who had arrived from Ireland eight years earlier, died of the disease. By the end of the month, there were nine additional cases in the city, with seven more deaths. All the victims were immigrants, with the exception of the infant daughter of an immigrant couple. Yet strangely all, with one exception, had been in the country for over two years. None could be linked to recently arrived immigrants.

Still, the brunt of the disease was borne by immigrants coming over in steamships and trapped in quarantine. Forty-four individuals, many of them Russian Jews, died of cholera in New York's quarantine stations, in addition to the seventy-six who had died of the disease at sea. As the cholera scare in Europe abated, it appeared that the quarantine had spared the city the worst of the outbreak. Isolating European immigrants had seemingly served its purpose, but with a price. Those destined for quarantine received poor medical treatment in overcrowded conditions. Many of those quarantined were Jews who were prevented from following kosher food regulations. Worse still, contrary to Jewish burial practices, many of the Jewish victims were cremated.

The quarantine policy lasted until February 1893. As a result, immigration from Europe fell off dramatically in late 1892 and early 1893. Steamship companies could not afford to keep their ships tied up in quarantine for twenty days without a serious dent in profits. The federal government also felt the financial pinch since the Immigration Service operated through a 50-cent head tax on every immigrant, paid by the steamship company. With so few immigrants, the coffers dried up. In the face of cost-cutting measures, Colonel Weber offered to resign. Some worried about the effects of the quarantine on the upcoming World Exposition in Chicago in 1893, leading the *Times* to call the quarantine an "opera bouffe order . . . a delayed April-fool proceeding."

The cholera scare provided Chandler with an opportunity to put forth a more restrictive immigration bill, something that he chose not to do after the typhus scare earlier in 1892. He introduced a bill in the Senate in January 1893 that would suspend all immigration for one year, which would give Congress more time to draft a more permanent law restricting immigration.

Weber called Chandler's bill "a senseless panic among our people." *The Nation* called it "a medieval admission of inability to take scientific precautions against cholera." Others noted that even with a moratorium on immigration, disease could still be transmitted by arriving cabin passengers such as tourists and businessmen, as well as imported cargo arriving in the port.

Despite the fears brought about by the outbreaks of two deadly and contagious diseases linked to immigrants, the effect on national policy was quite the opposite of what might have been expected. Few members of Congress were ready to join Chandler in calling for even a temporary suspension of immigration. Public opinion, despite worries over immigration, was not willing to jettison America's traditional vision of immigration. Chandler's bill went nowhere.

Instead, Congress passed the National Quarantine Act in early 1893, which strengthened the federal government's role in immigration and created new rules governing public health. As a sop to immigration restrictionists, the law formally gave the president the power to suspend immigration in case of an epidemic, a power never used by Harrison or any subsequent president.

The nation did get an immigration bill in 1893. There was no moratorium on immigration, and attempts to expand exclusionary categories to include anarchists and those not able to read and write in their native language failed to survive the bill's final draft. The pattern of near-absolute exclusion by race as set in the Chinese Exclusion Act would not be followed for white European immigrants. "Exclusion all along the lines is not to be thought of excepting under the pressure of some extraordinary urgency," one journalist wrote. "It were better for the government to recede from its Chinese policy instead of venturing to extend it."

What did pass was a series of administrative reforms that tightened the regulation of existing laws to weed out undesirables, as Chandler had proposed in his final report after his 1892 hearings. The system of bonding was allowed to continue, but with greater scrutiny. Bonds could now only be granted under the authority of the superintendent of immigration in Washington.

The new law created boards of special inquiry at each inspection station to hear the cases of all immigrants not clearly entitled to land. Immigrants who appeared before these boards would need to gain a majority vote from the board before being allowed to land. In addition, any dissenting board member had the right to appeal the admissions decision to the commissioner.

The bill called for a more detailed ship's manifest to be provided by the steamship companies to immigration officials at the port of entry. When Annie Moore landed in January 1892, her manifest listed passenger names and the answers to eight basic questions. Now, besides having to answer the basics—name, age, sex, and occupation—an immigrant would have to answer a total of nineteen questions. Among them were:

- "By whom was passage paid?"
- "Ever in prison or almshouse or supported by charity?"
- "Whether a polygamist"
- "Whether under contract, express or implied to labor in the United States"

Steamship officials would also have to make note of the mental and physical condition of immigrants on the manifest and list any deficiencies.

The new manifests would allow steamship companies to better sift immigrants. The detailed questions allowed for a more thorough cross-examination of immigrants. When immigrants arrived at Ellis Island, inspectors would ask them the same questions as appeared on the ship's manifest, whose answers would be in front of the inspector. If the answers did not align with the information on the ship's manifest or if the inspector felt there was something wrong with the answers, the immigrant would then be sent to a board of special inquiry hearing. In less than ten months of operation under the new law, the boards heard the cases of 7,367 immigrants, of whom 1,653 were excluded.

These boards of special inquiry were not courts, however. They were administrative hearings within the executive branch, and therefore not bound by the traditional rules of the courtroom. Immigrants who appeared before these boards were not accorded the guarantees of the Bill of Rights. Hearings were not open to the public and immigrants were not allowed to have counsel present. While appeal was an option, immigrants were not eligible for bail as their cases made their way to Washington. Board hearings could rely on informal evidence, such as letters, telegrams, telephone conversations, newspaper clippings, and hearsay. Although boards did attempt to use affidavits and witnesses sworn under oath, critics would soon refer to these as "star chamber" proceedings.

A process of extended grilling of immigrants, coupled with the boards of special inquiry, meant that Ellis Island officials now had more tools with which to exclude immigrants. American officials had now succeeded in erecting an obstacle course for potential immigrants that stretched from the ports of Europe to New York Harbor.

Health concerns helped drive the fear of immigrants. Consequently, a great deal of work at Ellis Island fell to the medical staff of the Marine-Hospital Service. Although it was also part of the Treasury Department, the medical staff at Ellis Island and other inspection stations was not part of the Immigration Service. While a civil service posting at Ellis Island was not exactly a prized position and did not necessarily attract the best doctors in the country, the Marine-Hospital Service— renamed the Public Health Service in 1912—strove toward professionalism. The service was organized along military lines and its doctors wore military-style uniforms, which frightened many immigrants who were raised to fear the military in their homelands. To add to the culture clash at Ellis Island, many of the doctors were Southern-born. The medical staff at Ellis Island was always small, beginning with six in 1892 and increasing to twenty-five by 1915.

Though understaffed, doctors at Ellis Island were faced with over 170 different medical ailments. Many were relatively minor, from cuts to burns to sprained ankles to poison ivy to mysterious itches. Some were simply cosmetic, such as those detained because of acne or warts. Measles, chicken pox, and diphtheria were found among children. The extent of the maladies shows the thoroughness, as well as the intrusiveness, of the medical inspection. Gonorrhea and syphilis, as well as abscesses on the breast, ulcer of the vulva, and ovarian tumors were all spotted by Ellis Island doctors. In 1899, one poor sap was even ordered deported for masturbation. Doctors also marked for further examination and treatment those immigrants they deemed "idiots" or those believed to be insane or merely depressed.

It is no surprise that medical officials also saw their share of death. Between 1893 and 1899, 244 unfortunate souls died at Ellis Island and other medical facilities for immigrants in New York. At the same time, many of the young women who came to New York Harbor for a chance of life in America arrived pregnant. In 1897, seven babies were born at Ellis Island.

Doctors at Ellis Island had a dual role. They were supposed to treat illnesses and disease as best as they could; but they were also supposed to certify immigrants whose medical condition could be considered loathsome or contagious, resulting in their being excluded from entry. Between 1893 and 1899, a relatively slow period of immigration, the immigration service at the Port of New York treated almost 9,000 individuals at the rather primitive and cramped medical facilities there. During those years, medical officials certified over 1,200 immigrants for deportation, although immigration officials made the final determination of exclusion. In fact, doctors would not allow themselves to sit on boards of special inquiry.

The conditions that most concerned officials were favus, a mildly contagious fungal scalp condition, and what doctors classified at first as conjunctivitis and later as trachoma, a contagious disease of the eye. In 1902, Commissioner-General of Immigration Terence V. Powderly noted that "until the tide of immigration swelled up, and began to flow in on us from the countries of southern Europe and the Orient, these diseases were not very prevalent" in the United States. Powderly believed that authorities needed to exclude immigrants with these diseases because, if in the "future we should have occasion to trace the cause why our people are hairless and sightless through Favus and Trachoma, we should have ourselves to blame." A majority of those ordered deported because of disease suffered from those two ailments.

As time passed, inspection methods would improve and immigration officials would be given more tools with which to inspect, and possibly exclude, immigrants. Public health officials had a duty to cure and heal, but they were also part of the ever-expanding obstacle course through which immigrants who arrived at America's gate had to pass.

NEW LAWS OR NOT, Senator Chandler would no longer have to worry about Colonel John Weber. Although Chandler may have been saddened that his fellow Republican, President Benjamin Harrison, lost to Democrat Grover Cleveland in the 1892 election, it also meant that Weber would soon lose his job at Ellis Island. Weber would be replaced by Joseph Senner, an editor of the German-language newspaper *New Yorker Staats-Zeitung.*

The good news was that by mid-1893, the typhus epidemic had been contained and the cholera scare had passed. The bad news was that the national economy had now plunged into a nasty depression, the worst economic downturn until the Great Depression of the 1930s.

The epidemic scares of 1892 had cut in half the numbers of immigrants arriving in New York. Now the economic depression cut that low number even further, and the downward trend continued into the mid-1890s. As many potential immigrants stayed in Europe, growing numbers of foreigners already in the United States decided to pack up and return to their homelands. Newspapers ran headlines such as "More Going Back Than Coming Over" and "Many Leaving the Country." Senner credited this trend to stricter enforcement of immigration laws, not bad economic times. He approvingly pronounced that "heavy immigration has been made practically an impossibility for the future"—a declaration that would have amused Senner's successors at Ellis Island.

Although over a million fewer immigrants came to the United States in the 1890s than in the previous decade, the decline masked a deeper and more enduring trend. During the 1880s, almost 3.8 million immigrants from northern and western Europe entered the United States, compared to 956,000 from southern and eastern Europe. By the 1890s, despite the overall decrease in immigration, southern and east European immigrants outnumbered northern and west Europeans by 1.9 million to 1.6 million. By the first decade of the twentieth century, there were three eastern and southern European immigrants for every one from northern and western Europe.

The top three countries of origin for immigrants during the 1880s were Germany, the United Kingdom, and Ireland. By the 1900s, it was Italy, Russia, and Austria-Hungary. In 1884, 13 percent of immigrants came from Italy, Austria-Hungary, and Russia-Poland. In 1891, the figure was 39 percent, and in 1898, 60 percent of all immigrants to America hailed from those regions.

These changes were not lost on most Americans. In 1892, before the depression struck, the *New York Times* noted that "an increasing proportion of the total volume is of immigration evidently undesirable. Americans are pretty well agreed that the immigration from Italy is very largely, and from Russia and its dependencies almost altogether, of a kind which we are better without." Just a few months later, the paper repeated its claims about these new immigrants:

> The New Yorker who goes to the Barge Office these days gets
> a good idea of the class of people now seeking homes in the
> United States. It needs only a glance to assure him that it
> is a most undesirable class. Ignorance and dirt are the chief
> characteristics of the average immigrant of to-day . . . it is
> plain that the United States would be better off if ignorant
> Russian Jews and Hungarians were denied a refuge here.

Henry Cabot Lodge captured these feelings when he bemoaned the fact that "the immigration of those races which had thus far built up the United States, and which are related to each other either by blood or language or both, was declining, while the immigration of races totally alien to them was increasing."

Such feelings extended to top officials at Ellis Island. Even though Colonel Weber was sympathetic to new immigrants, his former assistant was not. After he left his post, General James O'Beirne made a public plea for all Americans to make a pilgrimage to Ellis Island. Once there, the average American would gain "a full appreciation of the present and near impending dangers which seem to me to threaten the future stability of the Republic arising from immigration."

While the *Massilia* incident had raised fears about Russian Jews, concerns were soon raised about another new immigrant group also on board that same ship. If filth and disease were the negative traits associated with Jewish immigrants, criminality and violence were the supposed dysfunctional traits of Italian immigrants.

In May 1893, the *Times* discussed the exclusion of nine Italian immigrants at Ellis Island who admitted to having been in jail in Italy for such minor crimes as quarreling with a relative, throwing stones at a woman, and carrying a concealed weapon. The article described Italy as "the land of the vendetta, the mafia, and the bandit" and southern Italians as "bravos and cutthroats" who seek "to carry on their feuds and bloody quarrels in the United States."

If the crimes seemed minor, Ellis Island clerk Arthur Erdofy hoped to disabuse his readers of such thoughts. "Fighting . . . means something more among people of this class than what would be understood by the word here," explained Erdofy. "It means a pistol or knife being brought into play. All this 'beating a woman without injuring her' and 'hitting a man with a stick without hurting him' . . . is bosh, pure bosh. . . . We read between the lines and substitute knifing or shooting for quarreling."

As deportations increased, Italians struck back at authorities, inflaming public opinion and reinforcing negative stereotypes. Thomas Flynn, an official at Ellis Island and the son of a Democratic city alderman, was attacked in the doorway of his lower Manhattan home one night. He was struck in the head by a large rock, allegedly from the hand of an Italian immigrant. It is unclear how the newspaper knew the attackers were "revengeful Italians." Perhaps it was because, as the *Times* dutifully reported, the men had left behind a bag of beans and macaroni. Some time earlier, another official was attacked in Battery Park by a group of Italians who threw a stone at him, missing his head but knocking the cigar he had been smoking out of his mouth.

The anger of Italians soon spilled over at Ellis Island. More focus on undesirable Italian immigrants meant more detainees awaiting deportation and increased congestion. One night in April 1896, eight hundred immigrants were detained. That spring, the exclusion rate among immigrants had been 8 to 10 percent, much higher than usual. Four hundred Italians had been sent back in one week. These pressures were too much for both Ellis Island and the Italian immigrants, who one afternoon staged a mini-riot while cooped up in an outside temporary detention pen. A *New York Tribune* reporter described the detainees as a "forlorn-looking lot . . . restless, depressed, degraded and penniless."

The fear of Italian immigrants was not just confined to New York or Ellis Island. The *Boston Globe* asked seven prominent individuals: "Are Italians a Menace? Are They Desirable or Dangerous Additions to our Population?" The shortest response came from a representative of the Italian consulate, who complained: "I cannot answer a question of the kind that you put because I cannot accept the implication which it involves that my countrymen compare unfavorably with any other class or race." Guiseppe De Marco, editor of Boston's Italian-language newspaper, had a similar reaction: "It is quite a hard thing for an Italian who loves his country to discuss such a question."

Despite the impolite tone of the question, most of the responses were positive, if somewhat condescending. Unitarian minister Christopher Eliot compared Italian immigrants to "untrained children," yet argued that Americans "have less to fear than most people think," as long as native-born Americans helped assimilate, train, and "protect them from their own ignorance and inexperience."

The head of Boston's Central Labor Union, John F. O'Sullivan, argued against any "further attempt at restriction of immigration of any kind, unless it be the restriction of laborers under contract, criminals (other than political) and paupers or those likely to become public charges." Naturally enough, O'Sullivan thought that all Italians needed was to learn to support labor unions.

Yet two responders were decidedly less sympathetic. Prescott Hall and G. Loring Briggs used the forum to push for a literacy bill for all immigrants. Both men were affiliated with a new organization based in Boston dedicated to stemming the tide of immigration. Each took pains to deny any prejudice toward Italians specifically. Briggs wrote that "anyone who states that Italian immigration is necessarily a menace to this country simply because it is Italian is governed by narrow-minded prejudice, which is certainly unbecoming to an American." However, both Briggs and Hall noted that a majority of Italian immigrants were illiterate and therefore unfit for American citizenship.

Arguments over the suitability of Jewish and Italian immigrants continued throughout the 1890s. The party affiliations of Ellis Island's workforce may have changed, but the debate over immigration continued, as would the eternal, yet elusive, desire for that proper sieve that would neatly sort out immigrants—good from bad, desirable from undesirable, wheat from chaff.

In this debate, Bostonians like Prescott Hall would continue to lobby for stricter regulation of immigrants. For them, immigration was personal.

Chapter 5

Brahmins

———•·•———

*Let us welcome all immigrants who are sound mentally and
physically and intelligent, and let us protect the country from
those who tend to lower the average of health and intelligence.*
—Prescott Hall, 1907

*The Puritan is passed; the Anglo-Saxon is a joke; a newer and
better America is here.*
—James Michael Curley, 1916

BOSTON—THE "HUB OF THE UNIVERSE," THE "ATHENS
of America"—was America's most important city up to the midnine-
teenth century. At least it appeared that way to most Bostonians. This
was John Winthrop's City on a Hill that became the cradle of the Revolution and incubator of American democracy. By the 1800s, the Puritan
drive for perfection had morphed into the crusade for more temporal
reforms: William Lloyd Garrison's abolitionism, Dorothea Dix's work
with the mentally ill, and Julia Ward Howe's work with the blind.

Boston had helped create and nourish America's first truly home-
grown literature and culture, with Hawthorne, Emerson, Longfellow,
Thoreau, and Whittier. Its historians—Parkman, Adams, and Ban-
croft—wrote the first drafts of American history. Its magazines—*The
Atlantic* and the *North American Review*—shaped the nation's elite
opinion. And then there was Harvard University across the river in
Cambridge.

Boston had long stood at the apex of Anglo-American culture.

Yet by the end of the nineteenth century, that culture's foundations seemed on shaky ground. The 1880 Census showed that 63 percent of Bostonians were either immigrants or the children of immigrants. By 1877, Catholics accounted for more than three-quarters of all births in New England. Irish Catholics had already taken over the city's police and fire departments. Catholic parents increasingly abandoned the public schools for parochial schools. In 1884, Bostonians elected Hugh O'Brien as the city's first Irish Catholic mayor, and by 1890 Irish politicians had taken office in sixty-eight Massachusetts towns and cities.

It is no surprise that much of the agitation for immigration restriction should find its origin in New England. Francis A. Walker, Henry Cabot Lodge, and William Chandler all hailed from Yankee stock. When discussing the "masses of peasantry" from Italy, Hungary, Austria, and Russia in the 1890s, Walker expressed the combination of dismay, disdain, and deep pessimism that characterized New England's Anglo-Saxon mind.

> These people have no history behind them which is
> of a nature to give encouragement. They have none of
> the inherited instincts and tendencies which made it
> comparatively easy to deal with the immigration of olden
> time. They are beaten men from beaten races; representing
> the worst failures in the struggle for existence. Centuries are
> against them, as centuries were on the side of those who
> formerly came to us.

Perhaps the best expression of the insecure New England mind-set was Thomas Bailey Aldrich's 1895 poem "The Unguarded Gates." A native of New Hampshire and former editor of *The Atlantic*, Aldrich was more William Chandler than Henry Cabot Lodge, though he stood second to no one in his defense of the Boston Brahmin tradition. Aldrich described his poem as "misanthropic."

> *Wide open and unguarded stand our gates,*
> *And through them presses a wild motley throng . . .*
> *Flying the Old World's poverty and scorn;*
> *These bringing with them unknown gods and rites,*
> *Those, tiger passions, here to stretch their claws.*
> *In street and alley what strange tongues are loud,*

Accents of menace alien to our air,
Voices that once the Tower of Babel knew!

Aldrich ends his poem with the popular historical allusion to the invasion of the barbarians and the fall of Rome.

Not all of the voices coming out of Boston opposed immigration. In 1896, Congressman John F. Fitzgerald gave a rousing, hour-long July Fourth address at historic Faneuil Hall. Mixed in with traditional patriotic sentiments, the thirty-three-year-old, second-generation Irish-American defended "the down-trodden and oppressed of every land," who come to America "to mould in their own fashion the way to fortune and to favor in this, the land of their adoption." For Fitzgerald, the nation's strength and economic power was intimately tied to immigrants, and he spoke up for the foreigner and against any new restrictions on immigration, including the literacy test.

As the Fitzgeralds of Boston and other Irish Catholics rose to prominence, the Brahmins could see their power and influence waning. Boston had long ago ceded its dominance in trade to New York, with the hub of culture and communications to follow. "As Brahmins ceased to be the undisputed arbiters of the public good," wrote one historian, "they became less confident of the Americanization of the newcomers." Henry Adams, grandson and great-grandson of presidents, advised his brother to start writing their epitaphs, for the more he witnessed "the formation of the new society, I am more and more impressed with my own helplessness to deal with it." The intellectual arguments of the Brahmins carried progressively less weight, especially regarding immigration. Francis Walker noted:

> For myself, strongly as I feel the evils of the existing situation [immigration], I have little hope of their early correction by law. On one or two occasions, when I have been called to speak in public upon this theme, I have seen how much more taking is the appeal to sentiment than the address to reason, in this matter; how great is the controversial advantage of him who speaks in favor of the complete freedom of entrance which has characterized our career thus far; how strong is the instinctive dislike of an American audience for any schemes of restriction or exclusion, in the face of the clearest considerations of expediency and even of national safety.

On this issue, Walker, like Adams, seemed to be signaling defeat. Yet a new generation of Bostonians chose not to give up the fight.

At just twenty-five years old, Prescott Farnsworth Hall formed the Immigration Restriction League (IRL) in Boston in 1894 with his friends Charles Warren and Robert DeC. Ward. All three were members of Harvard's class of 1889 and possessed impeccable Brahmin credentials. Warren was descended from a famous colonial Boston family. Ward was a Brahmin Saltonstall on his mother's side; his father was a wealthy Boston merchant. Hall's father was also a wealthy merchant. Warren and Hall were lawyers and Ward was beginning his career as a professor of climatology at Harvard.

Both pride and insecurity fueled the three young Bostonians— a prideful defense of Anglo-Saxon traditions mixed with insecurity brought about by the Brahmins' increasing loss of influence. Members of the IRL were driven by a fear that American democracy, founded by Anglo-Saxon settlers using Anglo-Saxon law and government, could perish under the avalanche of exotic immigrants.

Prescott Hall, who would become one of the most passionate and active keepers of the Anglo-Saxon flame, articulated this fear best when he asked: "Is there, indeed, a danger that the race which has made our country great will pass away, and that the ideals and institutions which it has cherished will also pass?" To Hall, the warning signs were ominous. Decades of Irish and German immigration had produced vast alterations in the nation's fabric, such that in "many places the Continental Sunday, with its games and sports, its theatrical and musical performances, and its open bars, is taking the place of the Puritan Sabbath."

The IRL raised specific questions about American society and democracy. Was America great because of the hard work of successive waves of immigrants coming to the nation's shores looking for opportunity? Or, as Hall and his colleagues were suggesting, was its greatness a by-product of its Anglo-Saxon settlers?

Hall, who would be the driving force behind the IRL for over twenty-five years, looked more like an earnest country parson than a fire-breathing activist. Physically unprepossessing, Hall had trouble filling out his topcoat. His mild appearance and soft, elongated features were matched by a sentimental personality. Hall's wife noted that her husband possessed a "loving and a lovable nature. He hated moral prigs with a cordial hatred."

According to one description, Hall was a "gaunt, sunken-eyed figure" who suffered from insomnia and ill health for most of his life. His mother was forty-five years old when Prescott was born, and an invalid for most of her life. She raised her son in a protective cocoon. Hall's wife later described how her husband, as a child, "grew up a frail little hothouse plant, for he was never allowed to romp, to climb, and to be reckless as other boys were." One historian described Hall as "an unstable New Englander, contemplative, subject to depressions."

The deep depressions from which Hall suffered were not unusual for his era and social class. In fin de siècle America, well before the age of Prozac, doctors diagnosed an epidemic of what was then called neurasthenia. Many contemporary social critics and physicians noted a general "lowering of the mental nerve" among the northern urban middle class, who seemed increasingly plagued by self-doubt, paralysis of will, insomnia, and other neuroses.

Insecurity and melancholy went hand in hand with these New Englanders' fears of being displaced, in terms of absolute numbers as well as political power and cultural influence. By the late 1800s, Boston Brahmin society was in decline. An increase in divorces and suicides and a decrease in birth rates among native-born Protestants—especially when compared with large Irish Catholic families—only added to the sense of loss and pessimism. The new immigration from eastern and southern Europe provided the double whammy to the Brahmin psyche, reinforcing whatever gloom and insecurity was caused by their loss of control to the Irish.

Francis Walker provided the intellectual explanation for this phenomenon, blaming immigrants and the supposed degrading conditions they brought to America for the declining Protestant birth rates. Prescott Hall picked up the idea as just one rationale for immigration restriction. (Hall and his wife were childless.) At the dawn of the twentieth century, old-stock Americans saw grave national consequences in the declining birth rates among native-born white women and a seeming softening of the dwindling Anglo-Saxon stock, as exhibited by a prevalence of neurasthenics.

In response, the boisterous governor of New York in 1899 advocated what he termed "the strenuous life." Theodore Roosevelt was from an old Dutch New York family on his father's side, but he had a message for the Boston Brahmins and other native-born Americans.

"If we stand idly by, if we seek merely swollen, slothful ease and ig-noble peace, if we shrink from the hard contests where men must win at hazard of their lives and at the risk of all they hold dear, then the bolder and stronger peoples will pass us by, and will win for themselves the domination of the world," he warned. The words were as pertinent to a nation beginning to enlarge its role in the world as it was a warn-ing to Anglo-Saxons who risked being overtaken by more vigorous im-migrant groups. "New England of the future will belong, and ought to belong, to the descendants of the immigrants of yesterday and today," Roosevelt would predict in 1914, "because the descendants of the Pu-ritans have lacked the courage to live."

Despite his problems, Prescott Hall embodied a different form of the strenuous life. Through ill health and melancholy, Hall fought with his pen, badgering public officials and newspapermen, ever relentless in seeking to restrict immigration from undesirable groups. Rather than completely retreating into gloom or going into exile, Hall remained to fight his imperfect fight. As the years went by, Hall found history steadily drifting away from him. He grew increasingly bitter as his ideas lost whatever sliver of youthful sympathy they once had for the Ameri-can ideal of immigration.

Yet Hall's lifelong battle against immigration exhibited a simple irony. The IRL stood as defenders of Anglo-Saxon values, of which democracy was at the forefront, yet its members chose to eschew dem-ocratic politics and organizing. The people could not be trusted. As Walker believed, they were too easily swayed by sentiment to face up to the tough task of limiting immigration.

In fact, Hall embodied the last gasp of the old New England Federal-ist tradition. He opposed abstract universals in favor of what he termed "Nordic concreteness." To Hall, America's founding fathers used the universalist ideals of the Declaration of Independence to institute a type of aristocracy. By the early twentieth century, Hall saw that ideal in tatters. To remedy that, he argued for limiting voting rights to those Americans who paid a certain level of taxes, possessed a certain level of education, or owned a business of a certain size.

So it was no surprise that instead of working through politi-cal means, the IRL opted for an elite approach. A precursor to the modern think tank, the IRL focused on social science research, which it published in pamphlets and distributed to journalists, politicians, businessmen, and other community leaders. Between 1894 and 1897, in the wake of the typhus and cholera scares and the continuing debates regarding Ellis Island, the league printed some 140,000 copies of its pamphlets, with titles such as "Immigration: Its Effects upon the United States, Reasons for Further Restriction." The IRL bragged that over five hundred newspapers nationwide were receiving its pamphlets and some were even reprinting part or all of these reports as editorials.

Yet the organization would never approach a mass movement. After two years in existence, its membership totaled only 670 and IRL meetings rarely consisted of more than twelve members. No doubt embarrassed by so few members, Hall tried to fudge the issue in his testimony before a federal commission in 1899. He claimed that five thousand individuals who were not members received the League's materials and "for all practical purposes might be considered members," even if they did not pay dues.

The IRL's strength was not the size of its membership, which was perhaps too plebeian a yardstick, but rather its quality. The membership of the IRL consisted of a who's who of Boston Brahmins. As the years went by, prominent national figures added their names to its roster, including novelist (and close friend of Theodore Roosevelt) Owen Wister and publisher Henry Holt. Academia also added intellectual sheen to the group, notably in the form of Harvard president A. Lawrence Lowell, the presidents of Bowdoin College, Georgia School of Technology, and Stanford University, and University of Wisconsin professors John R. Commons and Edward A. Ross.

The IRL worked closely with Henry Cabot Lodge, who had moved over to the U.S. Senate by 1893 and would soon take over as chair of its immigration committee from Senator William Chandler. The IRL would provide specialized knowledge to opinion makers and lawmakers, giving a patina of intellectual respectability to the drive to limit immigration. This would lead *Lippincott's Monthly Magazine* to tell its readers that it was not "professional alarmists who are taking up the vital question of immigration and call for a halt; it is students of social science . . . who toll a warning bell."

As with Senator Chandler—who was not associated with the IRL—it might be easy to pigeonhole the Immigration Restriction League as a nativist organization and leave it at that. However, anti-immigrant feelings easily coexisted with more liberal ideals. Many of the Boston families associated with the IRL had been staunch abolitionists two generations earlier. At the end of the century, with the nation embroiled in a guerrilla war in the Philippines, many of those associated with the IRL, such as Ward, Joseph Lee, and Robert Treat Paine Jr., became vocal opponents of American imperialism.

The founding of the Immigration Restriction League was part of a national wave of reform during the 1890s, with organizations forming to push for temperance, ban prostitution, and protect the environment and consumers. Immigration regulation, rather than an aberration, was part of a national movement that turned its back on the laissez-faire philosophy of government and sought to transform American society and control the social changes roiling the country in the late 1800s. Two prominent patrician members of the IRL were better known for their support of other Progressive reforms. Joseph Lee earned his fame as the "father of America's playgrounds," while Robert Woods was a leader in Boston's settlement house movement.

The descendants—and beneficiaries—of Boston's merchant elite were now turning their collective backs on capitalism. The young founders of the IRL, according to one historian, were now "contemptuous of industrial profiteering." Francis Walker, the economist and son of a wealthy manufacturer, led the way in criticizing the excesses of big business. Immigration restrictionists carped at steamship companies and railroads, which made money off the immigration trade. One anti-immigrant writer could have been speaking for the Boston patricians when she asked: "Why should the American people suffer in this way through the selfish and unpatriotic greed of the steamship companies who are in league with the immigrants?"

The IRL's constitution laid out its main objectives: "to advocate and work for the further judicious restriction or stricter regulation of immigration. . . . It is not an object of this League to advocate the exclusion of laborers or other immigrants of such character and standards as fit them to become citizens." Its early advocacy was pointedly free of ethnic prejudice, as Ward wrote that the League did not believe that immigrants should be excluded "on the ground of race, religion, or creed." Yet they were unhappy with the current immigration laws. Even with the opening of Ellis Island and the expansion of excludable categories, the IRL thought the quality of immigrants was deteriorating. It demanded radical changes in the nation's immigration laws. However, the organization stayed away from calling for an end to immigration or from singling out any specific ethnicity or nationality for exclusion.

Among its proposals, the IRL lobbied for increasing the head tax from $1.00 per immigrant to at least $10, and possibly as high as $50; a consular certificate for each immigrant, acknowledging his or her character and desirability; and a mandate that every immigrant had to read and write in his or her own language. However, the IRL thought an education test in English would be unfair.

For a young man not yet thirty, holding no political office, and with no past accomplishments beyond his Harvard degree, Prescott Hall managed to receive a good deal of attention and deference from newspapers and government officials. Just a few months after the founding of the IRL, he received a written assurance from the superintendent of immigration, Herman Stump, that he was "determined to restrict immigration to the most desirable classes. You will observe this by the great number of those now arriving who are detained for special examination."

Like so many other Americans with an interest in immigration, Prescott Hall and the rest of the Immigration Restriction League saw Ellis Island as the focus of debate. The young reformers were allowed to visit Ellis Island on at least three occasions in 1895 and 1896, where they were given near carte blanche to conduct their own unofficial investigations. In April 1895, Hall visited Ellis Island and deemed its operation greatly improved over previous years, although he still saw too many illiterate, unskilled workers, especially Italians, during his visit. "As nearly as I could judge in the case of the Italians whom I saw at Ellis Island," Hall told the *Boston Herald*, "there was in general a close connection between illiteracy and a general undesirability."

In mid-December 1895, Charles Warren and Robert Treat Paine Jr. visited Ellis Island, bringing pamphlets in English and other languages. Once there, the young Bostonians were granted remarkable access as they handed out their pamphlets to immigrants who had already told officials they could read. According to Warren and Paine, 9 to 10 percent of those claiming to be literate were lying. Over three days, the two men examined immigrants from six separate ships, most hailing from the Austro-Hungarian Empire and Russia. All the Germans and Bohemians they interrogated could read and write. However, 48 percent of Russians, 37 percent of Hungarians, 62 percent of Galicians, and 45 percent of Croatians could not read. Despite their unusual access, the IRL deemed the investigation a failure since during the visit not a single Italian immigrant passed through Ellis Island, and no investigation would be complete without assessing the vast throng of illiterate and unskilled Italians pouring into the country.

So in April 1896, IRL members visited Ellis Island again immediately after the mini-riot of Italian immigrants. This time, Prescott Hall, Robert DeC. Ward, and George Loring Briggs came at the invitation of Commissioner Senner. The three men spent several days on the island. It must have been a sight to see the young Brahmins, holding their pamphlets as tightly as their prejudices, set out among the dazed and dirty crowd of immigrants. We have no record of how the immigrants felt about the well-dressed young men thrusting pamphlets in front of them, but we do know what Hall thought about the immigrants.

In its April 1896 investigation, the IRL committee examined 3,174 Italian immigrants at Ellis Island and found that 68 percent of them were illiterate. Yet, much to the dismay of the IRL team, only 197 of these Italians were excluded from entry. In just a few days, officials at Ellis Island had let in almost two thousand illiterate Italians. Overall, they found that only 4.5 percent of the immigrants from northwestern Europe coming through Ellis Island during their visit were illiterate, while nearly 48 percent of those from southern and eastern Europe could not read. To the Boston Brahmins, this alarming trend threatened America's moral and intellectual fabric.

The IRL members came away from their visit feeling surprisingly positive toward the enforcement of the current law—perhaps due to the great deference granted to them by Ellis Island officials. Yet they called current immigration laws "radically defective" in keeping out undesirables. A miniscule number of immigrants were actually debarred or deported. In 1892 and 1893, the first years of the new immigration law, the number was around 0.5 percent. With the institution of boards of special inquiry under the 1893 immigration law, the percentage doubled between 1894 and 1895 to around 1 percent. To rectify this situation, the IRL continued to press for a literacy test.

Such a test would separate desirable from undesirable immigrants, keep the nation true to its history of welcoming immigrants, and make exclusion based on individual characteristics, not race, religion, or nationality. Although a literacy test was theoretically race- and ethnicity-neutral, restrictionists rightly believed it would have a disparate effect on immigrants. Henry Cabot Lodge used the work of the IRL to push for the literacy test in the Senate and was quite explicit that the test would mostly affect eastern and southern Europeans.

Not all restrictionists were enamored with the literacy test. Although he died in January 1897 before Congress took up the idea, Francis A. Walker, for once applying some of his economist's skepticism, had earlier noted that the "anarchist, the criminal, the habitual drunkard would be far more likely to pass the ordeal of a reading and writing test than the pocket-book test."

Eventually, both the House and Senate passed a literacy test in early 1897. The test would consist of roughly twenty-five words from the U.S. Constitution translated in the immigrant's native language. However, both Joseph Senner and Herman Stump urged President Grover Cleveland to veto the idea. Back in 1893, Stump had been an ally of Senator Chandler in the investigation of Ellis Island and was highly critical of the new immigrants. Now, after four years at the Immigration Bureau, he modified his views.

Writing to the secretary of the Treasury, Stump agreed that the public demanded greater immigration restriction. However, he argued that any such laws "should be tempered with sympathy for our unfortunate fellow beings who are compelled by adversity to abandon their homes to seek an asylum in an unknown country." Making a familiar argument, Stump said that America needed unskilled labor to "construct railroads, macadamize our highways, build sewers, clear lands," thereby freeing up native-born Americans from jobs they found distasteful and allowing them "to engage in the higher and more remunerative trades and occupations."

Such arguments helped sway Cleveland who, in one of his final acts as president, vetoed the literacy bill. Congress was unable to override it. Cleveland's veto message was a defense of traditional views of immigration. "It is said, however, that the quality of recent immigration is undesirable," Cleveland stated. "The time is quite within recent memory when the same thing was said of immigrants who, with their descendants, are now numbered among our best citizens." Cleveland would rather "admit a hundred thousand immigrants who, though unable to read and write, seek among us only a home and opportunity to work, than to admit one of those unruly agitators and enemies of governmental control, who can not only read and write, but delights in arousing by inflammatory speech the illiterate and peacefully inclined to discontent and tumult."

For years, immigration restrictionists would speak of the perfidy of Cleveland's veto, never forgetting how close they had come to saving the Republic from tens of thousands of illiterate undesirables. Yet the IRL continued the fight for a literacy bill, single-mindedly working on the issue for twenty more years, publishing pamphlets and lobbying newspaper editors and elected officials.

While these Bostonians attempted to exert power at the elite level of the debate, another New Englander with very different ideas was perhaps the most powerful person in charge of immigration policy at Ellis Island in the late 1890s. For years, Edward McSweeney did his work quietly until scandal forced his name onto the pages of newspapers, again raising doubts about Ellis Island's ability to enforce the nation's immigration laws.

Chapter 6

Feud

I have not been understood by many.
—Terence V. Powderly

*McSweeney . . . is governed by his motives, resentment, and
inordinate desire for distinction. . . . [He] is now surrounded
by a lot of servile, obsequious flatterers.*
—Roman Dobler, Ellis Island Inspector, 1900

JUST AFTER MIDNIGHT ON JUNE 15, 1897, A FIRE BROKE
out in the northeast tower of the main building on Ellis Island. The
fire's location made it hard to reach with water hoses. The building,
constructed largely of Georgia pine, burned quickly; within a half
hour the roof had collapsed. Immigration records dating back to
the Castle Garden era, which had been held in half-buried stone and
concrete magazines from the island's former days as an ammunitions
depot, were completely burned. The fire quickly spread to other buildings on the island, with flames that lit the night sky. An official investigation into the cause of the fire would later fail to solve the mystery;
however, Victor Safford, a doctor at Ellis Island, thought that it was
set deliberately, probably by a disgruntled night watchman who should
long before have been declared insane.

Whatever its cause, the fire drove out nearly two hundred immigrants who were detained on the island. Most of them were Italians,
but the group also included several Hindus in colorful robes and bead
hats who had arrived as part of a traveling exhibition. In addition,

thirty-one workers, including guards, an apothecary, a cook, two doctors, and three nurses were stationed on the premises.

To some, it was a reminder of the flimsy nature of the original buildings. *Harper's Weekly* said the buildings had been "monuments of ugliness" and "wretched barns and architectural rubbish heaps." The *New York World* blamed the government for constructing "great piles of rosin-soaked lumber, admirably arranged for burning." Commissioner Joseph Senner condemned the buildings as firetraps and said that he had been haunted by the fear of fire for years. "Every day as I left the island during the past four years," Senner told the *Times*, "I gave a farewell look at the buildings, for I expected to return the next day and find them all in ashes." His prophecy finally came true.

Thankfully, no one was hurt in the fire, but officials had a bigger problem. An estimated seven thousand immigrants were already on ships in the Atlantic headed for New York, with over six hundred scheduled to arrive the day after the fire. As the ruins on Ellis Island continued to smolder, immigration officials set up a temporary inspection center on the piers at the Battery, on the tip of Manhattan. That first day after the fire, fifty-five immigrants were detained for further inspection.

Officials then moved into the old Barge Office in the southeastern section of the Battery, which had served as a temporary facility after Castle Garden's closing. The fanciful Venetian Renaissance gray stone building, with its tall, thin turret overlooking the harbor, would again serve as the nation's primary immigration depot for two and a half more years. The immigration service chartered the steamboat *Narragansett*, now docked at the island, to serve as a temporary, floating dormitory for as many as eight hundred immigrants who had not yet passed inspection at the Barge Office.

As talk began about rebuilding the facilities on Ellis Island, it became clear that the new facilities would have to be built of stone and steel, not wood. Still, one upstate newspaper in all apparent seriousness suggested that new wooden buildings would not be such a bad idea. An occasional fire on the island, the paper's editor reassured its readers, would kill off the germs and microbes carried over by immigrants.

The chaos that ensued from the fire and the resulting move into the Barge Office left the New York immigration service in disarray. A newly elected president—William McKinley—began replacing Democratic officeholders in the immigration service with Republicans. A month after the fire, McKinley nominated Thomas Fitchie to replace Senner as commissioner. Fitchie had been a loyal Brooklyn Republican officeholder, but at age sixty-two and with no prior experience with immigration, he could hardly have been counted on to be a vigorous leader in difficult times.

America was digging itself out of the deepest economic depression in its history. As a new century approached, immigration would again pick up. The business of regulating this influx would have to continue for the time being without Ellis Island. To make matters worse, over the next four years the New York immigration service would become mired in a swamp of bureaucratic pettiness and personal vendettas that showed the limits of patronage politics.

THIRTY-YEAR-OLD EDWARD F. MCSWEENEY, the second in command at Ellis Island, was a bulldog of a man, whose bullet-shaped head was topped by thinning black hair. Victor Safford remembered his lifelong friend as a "live wire."

Growing up in Marlborough, Massachusetts, about thirty miles west of Boston, McSweeney dropped out of school as a child and began working in a shoe factory. Though his early biography had the makings of a Dickensian novel of drudgery and exploitation, McSweeney was more Horatio Alger than Oliver Twist.

By the time he was nineteen, he had helped found the Lasters' Protective Union; two years later he became the union's president. Labor work led to political work, as McSweeney became active in the Massachusetts Democratic Party. As a reward for helping round up labor support for Grover Cleveland's successful presidential campaign, McSweeney was named assistant commissioner at Ellis Island in 1893.

Befitting someone from humble beginnings who clawed his way up through the industrial and political jungles of late-nineteenth-century America, McSweeney had an air of physicality about him. Referring to a Protestant missionary who spent much of his time proselytizing to Catholics at Ellis Island, McSweeney told Archbishop Michael Corrigan of New York that if "any good would come of it, I would be delighted to call him to account with a round turn." When an immigrant tried to bribe him with $5.00, McSweeney became indignant and

smacked the man in the face. Along with such eruptions, he also displayed widely recognized administrative skills and shrewd intelligence.
When McKinley became president, McSweeney, a partisan Democrat,
not only retained his position but also managed to become the de facto
boss at Ellis Island. Above all, McSweeney was a survivor.

McSweeney remained in a Republican administration thanks largely
to new civil service regulations. Patronage was the lifeblood of politics
and helped staff the small federal bureaucracy, but it also led to corruption and a tolerance for ineptitude. To deal with the problems of
an increasingly complex society, a more professional federal workforce
was needed. In 1896, President Cleveland placed Immigration Service
workers under civil service protection. Current federal workers were
not forced to take the civil service exam and were able to keep their
jobs. This meant that many patronage workers remained in the service,
but this time with the job protection that civil service offered. Mc-
Sweeney kept his position, although his salary was reduced.

Meanwhile, the McKinley administration searched for someone to
run the immigration office in Washington. The president finally settled
on Terence V. Powderly, one of the most famous Americans of the
late nineteenth century. The former Grand Master Workman of the
Knights of Labor—a fanciful title that befitted the utopian nature of
the organization—helped build the country's first major national labor
union, and in doing so became a celebrity whose "face and name graced
everything from chewing tobacco packages to haberdashers' trade
cards." His portrait hung inside humble homes, and a town just outside
of Birmingham, Alabama, was named in his honor. Powderly had also
served as mayor of Scranton, Pennsylvania.

On the surface, McSweeney and Powderly possessed many similarities. These two sons of Irish Catholic immigrants grew up in large
families—McSweeney was one of eight, Powderly one of twelve. Their
careers began in the labor movement, yet they were conservative by
temperament and opposed to socialism. Unions were their avenue into
partisan politics. Their backgrounds fed their interest in immigration.

Yet their differences outweighed their similarities. Whereas Mc-
Sweeney was a Democrat, Powderly, sixteen years older than his soon
to be nemesis, was a Republican. McSweeney played the political game
with aplomb, cultivating influential and powerful people throughout
society. Powderly, on the other hand, had a knack for angering both

subordinates and superiors wherever he went. McSweeney was slick, while Powderly could be moody and abrasive. McSweeney retained strong ties to labor and the Catholic Church throughout his life; Powderly became estranged from both. Though both men supported the current immigration laws, McSweeney was sympathetic toward immigrants, while Powderly's views were decidedly more negative.

McSweeney seemed to be born for political life, but Powderly was miscast in the profession. A slender, almost frail man, with a long droopy mustache and pale blue eyes, Powderly had the look, according to one contemporary journalist, that some mistook for "poets, gondola scullers, philosophers, and heroes crossed in love." He was not, in appearance at least, a typical union man, and his looks suggested other character flaws: indecisiveness, moodiness, thin skin, and a querulous nature.

One historian described him as "a vain, pigheaded, unyielding, difficult man," who was hard to like even from the "safe distance of an archive one hundred years" later. He had a tendency to quarrel with friends and foes alike. Recalling his days as leader of the Knights of Labor, Powderly noted: "I cannot forget either that I had been the recipient of a much larger share of unstinted censure, condemnation, denunciation, and abuse from those I had worked for as well as from those I had opposed." By the early 1890s, the Knights had gone into decline, wracked with dissension, and Powderly was looking for other opportunities. He later claimed that when he left the Knights, he was "broken in health and spirits" and doctors had given him only months to live.

Powderly somehow managed to survive, and in 1896 he supported Ohio's Republican governor, William McKinley, for president. He became McKinley's main adviser on labor issues. The Immigration Bureau was to be Powderly's reward.

Over the course of his career, Powderly made many enemies, a dubious skill he would soon put to use in his new job. Some of those enemies pressured the Senate to block Powderly's confirmation, forcing McKinley to make a recess appointment. Even the Knights of Labor's official newspaper came out against its former chief. Many of them distrusted Powderly's Republican friends and criticized Powderly for dropping his opposition to the gold standard to align with McKinley's views.

Another critic was Samuel Gompers, president of the American Federation of Labor (AFL), who called Powderly's selection "an affront to labor." Gompers's AFL succeeded the Knights as the country's leading labor union, and the two men clashed repeatedly over the years.

Powderly fought back, however, and McKinley stuck with him. Powderly went so far as to elicit the support of Edward McSweeney, his soon to be subordinate, to lobby Gompers to lift his opposition to Powderly's appointment. Gompers didn't budge, telling McSweeney that he opposed Powderly because his reputation "had been to break down and disrupt, and that he had used his position for unworthy ends." Even without the support of Gompers, Powderly eventually received Senate confirmation in March 1898.

Powderly's brother Joseph had been part of John Weber's 1891 commission on European immigration. For the brothers, immigration was a personal issue. Terence accused new immigrants of coming to America "to compete in the struggle for food with the American workman." He had gone to Castle Garden years earlier and saw what he called "agents of corporations" waiting for immigrants to arrive. Powderly recognized one of the men, who then arranged for some newcomers to travel to Pennsylvania, where they displaced native-born workers, many of whom Powderly knew personally.

Powderly did not stop with his economic arguments. He went on to call the new immigrants "semi-barbarous." His views of immigration were somewhat ironic considering his background. As one of his many critics noted, if the laws Powderly wanted enforced had been applied to his Irish immigrant parents, Powderly might "be carrying turf, in an Irish bog, instead of being able, from the influential position he enjoys among Americans, to warn off later comers." It was an irony not lost on Powderly, whose father had been arrested as a youth in Ireland for trespassing on a gentleman's estate with a gun and killing a rabbit. For the offense, the elder Powderly spent three weeks in jail—a fact that would now have excluded him from entry to America.

Powderly was now in charge of enforcing the nation's immigration laws. One of the biggest problems he had to deal with was the worsening situation in New York. As construction of the new buildings on Ellis Island continued, immigration officials were forced to conduct their business in the much more cramped quarters of the Barge Office. While immigration had been cut in half during the depression, better economic times now lured more immigrants to the country. More immigrants coming through the inadequate facilities at the Barge Office spelled trouble.

That trouble would spark a growing rift between Powderly and Edward McSweeney. It is difficult to pinpoint just when things began to go wrong. Upon taking office, Powderly had learned that there were problems with the immigration station in New York. "Ill treatment of arriving aliens, impositions practiced on steamship companies, and discourtesy to those who called to meet their friends on landing were frequent," wrote Powderly. Eager to ingratiate himself with his new boss, McSweeney told Powderly that he could see "some rocks ahead" and offered to put his boss "in the way of escaping them." He cryptically warned Powderly that the Barge Office was "a peculiar Service and some peculiar practices and precedents have come into vogue."

Powderly made a surprise visit to the Barge Office in March 1899. He arrived with a stenographer and was given his own room for a number of days to investigate and interview Barge Office employees. Powderly quickly discovered that McSweeney was the real chief of the immigration station and Fitchie, the nominal head, was "almost unknown to most of the employees at the station although he had been in office two years." Powderly found some small irregularities, but decided to take no further action. Still, McSweeney took Powderly's investigation as a personal affront.

Powderly's views on immigration—and his zealous pursuit of those goals—also added to the growing divide between the two officials. Powderly was busy shaking up immigration enforcement across the country, from New York to California to the Canadian border. To his credit, he was no mere political hack. Powderly was determined to enforce more strictly the laws against contract workers and Chinese immigrants—both traditional bugbears of labor. Powderly proudly noted that in 1899, 741 illegal contract laborers had been excluded, nearly double the number from the previous year.

Yet his strict enforcement of contract-labor laws ran into predictable opposition in the Treasury Department. Powderly's complaints about cheap immigrant labor did not warm the hearts of his pro-business Republican superiors. In 1899, a large group of Croatian immigrants arrived in Baltimore and were detained on suspicion of being contract laborers. When they appealed their case to Washington,

Powderly ordered their deportation. He claimed the men were heading to a Chicago address of a "man whose name is a stench in the nostrils of organized labor."

Powderly's decision revealed the weakness of the contract-labor law. By 1899, most employers were careful not to make any contracts for incoming immigrants, who themselves were careful not to tell immigration officials that they were arriving in the country to work on a specific job. Powderly admitted that the evidence in the case "was not such as would warrant a conviction in a criminal court." Yet, believing deep in his heart that these men were violating the law, he ordered their exclusion anyway.

Such reasoning did not wash with Powderly's boss, Treasury Secretary Lyman Gage, who overturned his decision and allowed the Croatians to proceed to Chicago. Suspicion without evidence, Gage argued, was not a sufficient reason to bar immigrants. Powderly, seemingly incapable of restraining his anger at the decision, told the German-language *New Yorker Staats-Zeitung* that Gage "has no sympathies for the [native-born] laborers." He later complained about being misquoted, but the quote captures both Powderly's anger and his perennial inability to bite his tongue.

Such impolitic behavior won Powderly few friends in the Treasury Department. Not only would Powderly battle his subordinates in New York, but he would also find himself fighting with his bosses in Washington. Powderly especially ran afoul of Horace Taylor, the assistant secretary of the Treasury, who took office in early 1899 and often referred to Powderly as "that labor crank." Their mutual disregard for Powderly would lead McSweeney and Taylor to become close allies in the coming bureaucratic struggle.

The decision on the Croatian laborers fed Powderly's suspicions—or paranoia—that his colleagues were uninterested in enforcing the contract-labor law. He seemed to have his opinions confirmed in April 1900 when Fitchie and McSweeney enacted a minor reform at the Barge Office. Since Congress passed the contract-labor law in 1885, a separate group of inspectors had existed who only dealt with suspected immigrant contract laborers. In the late 1890s, Fitchie and McSweeney, in the interest of efficiency, decided to merge the contract-labor inspectors with the regular inspectors into one inspector class. Assistant Secretary Taylor approved the plan without consulting Powderly, who bitterly opposed the idea. Though it appeared to be a rational bureaucratic reform, it had the effect of reducing the number of immigrant exclusions in New York based on contract-labor violations by almost 90 percent, according to Powderly.

Powderly was in an awkward position. He was a labor man opposed to cheap immigrant labor, yet he worked for a pro-business Republican administration. Even worse, he had also alienated many people in the labor movement. Recognizing this situation, he worked hard to stay in the good graces of McKinley, without whose support Powderly would have found himself out of a job.

Perhaps that insecurity led Powderly to ask for McSweeney's help with the 1898 gubernatorial race in Connecticut. John Addison Porter, the personal secretary to McKinley, was running for the Republican nomination. Powderly wanted Fitchie to ask McSweeney to "run over and get some of his Democratic friends to get into the caucuses and help our friends out." There is no evidence that McSweeney agreed to the request, and Porter failed in his bid to become governor.

Just a few months before Powderly made his request, New York senator Thomas C. Platt, the longtime Republican boss of the state, complained about the "extreme partisan conduct" of McSweeney. "Is there not some way that he can be removed and a good Republican put in his place," Platt asked Thomas Fitchie. Platt was angered less by McSweeney's Democratic affiliation and more by the fact that he had run for New York sheriff in 1897 on Seth Low's reform Citizens' Union ticket. McSweeney was trying to prove his Republican bona fides by supporting Low, but the Citizens' Union ticket consisted of reform Republicans opposed to Boss Platt.

Despite Platt's urgings, McSweeney remained in office. Perhaps Fitchie recognized that the immigration service in New York could not run without McSweeney's administrative talents. There is another possible explanation. When the bitterness between McSweeney and Powderly broke out into open warfare a few years later, Powderly would accuse McSweeney of delaying the stay of immigrants at Ellis Island "for the purpose of swelling the receipts of Mr. Hess who has the contract for providing food for immigrants at Ellis Island." Charlie Hess also happened to be a loyal member of Senator Platt's Republican machine. Powderly claimed that McSweeney told him: "I can rely upon Senator Platt to do the right thing by me." So it is not beyond the realm of possibility that McSweeney had made his peace with Platt, a man more interested in patronage than partisanship.

The accusation that McSweeney was involved in unethical conduct was part of a larger problem at the Barge Office. While Ellis Island had put the buffer of New York Harbor between immigrants and those who prowled the waterfront looking to take advantage of greenhorns, the Barge Office provided no such luxury. McSweeney himself explained that all of the problems that had once existed at Castle Garden were reappearing at the Barge Office.

More complaints emerged about the Barge Office. Words like "listless," "inexcusably insolent," and "inefficient" were thrown about to describe the staff. Victor Safford spoke of one worker, a German immigrant with a bushy beard, whose sole duty seemed to be to march around with great pomp dressed in naval cap and double-breasted coat with brass buttons. The man was obviously a political appointee, and Safford could never figure out what the man did.

By the end of 1899, word reached Washington of serious problems at the Barge Office, prompting Secretary Gage to appoint a committee to investigate, led by John Rodgers, commissioner of immigration at Philadelphia, and Richard K. Campbell, from the Washington office. Rodgers and Campbell conducted two months of hearings in lower Manhattan in early 1900, collecting over two thousand pages of testimony.

Much as Powderly had found earlier, the Campbell-Rodgers report concluded that McSweeney was the real power at the Barge Office. It laid out in detail charges of cruelty, corruption, and the abuse of immigrants "of such a pronounced and inexcusable character." The report concluded that McSweeney "countenanced extreme cruelty and impropriety in the methods of inspection in the registry division" and recommended the firing of a dozen employees at the Barge Office, including McSweeney.

One form of corruption occurred in the Boarding Division. When ships reached the docks, American citizens were separated out from immigrants and allowed to pass. Albert Wank, an assistant officer in the Boarding Division, reportedly took cash payoffs to let immigrants through, thereby avoiding inspection. A clerk for a French steamship line testified that it was common for immigrants to pay Wank $1 or $2 to get out of the inspection line. Those immigrants not paying the bribe would then often pass by Emil Auspitz, the gateman in charge of the entrance to the registry room. Auspitz was accused of treating immigrants roughly and using foul language.

The most serious charges were leveled against John Lederhilger, the chief of the Registry Division and one of McSweeney's closest allies. "Mr. Lederhilger is insolent, overbearing, dictatorial and cruel to his subordinate officers," Campbell and Rodgers concluded, "and is jealous and resentful in his bearing toward those over whom he cannot legitimately exercise control." More specifically, witnesses accused Lederhilger of being a letch obsessed with the sexual behavior of young female immigrants. One Barge Office worker told the committee: "Every good looking young woman has been put to what they call the 3rd Degree." Lederhilger often used indecent language with young women because, according to the witness, "he cannot help himself; he is a brutal man."

An interpreter at the Barge Office testified that Lederhilger was in the habit of asking women about their sexual activity. Sometimes the interpreter, who was forced to translate these questions, had to clean up his language. Another interpreter complained that Lederhilger's interviews of French girls were obscene. The interpreter refused to interpret for him on a number of occasions when he wanted the following question asked: "Who fucked her on board the ship?"

The report also blamed Lederhilger for the suicide of one Italian woman, who suffered "under the mortification and distress incident to her being held and examined as a procuress [madam]." If Lederhilger thought a woman was possibly a prostitute, it gave him license to molest her physically. Pointing to a woman's breast, Lederhilger allegedly said: "Open that dress and see if you have anything in that pocket." Another witness claimed he saw Lederhilger and other officers open the clothing of women and "thrust their hands in their bosoms and in other ways improperly handle their person."

Treasury Department officials sat on the report for two months. Meanwhile, Powderly drew up formal charges against thirteen individuals, including McSweeney. His superiors quashed the charges, leading Powderly to accuse McSweeney's friends in Treasury of protecting him. By September, McSweeney felt confident enough to write to Archbishop Michael Corrigan that although unscrupulous persons had attempted to discredit his work, his bosses in the Treasury Department had foiled the plot.

The report was certainly slanted in Powderly's favor. Both Rodgers and Campbell were Powderly allies and most witnesses were Powderly's friends at the Barge Office. McSweeney was the main target of the report, and he called the investigation "a persecution, of which I was the proposed victim," while Fitchie said it was "conceived in iniquity and born in sin." Yet it is hard to believe that all of the charges and testimony in the massive report were simply fabricated to frame Mc-Sweeney.

Edward Steiner, a Grinnell College professor and immigrant from the Austro-Hungarian Empire, traveled a number of times across the Atlantic Ocean in steerage collecting material for his books on immigration. "Roughness, cursing, intimidation and a mild form of blackmail prevailed to such a degree as to be common," Steiner noted at this time. On one trip, an inspector approached Steiner and hinted that he might have difficulties getting through inspection. A little money, the inspector intimated, might make the problem go away. A Czech girl told Steiner in tears that an inspector promised to pass her through inspection if she agreed to meet him later at a hotel. "Do I look like that," she asked Steiner through her embarrassment. The inescapable conclusion is that, even accounting for the personal vendetta between Powderly and McSweeney, a lot of petty corruption and abusive behavior was being tolerated at the Barge Office.

Even as the Treasury Department tried to bury the report, excerpts were leaked to the press. Now that the charges were aired publicly, Washington needed to act. In a classic case of creating scapegoats to protect higher-ups, officials fired a handful of minor Barge Office workers, gatemen, and messengers, charging them with taking bribes and treating immigrants roughly. In a tragic footnote, one of those dismissed was a fifty-five-year-old black messenger named Jordan R. Stewart. In addition to bribery, Fitchie also accused Stewart of being repeatedly drunk on the job.

Stewart had been born a slave and served as a lieutenant in the 73rd U.S. Colored Infantry in the Civil War. During Reconstruction, he represented Tensas Parish in the Louisiana state legislature. In addition, he had been a businessman, a deputy sheriff, and a watchman at the New Orleans customs house. By the 1890s, with increasing violence against blacks in the South, Stewart found himself in New York City, where he no doubt used his Republican political connections to land a patronage position in the New York immigration service. Now he was out of a job.

While men like Stewart took the fall, McSweeney and his allies, including John Lederhilger, dodged a bullet and resumed their jobs. Powderly had been foiled and so had his attempt to use the immigration service on behalf of McKinley and the Republican Party. Just after the conclusion of the Campbell-Rodgers investigation, Powderly wrote to an ally that if only he could control the Immigration Bureau without meddling from superiors, he could "pave the way for Republican success in many a doubtful place, and do it without detracting from the usefulness of the Bureau." He promised McKinley that if only he were allowed a free rein, he could strictly enforce the immigration laws and win more support for the president from labor men, since Powderly's Immigration Service would be looking after their interests regarding contract labor.

Powderly wanted to help both American workingmen and McKinley, but he believed that personal enemies stymied his mission at every turn. The reason, he felt, was that the Immigration Service was filled with Democrats and the Treasury Department was rife with anti-labor men. With McKinley up for reelection in 1900, Powderly became obsessed with the belief that McSweeney and his allies were working for a Bryan victory.

Some of Powderly's friends ventured close to paranoia. James "Skin the Goat" Fitzharris and Joseph Mullet arrived in New York in May 1900, having left Queenstown, Ireland. They had been part of a group called the Invincibles, Irish Republicans who carried out the infamous 1882 Phoenix Park murders of Lord Cavendish and Thomas Henry Burke in Dublin. Having served eighteen years in prison, the two men were now free and headed to the United States for a visit. The sixty-year-old Fitzharris, dapperly attired in a blue serge suit and green scarf with a pin bearing the face of Irish hero Robert Emmett, and the younger Mullet, a hunchback, were quickly detained. Their case clearly came within the 1891 law barring the admission of criminals; the only question was whether their crime was of a political nature or not.

A Powderly ally named A. J. You believed that the detention of these two men had McSweeney's fingerprints all over them. "You can readily see what an alarm will be sounded by the Irish people if these parties are held for investigation by our force and the hellish purpose conceived by the Deputy Commissioner [McSweeney] in having this order issued over the signature of the Commissioner," You fretted. "How easily the holding up of the Irish immigrants or foreigners can be turned with a free hand against us and especially directed against yourself as the head of the Immigration Services."

When the Treasury Department finally decided that the crimes of the two men were not politically motivated, "Skin the Goat" and Mullet were sent back to Ireland, but only after they spent an unhappy month in detention. Mullet wrote to Commissioner Fitchie to complain about their treatment, calling their month in detention worse than their eighteen years in a British jail. In the latter, at least, the Irishmen were kept apart from the other convicts and treated like political prisoners, while in New York, Mullet and Fitzharris were forced to "mix with the scum of Europe."

A Democrat who had kept his job under a Republican administration thanks to new civil service regulations, McSweeney knew that his civil service classification could be overturned at any time, so he went out of his way to ingratiate himself with New York Republicans.

In what was probably as shrewd an assessment as he ever made, Powderly noted that McSweeney was known always to be "a most ardent McSweeney man." Whatever the case, McSweeney proved himself a consummate survivor and political operator. Powderly could have learned a few lessons from him.

In MID-DECEMBER 1900, TWO and a half years after the fire that devastated the first immigration station at Ellis Island, new facilities were finally completed and open for business. On December 17, Fitchie, McSweeney, and their entire staff welcomed the first boatload of steerage immigrants to Ellis Island. The *Kaiser Wilhelm II* brought 654 immigrants, the first of 2,252 who would pass through Ellis Island that first day. There was no pomp and circumstance as there had been in January 1892; the only celebration was a good-luck horseshoe of flowers presented to Thomas Fitchie by his friends. The first immigrant off the boat was a young, laughing, red-headed Italian girl named Carmina di Simona, "so much inclined to rotundity that it was a question whether her greater dimension was length or breadth."

There was no Annie Moore treatment for Carmina, no ten-dollar gold piece or front-page articles. Americans may have been happy about the new facilities, but they seemed less inclined to celebrate the new immigrants.

Officials claimed that the new reception building could accommodate more than seven thousand immigrants in a single day. It was not designed in the neoclassical, white marble, Beaux Arts style then fashionable for public buildings. Instead, it was a steel-frame structure covered with red brick laid in Flemish bond with limestone trimmings. Four 100-foot, copper-covered, bulbous towers crowned each corner, giving the building a vaguely Byzantine feel. Massive arches with moldings of eagles and shields capped many of the windows. There were new offices, dining facilities, hearing rooms for the boards of special inquiry, shower rooms, and a roof deck for entertainment.

The centerpiece of the main building was the second-floor registry room. Measuring 200 feet by 100 feet and with a 56-foot ceiling, this large airy space was divided into narrow aisles by iron railings for immigrants to pass through on their way to the registry clerk holding the ship's manifest. Unlike the previous shabby wooden quarters, all the new buildings were fireproof. Even secondary buildings like the hospital and the power plant exhibited a stolid dignity.

Ellis Island now consisted of an imposing set of structures that announced to immigrants the grandeur of their adopted country. Inspection would again be cloistered away from the hubbub, distractions, and immigrant sharks at the Barge Office in the Battery. "The crowd of foreigners who besiege the present quarters every day making life hideous with their quarrels or cursing the guards and gatemen in a babel of tongues will be a thing of the past," rejoiced the *Times*.

Yet fancy new buildings did nothing to improve the quality of inspection, reduce corruption, stem the abuses of immigrants, or quell the increasingly vicious infighting between the McSweeney and Powderly camps. Washington had created Ellis Island and an immigration service to run it. However, those who worked in this infant bureaucracy were still mired in the political patronage that defined an older period of history. A stronger federal government was needed to deal with the problems of a modern industrial and urbanized society, but a more professional staff to run that government was also needed. Turn-of-the-century Ellis Island embodied that clash between traditional political patronage and the more strenuous demands being placed on government to regulate an increasingly complex society.

Ensconced in Washington, Powderly received regular updates from Ellis Island officials loyal to him. One of them called McSweeney a "Dr. Jekell [sic] and Mr. Hyde character. . . . He is so bigoted, partisan, spiteful and malevolent. It is terrible." Powderly himself referred to his nemesis as "McSwine." His friends intercepted letters from McSweeney to his allies in the Treasury Department, which were dutifully copied and sent to Powderly.

Not to be outdone, McSweeney recruited Powderly's confidential secretary to spy on Powderly and report on his actions. Powderly discovered this and fired the clerk. Powderly's allies at Ellis Island accused McSweeney of harassment, while McSweeney played the martyr for his superiors at the Treasury Department. "I am free to admit that it has been pretty hard to come into contact day after day with men who are trying to cut your throat," he complained to Assistant Secretary Taylor.

When not bogged down with his battles with McSweeney, Powderly continued to think about the effects of the inspection process. Though Powderly was known as a restrictionist, his views were being tempered by political reality. "Italian, Hungarian, Polish and Oriental immigrants passing through Ellis Island should be treated kindly," he wrote to McKinley. "Such immigrants in time become citizens and their influence among their compatriots will play no insignificant part in the politics of the future." When future Republican politicians asked these new Americans for their votes, Powderly worried that they would ask: "Is this the party that was in power at Ellis Island when I landed?"

Powderly understood that while Ellis Island symbolized the nation's vigilance toward immigration regulation for native-born Americans, it was also becoming a symbol for first-generation Americans. Calls for stricter regulation of immigrants would have to be balanced by the concerns of new immigrant communities.

The new buildings at Ellis Island stood as a testament to a nation entering a new century determined for greater power and glory. The main building impressed upon immigrants that America was a substantive and wondrous land; the power of the federal government and the American nation made their stamp on the immigrant immediately. This same government would soon compel some of those who entered Ellis Island or their children back to Europe as soldiers in World War I. The same government would assist many of these immigrants in their old age with Social Security many years later. Just as every immigrant would feel the force of the federal government at this most important point in their lives, it would be only a matter of time before native-born Americans would experience the presence of the state in their own lives. Immigrants at Ellis Island were just a little ahead of the curve.

The government owed the American people a fitting structure to enforce the law, and it owed immigrants a building that would welcome as well as awe. Unfortunately, the new façade could not mask the disarray and corruption that took place inside its walls.

By the summer of 1901, Inspector Roman Dobler, a Powderly informant, talked about "a bellicose spirit" pervading Ellis Island. As proof, he told the story of Helen Taylor, a twenty-six-year-old assistant matron, who had gotten into an altercation with an inspector named Augustus Theiss, a McSweeney ally. When Theiss passed through an immigrant whose entire family, including wife and two daughters, Miss Taylor had marked for special inspection, an indignant Taylor delivered a "stinging slap across the face" of the short and doughy Theiss. "Do you mean to say I am a liar," she asked him.

From Washington to New York, tempers had reached a boiling point. Scandals, squabbles, and pettiness reigned. Americans like Henry Cabot Lodge feared a peril at the portals and wanted a gate to guard against the wrong kind of immigrants. The dawn of the twentieth century found Americans still raising questions about not only who was entering through those portals, but also who was guarding those gates.

A restless nation—and more importantly a restless new president— would try to remedy this unhappy situation.

Part III

REFORM AND REGULATION

Chapter 7

Cleaning House

———•◦•———

It does seem to me that mental and physical inferiority are the highest recommendations for promotion at this station.
    —Roman Dobler, Ellis Island Inspector, 1900

Ellis Island has been a place for the harboring of vultures who preyed upon the immigrants and people began to look upon it as the hell hole of America.
    —Frank Sargent, Commissioner-General of Immigration, 1903

LEON CZOLGOSZ.

It was not a name that rolled off the tongues of native-born Americans. With great authority, the *Journal of the American Medical Association* informed its learned readers that the man who fired two shots into President William McKinley on September 6, 1901, "bears a name that can not be mistaken for that of an American."

To make matters worse, the press reported that Czolgosz was an anarchist. To many Americans already unsettled by large numbers of immigrants from strange lands, the shooting reinforced the connection between foreignness, criminality, and radicalism.

Yet there was one problem: Leon Czolgosz was an American citizen, native-born in Michigan to Polish Catholic parents who had fled Prussia. Despite this inconvenient fact, McKinley's assassination again stoked America's fear of immigrants. Yet Congress took its time in reacting to the tragedy, waiting almost two years before it added anarchism to the list of offenses for which immigrants could be excluded. While it was at it, Congress also added prostitutes, epileptics, and professional beggars.

Theodore Roosevelt had little use for anarchists, calling them treasonous criminals who "prefer confusion and chaos to the most beneficent form of social order" and arguing that their philosophy was "no more an expression of 'social discontent' than picking pockets or wife beating." Yet for Roosevelt, out of the tragic murder of William McKinley came the fulfillment of his own ambition as he was now catapulted into the White House. Within three short years, Roosevelt had gone from war hero to governor to vice president to president.

The bullets that ended McKinley's life also put a close to nineteenth-century America. Roosevelt seemed different from his predecessors in almost every way. He approached the presidency with the vim and vigor he had approached everything else in his life. He possessed a restless and curious mind. His speeches pulsed with energy, with little of the florid and flabby rhetoric of his predecessors. Instead, he spoke the language of action, urging Americans toward the strenuous life. He wrote in 1894: "We Americans have many grave problems to solve, many threatening evils to fight, and many deeds to do, if, as we hope and believe, we have the wisdom, the strength, the courage, and the virtue to do them." And by 1901 Roosevelt saw that there was still much to do.

In the previous decade or so, the pieces had been put in place for a strong national government at home and abroad. Roosevelt wanted to use that national government for solving problems and fighting evils. "I did not care a rap for the mere form and show of power," he wrote in his autobiography, "I cared immensely for the use that could be made of the substance."

Roosevelt is often associated with trust busting and conservation, but he was just as interested in immigration. If Washington was the father of the country and Lincoln the savior of the union, then Theodore Roosevelt was the philosopher of the modern nation. He believed that immigration was central to the question of American identity.

Roosevelt was no newcomer to the issue. Back in 1887, he delivered a blistering, red-meat political speech in front of the cream of New York's elite gathered for a feast at Delmonico's, in which he lashed into Governor Grover Cleveland for allowing the admission of "moral paupers and lunatics" at Castle Garden. In 1897, while serving as New York police commissioner, Roosevelt expressed his horror that a local news-paper had said he was opposed to immigration restriction. Roosevelt had the paper quickly correct the error. When President Cleveland later vetoed the literacy bill, Roosevelt "took a kind of grim satisfaction in Cleveland's winding up his career by this action, so that his last stroke was given to injure the country as much as he possibly could."

Roosevelt worried about the negative effects of unrestricted im-migration. The young patrician criticized businessmen who demanded cheap immigrant labor, saying they were "committing a peculiarly contemptible species of treason." While they might benefit from im-migration in the short term, Roosevelt argued, their "children and grandchildren may have to pay dearly for their ancestors' selfish greed, when the descendants of the brutalized men whom we imported have grown to be a power in the land, and have cast off the old-world shack-les without learning the new-world capacity for self-restraint and self-government." To protect the wages of workingmen and the future of American self-government, Roosevelt wanted laws that would let in "really good immigrants" and sift out the "very unhealthy elements."

The relationship between immigration and national character was never far from Roosevelt's mind. Postulating the definition of "True Americanism," he gave a rousing, if somewhat vague, definition of American identity and defended American exceptionalism. While Roosevelt's America accepted European immigrants, it also needed to Americanize them. Roosevelt wrote: "We welcome the German or the Irishman who becomes an American. We have no use for the German or Irishman who remains such . . . we want only Americans, and, pro-vided they are such, we do not care whether they are of native or of Irish or of German ancestry." Roosevelt noted that the "mighty tide of immigration to our shores has brought in its train much of good and much of evil," and therefore the nation needed to regulate immigration more strictly.

Roosevelt was a rare individual; a trust-fund patrician broadly read in history and literature, yet one whose curiosity led him to learn first-hand about social conditions. Roosevelt got an education from his friend Jacob Riis, who led him through the teeming slums of lower Manhattan that Riis was about to immortalize in his book *How the Other Half Lives*. Wanting to see more, Roosevelt, then police commis-sioner, hopped a ferry to Ellis Island in 1896 to witness the sifting of immigrants firsthand. With characteristic zeal, he eagerly nosed his way around the old facilities, paying careful attention to both the inspectors and the inspected.

A young inspector named Robert Watchorn remembered Roosevelt's visit. Roosevelt also remembered Watchorn and a decade later would name him commissioner of Ellis Island. Watchorn recalled seeing Roosevelt at a board of special inquiry hearing for a "stalwart, brawny young Swedish stowaway," as the future president paid rapt attention to the proceedings, noting that Roosevelt "probably regretted that he was powerless to decide the matter at once." The stowaway, despite his illegal entry into the country and his lack of money, family, or destination, represented the right kind of immigrant to Roosevelt. "I like the looks of that young fellow," Roosevelt told Watchorn, applauding the decision of the board to allow the stowaway to remain. "We need lots of good, vigorous, healthy blood to mingle with the national stream."

William McKinley, on the other hand, seemed uninterested in immigration. During his first presidential campaign, he supported a literacy test for immigrants and spoke of the need to prevent the importation of cheap labor. When the campaign ended, McKinley said little more about immigration and paid no attention to what was happening at Ellis Island and the Barge Office, allowing the troubles there to become festering sores. During his four-plus years in office, McKinley thought that silence was the best policy on immigration.

Roosevelt could not have been more different—or so it seemed. While McKinley was solidly middle American in background and outlook, Roosevelt was part of the nation's urban gentry, a Harvard-educated New Yorker, a politician and scholar with multivolume histories already under his belt. McKinley was the last of the Civil War veteran presidents, while the forty-two-year-old Roosevelt was the nation's youngest president.

Roosevelt exists in historical memory as a man of bluster, a straight-talking reformer, yet the reality is more complex. The Rough Rider with overseas expansion on his mind was also noted for his quiet diplomacy. He would see his antitrust record dwarfed by that of his much-maligned successor, William Howard Taft. On issues from his handling of the 1902 miners strike to his shepherding of the meat inspection bill to his relations with the Republican old guard, Roosevelt was more of a deft accommodationist than a take-no-prisoners reformer. Nowhere does that become more apparent than in his handling of immigration. In some ways Roosevelt was not that different from McKinley.

LESS THAN A MONTH after taking office, Roosevelt busied himself with affairs of state. As expected, the new president had his fingers in many pots. There was much to think about—appointments, bills, politics.

One area in particular focused Roosevelt's mind: the Immigration Service. Through friends in New York, he was already aware of the situation at Ellis Island. Three weeks after taking office, he confided to his close friend Nicholas Murray Butler that he was "more anxious to get this office straight than almost any other."

As a new boss entered the scene, people quickly calibrated how they would fare under the new order. For Mark Hanna, the brains behind the McKinley presidency and leader of the Republican establishment, the elevation of Roosevelt to the presidency was not good news. "That damned cowboy is president of the United States," he is reported to have said in a not entirely positive tone. Other Republicans were not sure what to make of the notoriously unpredictable Roosevelt.

Similar thoughts ran through the minds of those who worked on immigration. For Powderly, the death of McKinley was a huge blow. McKinley had been his biggest—and, increasingly, his only—supporter, the object of Powderly's near–hero worship. With Roosevelt, Powderly had no such relationship. While Roosevelt once applauded an anti-immigrant article Powderly had written, that was almost fifteen years earlier. Powderly feared that Roosevelt still remembered that he had supported Henry George in 1886, when both George and Roosevelt unsuccessfully sought the mayoralty of New York City.

Even with a new boss to impress, Powderly showed no sign of trimming his sails. Just as Roosevelt was settling into the White House, Powderly was trying to force the deportation of sixteen immigrants from Transylvania headed to Hubbard, Ohio. Charged with violating the contract-labor laws, the men had spent two weeks in detention at Ellis Island, but Powderly's superiors at Treasury found no reason to detain them any further and released them over his strenuous objections. Powderly, no doubt, had all his fears confirmed once again that his bosses had little interest in protecting the American worker from cheap immigrant labor. Treasury Department officials had their belief confirmed that Powderly was insufferably stubborn and not a team player.

Edward McSweeney had more reasons to be optimistic about the new chief executive. Despite McSweeney's background as a partisan Democrat, he had run for city office in 1897 on a reform ticket with Seth Low, which helped him ingratiate himself with a number of prominent New Yorkers who just happened to be good friends with Roosevelt. His new friends included soon to be president of Columbia University Nicholas Murray Butler and reformer Jacob Riis. When Senator Thomas Platt, New York State's Republican boss, tried to get McKinley to replace McSweeney, his new friends sent a letter to Washington praising McSweeney. One of the signers was Theodore Roosevelt. Though he had never met McSweeney, Roosevelt signed the letter on the recommendation of Butler.

For Prescott Hall, Roosevelt's elevation to the presidency must have seemed like a godsend. Although a New Yorker, the new president had strong ties to the Boston Brahmins. A Harvard graduate, Roosevelt was a close friend of Henry Cabot Lodge. The president's first wife, Alice, hailed from Boston's blue-blood Lee family.

Roosevelt's views on immigration, at first glance, appeared in sync with those of Hall and his fellow restrictionists. The new president was already on record condemning unrestricted immigration and castigating big business for its role in promoting it. During the cholera scare of 1892, Roosevelt told Lodge that he hoped the crisis would lead to a "permanent quarantine against most immigrants." By background, friendships, and temperament, Roosevelt imbibed a decided skepticism about new immigrants.

Yet immigrant defenders had reason to be optimistic as well. As New York police commissioner, Roosevelt had once assigned a group of Jewish policemen to protect an anti-Semitic German preacher in town to give a speech. His calls for immigration regulation were always coupled with strong denunciations of know-nothingism and pleas to treat immigrants with decency. Roosevelt himself was a mixture of Dutch, English, French, Welsh, German, and Scottish blood and possessed an optimism about America that somehow eluded many of his friends. "I am a firm believer that the future will somehow bring things right in the end of our land," Roosevelt wrote the notoriously dour Brahmin historian Francis Parkman.

Whatever may have been his true beliefs, Roosevelt first had to clean up the mess in the immigration service. A month after taking office, he met with Powderly. Though Powderly was willing to resign his post, the president said he had no intention of removing him from office. Just after the meeting, Roosevelt wrote Butler that "our people have been united in telling me that Powderly was a good man." Even more good news for Powderly was that Roosevelt told him that every "good man whom I have met who knows anything about that office has agreed in believing McSweeney to be corrupt."

Powderly left the meeting confident that he would be retained and that perhaps he would triumph over his enemies both at Ellis Island and in the Treasury Department. Even when rumors leaked out in the coming months that Powderly might lose his job, the old labor man held on to the president's personal reassurance like a life preserver. "From all I knew of Mr. Roosevelt that simple declaration was equivalent to another man's oath," Powderly later reminisced.

Whatever his feelings for Powderly, Roosevelt felt no sympathy for Thomas Fitchie, the nominal head of Ellis Island. Nicholas Murray Butler called Fitchie "an old man with weak will" and Roosevelt considered him "absolutely incompetent." Though Roosevelt did not know Fitchie personally, he certainly knew his type. He had been battling the New York Republican machine, of which Fitchie was a proud member, his entire political career. Although not personally corrupt, Fitchie was a time server who squandered the power he was given.

As the months wore on, nothing was done. As late as April 1902, more than seven months after Roosevelt took office, Fitchie, McSweeney, and Powderly all remained in office. Why had Roosevelt procrastinated? First, contrary to his blustering image, Roosevelt was a deliberate politician. Second, Roosevelt had trouble finding someone to run Ellis Island. "As for Fitchie's successor, all I want to do is to get the best possible man in the country," Roosevelt wrote Butler, setting the bar a bit high.

There was still a third reason. Despite his personal reassurances to Powderly and his initial negative impression of McSweeney, Roosevelt still remained torn as to who was at fault in the running battles in the Immigration Service. Depending on whom he last spoke with, his opinion about the two men could change from week to week.

Even with the charges of abuse and corruption swirling around Ellis

Island, McSweeney was highly regarded for his administrative skills—even by his enemies. No less a person than Terence Powderly noted that no one else "so thoroughly understands the immigration service at the Port of New York as Mr. McSweeney." McSweeney devoted himself to learning Italian so as to better handle the waves of Italian immigrants. He had become a leading national authority on immigration issues, writing articles and giving talks to academic audiences. McSweeney was the "ablest man in the whole immigration service," Butler confidently told Roosevelt.

"Nicholas Miraculous" Butler was one of McSweeney's biggest defenders. He was extremely close to Roosevelt (although they would later have a nasty falling out) and encouraged Roosevelt to investigate Powderly and keep McSweeney on. "I do not believe the rumors in circulation about his [McSweeney's] integrity, and I feel pretty confident that an investigation instituted by you would confirm this belief," Butler wrote Roosevelt. On top of that, he sent Roosevelt the 1898 letter from Powderly requesting to have McSweeney help out with the governor's race in Connecticut. "This is about as low a grade of political morality as we ordinarily come across," Butler wrote to Roosevelt. The president seemed disgusted with the letter, which succeeded in tarnishing Powderly's reputation in his eyes.

Jacob Riis called Powderly "a wart that should be removed" and praised McSweeney as "clean and straight." Even Henry Cabot Lodge supported McSweeney. The alliance between the Boston Brahmin Republican and the Irish Catholic Democrat was an odd pairing, but Massachusetts Republicans worried that a dismissal of McSweeney might hurt Republicans among the state's Irish voters. "McSweeney has most industriously worked up every kind of influence, political, charitable, and religious, especially Catholic," Roosevelt complained to Lodge.

What Roosevelt really wanted at Ellis Island, he wrote to a friend, was someone he could trust and "not some man about whom after hearing all the evidence I could be doubtful as to whether I ought to feel distrust." The word Roosevelt kept getting was that the inspection center was badly run. "Either McSweeney is absolutely incompetent or else he is more responsible than any other one man for these evils." Despite all of the positive words about McSweeney that he received from his friends, Roosevelt increasingly leaned toward the latter explanation.

Finally, in the spring of 1902, Roosevelt made the only decision that made any sense: he would get rid of the whole lot. He summoned Fitchie and McSweeney to Washington to inform them they would be replaced. Fitchie begged Roosevelt to rethink his actions and send a committee to visit Ellis Island so they could see that the charges were unfounded. Fitchie found the president adamant in his decision. A clean sweep was what he wanted.

Despite his earlier pledge to Powderly, Roosevelt had also concluded that the old labor leader would have to leave his post in Washington. "I believe the jig is up and that I have to go," a resentful Powderly wrote to his loyal ally Robert Watchorn. Powderly was convinced that his letter asking for McSweeney's help in the Connecticut political campaign was the main reason for his dismissal.

Powderly demanded to see the president. As humiliated as he was at getting fired, it particularly galled Powderly that he was being "coupled, before the public, with a man [McSweeney] who had, to my knowledge, brought the service beneath the rule of dishonest men." Roosevelt told Powderly that he was removing everyone who had brought the problems of the Immigration Service into the public eye. Frank Sargent, another Republican labor man and the former head of the Brotherhood of Locomotive Firemen, would replace Powderly.

The sheer number of enemies that Powderly had acquired over the years did him in. One was Archbishop Michael Corrigan, who had personally protested Powderly's behavior to the president. The Catholic Church had been concerned that the Knights of Labor was a secret organization, with its own rituals and vows, which might conflict with Catholic doctrine. Though Powderly was a Catholic, these concerns caused a number of run-ins with his local bishop in Pennsylvania and led to his estrangement from the Church.

McSweeney took advantage of his fellow Irish Catholic's difficulties. A regular churchgoer who had been president of the Marlborough Catholic Lyceum's debating society in his youth, McSweeney quickly allied himself with the Treasury Department's solicitor, Maurice O'Connell, who made it a point to visit with McSweeney during his trips to New York. Both men were members of the Knights of Columbus, and O'Connell proved useful to McSweeney, helping to squash the Campbell-Rodgers report and foiling Powderly's attempts to keep out contract laborers.

McSweeney found an even more important ally in Archbishop Corrigan. Dating back to the days of Castle Garden, the Catholic Church had taken an interest in the treatment of immigrants in New York. McSweeney kept the archbishop updated on Catholic immigrants entering Ellis Island. One problem was the presence of Protestant missionaries looking to make converts out of unsuspecting immigrants. For example, the American Tract Society handed out pamphlets at Ellis Island to Italian Catholics in their native language and Yiddish-language pamphlets entitled "Jesus of Nazareth the True Messiah" for Jewish immigrants. Protestant missions to Italian immigrants popped up all over Greenwich Village and Little Italy. Having McSweeney keep an eye on these Protestant missionaries was an invaluable service to the Archbishop.

All of these behind-the-scenes machinations were now over, and Roosevelt needed to find someone to take on the duties at Ellis Island. After a long search, he finally found a man who fit his exacting criteria. William Williams was a thirty-nine-year-old Wall Street lawyer, a loyal Republican with a reform bent, a former quartermaster officer in the army during the Spanish-American War, and a Yale man who belonged to the right clubs, including the University Club where the bachelor lawyer lived.

The son of a New London, Connecticut, merchant, Williams came from a family that was deeply intertwined with the history of early America. On his mother's side, he was the great-great-great-grandson of the famed preacher Jonathan Edwards. On his father's side, he was descended from Robert Williams, a Puritan settler who helped found Deerfield, Massachusetts. He was also a direct descendant of William Williams, a signer of the Declaration of Independence from Connecticut. The history of the nation's British settlers weighed heavily on William Williams's shoulders as he came reluctantly to Ellis Island.

Williams and Roosevelt had not previously met, but Williams came highly recommended. A Roosevelt friend praised him in words designed to tug at the president's conception of manhood and public service: "No more ruggedly honest man lives and few who have a keener desire to make their lives useful. . . . He would accept it as a most solemn trust although at a great personal sacrifice." This was just the kind of man who warmed Roosevelt's heart: wealthy, yet willing to sacrifice for the common good, with a résumé that spoke of both good breeding and public service.

After Fitchie, McSweeney, and Powderly had been told of their dismissals, Williams received a telegram from Roosevelt inviting him to lunch at the White House. It took Williams by surprise. Not only did he not know the president, but he was also not actively seeking any political office. A private man of independent means, he enjoyed his law work and had little ambition beyond that. But Roosevelt could be persuasive.

At lunch, he sat Williams down directly to his right and proceeded to talk his ear off for half an hour. It was vintage Roosevelt, but he had not entirely persuaded Williams. The president wanted him to take the offer immediately; Williams wanted to go back to New York and think about it. When Williams asked the president why he should take the job, Roosevelt responded by calling it "the most interesting office in my gift." Immigrants were being mistreated and something needed to be done about it. Upon his return to New York, Williams read up on immigration law and finally accepted the president's offer. He would be at his new desk on Ellis Island by the end of April.

Roosevelt had already chosen McSweeney's successor. Joseph Murray had been dubbed "the man who discovered Roosevelt." That had about as much truth as the statement that Columbus discovered America. The only thing that the older machine politician did was provide a little push to an ambition that already burned deep in Roosevelt's soul. In 1881, Murray, who came to America from Ireland as an infant and served as a drummer boy in the Union Army during the Civil War, had nominated the twenty-three-year-old Roosevelt for a seat in the New York Assembly.

Roosevelt always had a soft spot for the earthy Murray, despite their different backgrounds. Now it was time for the president to return the favor. It is not that Murray had not been adequately compensated for his political work. One historian noted that Murray's "good luck in picking a winner permitted him to reach offices beyond the limits of his capacity," which included a series of patronage jobs such as running the food counter at Castle Garden during the 1880s.

Roosevelt felt in the older man's debt, praising him in his autobiography "as fearless and as staunchly loyal as any one whom I have ever met, a man to be trusted in any position demanding courage, integrity, and good faith." Roosevelt noted that his friendship with the Irish Catholic politico helped broaden his understanding of other ethnic and religious groups. The only issue on which the two men disagreed was civil service reform: Roosevelt a supporter and Murray most certainly not. Now, his loyalty to Murray was going to force Roosevelt to go against the civil service rules that he so staunchly championed.

Roosevelt not only forced out McSweeney, despite civil service protections; he installed Murray into the spot, circumventing civil service rules. Roosevelt worried that Murray's previous service in the patronage-ridden Castle Garden might cause a problem. Before appointing Murray, Roosevelt asked him if he had ever been investigated. Satisfied with the answer, the president went ahead with the nomination.

William Williams soon learned that Roosevelt, despite his public persona, played the patronage game almost as well as the Tammany Hall politicians both men despised. Shortly after Williams took office, Roosevelt sent a man named Marcus Braun to meet with him about jobs at Ellis Island. Braun was the leader of a small Hungarian Republican political club in New York. A native-born Hungarian, he was a classic American archetype: the ethnic political entrepreneur. Braun leveraged his ethnicity for patronage jobs for himself and a few friends, in turn giving politicians real or imagined access to ethnic communities and their precious votes. Even a marginal figure like Braun could translate such access into power and prestige.

Williams informed Roosevelt he could appoint one of Braun's men as a laborer at $2 a day, but that under civil service rules he could not appoint another Braun colleague to a $1,800-a-year job. Williams could get the man a lower-paying job if the other candidates failed their civil service test. Not forgetting about his own needs, Braun wanted a job as supervising inspector at Ellis Island, a job for which Williams believed Braun was not eligible. Despite this, Braun managed to get named a special immigrant inspector at Ellis Island, a move that Roosevelt would later come to regret.

The cases of Murray and Braun show that Roosevelt was never too much of a reformer to play the patronage game. Bending the law was not outside of his comfort zone. Years later, lawyer James Sheffield wrote to Williams:

> The extraordinary part of a man like Roosevelt is that he finally comes to the conclusion that anything HE does is right, because HE does it. HE could beat the Civil Service rules on behalf of an utterly incompetent man and because his motives were to serve a friend, no one must criticize him for it. . . . It is strange how the country still believes in the Roosevelt brand

of righteousness as against the evidence of his constant use of the very men and methods he denounces in others.

Such behavior was not lost on Terence Powderly, who noted that although McSweeney had been fired effective May 1, he was given an additional thirty days' paid leave of absence, during which time Murray began work. Since the two men could not be paid for the same job, Powderly was ordered to name Murray as an immigrant inspector for thirty days at a salary of $10 a day. The lame-duck Powderly, still on the job for a few more weeks, expressed his pique with Roosevelt by refusing the order. Powderly's superiors at Treasury went ahead with the temporary appointment anyway.

Murray replaced McSweeney, but was not forced to take the civil service exam as required by law. The Civil Service Commission, no doubt influenced by the president, argued that the chaos at Ellis Island allowed for greater discretion of appointments. As soon as Murray took office, he was immediately put under normal civil service protections. An angry Powderly, in a letter to Robert Watchorn, clearly saw the irony of the situation: "Mark the consistency of dismissing me for writing a letter such as I wrote and then, in order to make room for a friend, he violates the civil service law himself by knocking it galley west and makes a place for a favorite." On this issue, Murray's view of civil service reform won the day, but Roosevelt's ethical flexibility would soon clash with the reform sensibilities of his new Ellis Island commissioner.

THEODORE ROOSEVELT HOPED WILLIAM Williams would reinvigorate Ellis Island, end the abusive treatment of immigrants, clean out the patronage dump, and strictly enforce the law. There would be no worries about meddling from Washington; Commissioner-General of Immigration Frank Sargent, who had replaced Powderly, was on the same page with his views of immigration, and the union man would prove exceedingly deferential to his underling, the wealthy Wall Street lawyer. William Williams wasted no time in getting to work, and not a minute too soon.

In 1902, more immigrants arrived than in any other year since 1881. More than 25,000 immigrants arrived in Williams's first week on the job. The island's sleeping quarters, which could accommodate as many 1,300 people, were bursting at the seams.

Williams let nothing escape his critical eye. In his first Annual Report, written just two months after he took the reins at Ellis Island, Williams talked about the lax enforcement and corrupt practices he found there. The inspection process was marked by a large degree of arbitrariness. Williams accused officials of placing "holds" on immigrants based not on an actual inspection, but rather on information on the ship's manifest. "The fact that most of those marked were able-bodied people with large amounts of money are points not without interest," Williams wrote, slyly implying a shakedown racket.

Angry at the sloppiness, corruption, and lack of professionalism among the Ellis Island staff, Williams continued to weed out workers who had given the place a bad name. By late September, he fired the accused serial groper John Lederhilger. By year's end, all officials named in the Campbell-Rodgers report had been pushed out.

Others also felt Williams's wrath. Emile Schamcham, a Syrian interpreter, was dismissed from his job for trying to get a date with an immigrant girl. While this woman was waiting at Ellis Island for a friend to meet her, Schamcham slipped her a note with the address of his boardinghouse. Another Williams target was a clerk named James Fraser, who had been away from his post for four straight days on an alcoholic bender—and apparently not for the first time. He told Williams he had contracted a disease during the Civil War that forced him to use alcohol as a stimulant. Under the new regime, such excuses would not be tolerated. Fraser was fired. Malingerers were no longer wanted and could no longer take cover under civil service rules or the protection of political patrons. When Senator Platt asked the new boss of Ellis Island to promote Samuel Samsom from gateman to inspector, Williams brusquely wrote back that Samsom "is not fitted either by temperament or training for a position much above that held by him now."

Nor would the abusive treatment of immigrants be tolerated. Six weeks into his administration, Williams posted the following notice throughout the main building at Ellis Island:

> Immigrants must be treated with kindness and consideration.
> Any Government official violating the terms of this notice
> will be recommended for dismissal from the Service. Any
> other person so doing will be forthwith required to leave Ellis
> Island. It is earnestly requested that any violation hereof, or
> any instance of any kind of improper treatment of immigrants

at Ellis Island, or before they leave the Barge Office, be
promptly brought to the attention of the Commissioner.

Williams was dead serious about enforcing his edict. He wrote to
one employee: "I was very much displeased at the rough and unkind
manner in which I heard you address two immigrants in the Discharg-
ing Bureau this afternoon. Do not let this occur again." Williams suspended a gateman named John Bell for two weeks without pay for using
"vulgar and abusive language" with an immigrant.

No area of immigration escaped Williams's attention. He kept a
close eye on steamship companies, fearing that they were not doing a
proper job of inspecting immigrants in European ports. On his fifth
day on the job, he fired off a letter to the French Line complaining
that while its manifests listed all immigrants as being in sound physical
condition, Ellis Island doctors found a number afflicted with various
ailments, such as hernias, blindness, and clubfeet. One immigrant had
only one leg, another had one leg shorter than the other, and a third
was a hunchback. Williams fined steamship companies for failing to inspect immigrants properly. Between May 1902 and May 1903, Williams
collected $6,560 in fines from steamship companies.

Next, Williams aimed his fire at those missionaries at Ellis Island
whom he believed were runners in disguise, suckering unwitting immigrants to their rooming houses and taking advantage of them. He barred
a German Lutheran minister, a man who ran the Home for Scandinavian
Emigrants, and members of the Austro-Hungarian Society for swindling
immigrants and keeping an unsanitary and unsafe boardinghouse.

To protect immigrants from falling prey to swindlers, Williams took
on the concessions at Ellis Island—the money exchange, baggage transfer contract, and food services. Herbert Parsons, a Republican leader
in the city, warned Williams about the food concession. Though the
contract was in the name of Schwab & Co., the business was really
run by Charles Hess, a local Republican leader connected to the Platt
machine. According to Parsons, Hess was "one of the most unmiti-
gated scoundrels in this city." McSweeney had shielded Hess under the
previous administration, but that would change. "I witnessed with my
own eyes the fact that immigrants were often fed without knives, forks,
or spoons and I saw them extract boiled beef from their bowls of soup
with their fingers," Williams reported.

New bids were put out and new contracts were awarded for the food, baggage, and money exchange privileges. Though the opening of Ellis Island and the federalization of immigration regulation were supposed to have eliminated the kinds of corruption that had existed at Castle Garden, the present state of these privileges showed that little had changed. The owner of the baggage contract had held it since Castle Garden, while the money exchange was in the hands of the nephew of the man who held it at Castle Garden. "This office has been run in the past largely in the interests of the restaurant privilege holder, and partly in the interest of some steamship companies, who have been violating our statutes with impunity," Williams wrote triumphantly to Roosevelt after the new contract had been awarded. "This office is now being run in the Government's interest."

Williams even tackled the landscaping of Ellis Island. While the *Times* noted that before Williams "there was not a flower or a bush of any kind on the island," by the summer of 1903 the island was taking the appearance of a "well-regulated and unusually prettily decorated park." And from the front door of the main building to the dock where the barges dropped off immigrants, a new steel canopy with a glass roof was erected to shelter immigrants as they began their inspection ritual.

Then there was the case of Edward McSweeney. Although McSweeney had been dismissed by Roosevelt, controversy still surrounded him. Williams and McSweeney overlapped for three days at the end of April, enough time for the streetwise McSweeney to sell his library of books and periodicals to Williams for the exorbitant price of $100. When Williams took office, not only was the entire service at Ellis Island a mess—from the quality of the inspectors to the quality of the food to the cleanliness of the buildings—but the records and files were also in disarray.

McSweeney had asked Williams if he could store five large boxes at Ellis Island until he could bring them up to Boston, where he was moving. The boxes, he told Williams, contained personal papers and materials. When someone told Williams that McSweeney had placed official documents in the boxes, he referred the matter to his superiors, who then dispatched a Secret Service investigator to New York.

The agent opened the boxes and found inside thousands of documents—4,292 to be exact—relating to official work at Ellis Island. There were letters, special reports, and minutes of boards of special inquiry.

When McSweeney wrote to Williams in August asking him to forward the boxes to Boston, he was informed they were being held by order of the secretary of the Treasury. Williams did pack up two small boxes of personal items from the larger boxes and shipped them to Boston.

In addition to the five large boxes, Williams was told that Mc-Sweeney had ordered a large cedar chest made at government expense. A Secret Service agent managed to track down the box to a storage facility in Manhattan, but could not open it. Government officials asked McSweeney, through his lawyers, to open the box. After some days of stalling, government officials were allowed to open the chest and found only bed linens. Clearly the chest had been cleaned out. While the original contents of the cedar chest will forever remain a mystery, William Williams made certain to keep a record of what was found in the boxes held at Ellis Island.

According to Williams, McSweeney was in a state of mental anguish when he discovered that the boxes had been opened. One reason was the strange accusation that emerged from those boxes concerning the case of two teenage girls. The Eloy sisters had been caught showing a "filthy and obscene photograph" to other immigrants awaiting inspection. The girls were then brought before McSweeney, who allowed the girls to land. McSweeney later told investigators that the photo in question had mysteriously disappeared. However, all of the material relating to the case, including the photo, was carefully filed away by McSweeney in a small manila folder marked "Eloy girls."

None of this seemed to slow down the indefatigable McSweeney. While Terence Powderly was at home in Washington sulking, McSweeney was back home in Massachusetts running the campaign of William A. Gaston, the Democratic candidate for governor. Gaston had been Roosevelt's classmate at Harvard and gave the president his personal assurance that McSweeney could explain the documents if given a chance.

Roosevelt then ordered Henry Burnett, the U.S. attorney in New York, to interview McSweeney. In an interview that lasted almost two days, McSweeney said that he never intended to take away the documents but instead wanted to put them aside to help William Williams, whom he had hoped would call upon him for advice. Williams called the testimony "confused and contradictory." He had no doubt as to McSweeney's dishonesty, writing Roosevelt that if he were wrong about McSweeney, then "I am so lacking in intelligence that I am not fit to hold this office one day longer."

It was not until the summer of 1903 that the president and Burnett agreed to file charges against McSweeney for purloining government documents. McSweeney's lawyers claimed that the charges were trivial, that he never meant to take official documents, and that the boxes never left Ellis Island. If they were so valuable to McSweeney, his lawyers argued, why didn't he take them with him immediately?

Because of an electoral fluke, William Gaston was running again for governor of Massachusetts in 1903 and McSweeney was again running the campaign. Gaston staunchly defended McSweeney, claiming that the charge was "technical" and the papers had "no earthly importance."

The case took on political overtones as Roosevelt took an interest in Gaston's campaign and the prosecution of McSweeney, calling the latter a "dog" and an "indicted scoundrel." The president suggested to the Republican candidate for lieutenant governor that he should use Gaston's employment of McSweeney against him in the campaign.

The case remained in limbo until a former clerk at Ellis Island named John Steele testified on behalf of McSweeney. He said that his old boss had ordered him to pack up his personal papers. In doing so, he emptied all the drawers in McSweeney's desk, mixing personal papers with official papers and then nailing the boxes shut. He claimed that McSweeney was not present when he did this. Thomas Fitchie also testified on behalf of his former assistant, claiming that in the move from the Barge Office back to Ellis Island, there was a mixup in the department's filing system.

Though McSweeney tried to remove government documents that would have made him look bad, Steele's testimony, combined with the fact that the papers never left the island, weakened the government's case. As William Williams noted, the question of McSweeney's guilt hinged on whether there was real motive surrounding the keeping of the documents. The *Boston Herald*, a staunch defender of McSweeney, argued that it was "utter frivolity . . . to accuse a man of having the criminal intent to steal papers which he voluntarily leaves in his accuser's possession." Nor could McSweeney be prosecuted for the mysterious cedar box that contained only bed linens. In June 1904, more than two years after McSweeney left office, all charges against him were finally dropped.

McSweeney was more a typical late-nineteenth-century political operative than a true villain. He cut corners, bent rules, ingratiated himself to powerful people, fought against real and perceived enemies, and too often put his personal survival ahead of public service.

Yet he was a deeply complex man. To his credit, he refused to let the indictment tarnish his career. Back in Boston, McSweeney reestablished himself as a prominent citizen. Besides running Gaston's two unsuccessful gubernatorial campaigns, McSweeney also became the editor of the *Boston Traveler*, where he led a campaign against tuberculosis. He fought for a workmen's compensation bill and was a member of the Massachusetts Industrial Accident Board. He was later put in charge of the Port of Boston. He continued writing and speaking on immigration, defending both the federal regulations enforced at Ellis Island, as well as the positive benefits of immigration.

As McSweeney was re-creating himself in Massachusetts, William Williams was taking firm control at Ellis Island, cleaning out McSweeney's allies and putting order to what had once been a dumping ground of political patronage. According to one Roosevelt biographer, at the new Ellis Island "a political snug harbor was swept, garnished, and set in running order on a strict merit basis."

Every move that Williams made seemed to vindicate Powderly. This was cold comfort for the ousted official who was now sitting in his home in the Petworth section of Washington tending to his vegetable and rose gardens, looking out across the street to the verdant grounds of the Old Soldiers' Home.

Shortly after his dismissal, Powderly found himself a speaker at the annual convention of the Brotherhood of Locomotive Firemen, the union that his successor, Frank Sargent, had previously led. Also in attendance that day was Theodore Roosevelt, who was being named a life member of the firemen's union. For Powderly, the event must have been difficult. "It is a great honor to a Labor Union to enroll the name of the nation's President on its roster," Powderly declared in his speech, "but I fear he has not the making of a good fireman, at least I don't like the way he 'fired' me."

These were difficult times for Powderly. Writing to his friend Robert Watchorn, Powderly said he was "feeling very blue and lonesome, and am also suffering from an attack of cholera morbus or something akin thereto." A few days later, he told another friend that he had "never felt so humiliated in all my life as on being turned out of a position that I did everything in my power to make respectable and dignified."

Powderly's depression deepened as tragedy continued to shadow his life. On top of being fired, he was still dealing with the grief of his wife's death in October 1901. In May 1903, his brother Joseph died suddenly. Robert Watchorn had visited Powderly to cheer him up, but he feared that when he left, his friend would "relapse into his morbid and apprehensive mood." Powderly, whose image once adorned the homes of the nation's working families, now felt abandoned and forgotten.

But Roosevelt had not forgotten Powderly. In the spring of 1903, less than a year after firing him, the president summoned him to the White House. Roosevelt admitted that he had been wrong to dismiss Powderly, and he wanted to reinstate him elsewhere in government. As plans to prosecute McSweeney were in motion, Roosevelt tried to get Powderly a job in the Justice Department. The president explained to Attorney General Philander Chase Knox that the more he looked into the immigration affair, "the more satisfied I am that Powderly was fundamentally right in his attitude." Roosevelt alluded to Powderly's 1898 letter as a mistake for which he had "amply atoned" with his time out of office. Roosevelt told a friend, "my conscience does not approve the action taken in Mr. Powderly's case, and the more I look into this matter, the more I am convinced that he was wronged, and I was misled."

Nothing came of the Justice Department job and Powderly sank further into gloom. While McSweeney moved on with his life in Boston, Powderly could not shake the embarrassment of his dismissal. He came to deeply resent Roosevelt. Powderly would vote for him in the 1904 election, despite the fact that he found "no reason to admire him" and felt good reason to dislike him. Powderly still hoped that after the election, the stain on his record would be wiped out and he would be returned to government service.

He would have to wait two more years for that moment to arrive.

LEON CZOLGOSZ'S SIMPLE ACT of murder had elevated one man to the presidency, but it also indirectly led another man to a basement prison at Ellis Island.

On October 23, 1903, seven months after anarchists were legally banned from the country, a contingent of Ellis Island inspectors, Secret

Service agents, and New York City policemen raided the Murray Hill Lyceum in Manhattan. They brought with them an arrest warrant for John Turner for espousing anarchist beliefs. A British citizen who had arrived in the United States days earlier, Turner had been invited by anarchist Emma Goldman to give a series of lectures. Now he was being taken to a small cutter waiting to ferry him to Ellis Island. Once there, he would be imprisoned in one of three nine-by-six steel-bar cells in the basement of the main building.

Goldman called Turner's new home a "fetid dungeon," not knowing that sixteen years later she too would become a prisoner of Ellis Island. A "philosophical anarchist," as the papers called him, Turner had the entire basement jail to himself, with the exception of two guards. While in public Goldman railed against Turner's situation, in private she noted that Turner had gained twenty pounds while at Ellis Island and was "wonderfully evenly balanced and easy going as only an Englishman can be." Despite this, Turner's plight proved useful fodder for anarchists like Goldman in their battle against what they saw as a reactionary government and established authority.

Writing from his Ellis Island jail, Turner noted how he was being held and threatened with expulsion because "the law imposes certain standards of opinion, of beliefs and disbeliefs." Steamship companies at European ports were now asking every prospective passenger to America whether he or she was an anarchist. If they answered yes, they were refused passage. Turner had kind words for William Williams, calling him "keen, businesslike, yet always courteous," but was baffled by what he called the "strange procedure" of the board of special inquiry that heard his case.

Clarence Darrow soon took up Turner's case, joined by the poet Edgar Lee Masters. While the case made its way to the Supreme Court, Turner was released on bond in March 1904 after four and a half months in jail at Ellis Island. He continued on his lecture tour, speaking out in favor of the merits of a general strike, by which workers would grind the capitalist system to a halt and bring freedom to the oppressed.

In May 1904, the Supreme Court ruled in favor of Turner's deportation. Perhaps sensing that his legal case was doomed, Turner beat the authorities to the punch and left the country of his own volition two weeks before the decision.

Though Turner was technically never deported, his was the first case of an alien to be ordered deported from the United States because of his political beliefs. In upholding Turner's arrest and eventual deportation, the Supreme Court once again validated the unique status of the administrative rules in place at Ellis Island. Darrow had turned to the First Amendment to defend his client, but the Court unanimously declared they were unable to understand how immigration law violated the First Amendment.

Turner certainly had his right to speak abridged, but that was merely a function of his exclusion under immigration law. "To appeal to the Constitution is to concede that this is a land governed by that supreme law, and as under it the power to exclude has been determined to exist, those who are excluded cannot assert the rights in general obtaining in a land to which they do not belong as citizens or otherwise," wrote the Court. In other words, since John Turner did not belong in the United States, the constitutional protections of personal freedoms did not apply to him. It was as if he had never entered America in the first place. As Emma Goldman would discover years later, the precedent of the Turner case would continue to reverberate at Ellis Island.

There was another quirk to the Turner case: he was not an immigrant. Turner had merely come to the United States to lecture and then planned to return to his job in England. Ellis Island was no longer just regulating immigrants, but rather any alien headed to U.S. shores even for the shortest visit. Even a casual tourist could find himself stewing in detention at Ellis Island while authorities decided whether he fell into one of the categories for exclusion.

When Turner's case made it to the Supreme Court, the named defendant was William Williams. The two men are forever linked in legal history. The commissioner of Ellis Island was not just content to clean up the patronage mess at the immigration station. His real goal was a stricter enforcement of immigration laws that would keep undesirables like John Turner from entering the country. That goal would prove far more controversial than making sure inspectors spoke politely to immigrants.

Chapter 8

Fighting Back

———••———

*The fact is that a reformer can't last in politics. He can make a
show for a while, but he always comes down like a rocket. . . .
He hasn't been brought up in the difficult business of politics
and he makes a mess of it every time.*
—George Washington Plunkitt, 1905

GEORGE WASHINGTON PLUNKITT LIVED JUST THREE LONG
city blocks west and three short city blocks south of William Williams's bachelor accommodations at the upper-crust University Club.
But those six blocks were a gulf as wide as any ocean. Plunkitt, who
served as a New York state senator while Williams was at Ellis Island,
was the epitome of the Tammany Hall ward boss. Working-class Irish
Catholics like Plunkitt entered politics as a profession looking for
profit, not as a service to the public interest.

Reformers like Williams had little respect for the Tammany bosses
and the feeling was mutual. To Plunkitt, men like Williams were amateur dabblers with no real understanding of the messiness of democracy and a disdain for the average citizen. They put ideals and morals
ahead of practicality. True to Plunkitt's maxim, Williams would make
a good show for a while, but would soon come down like a rocket.

Theodore Roosevelt was a reformer too—at least he styled himself
that way. He was equal parts zealotry and flexibility, a combination that
served him well in public life. It was a style that even Plunkitt probably
could appreciate.

William Williams, on the other hand, was all zealotry and no flexibility. Starchy as the high, white collars favored by men of that era, Williams was convinced of his utter correctness in all matters and had little use for those who might differ with him. That Williams had done yeoman's work in cleaning up Ellis Island made his attitude even more unfortunate.

Having first moved swiftly to clean up the immigration service in New York, Williams proceeded to tackle what he felt was an even more vital part of his job: a rigid enforcement of the immigration laws.

A year in office at Ellis Island confirmed Williams's low opinion of America's new immigrants. What he had seen in his first year on the job was "a particularly undesirable stream of immigration." In response, Williams stepped up the exclusion of immigrants, keeping an especially close eye on those he considered paupers or likely to become public charges.

Williams's appointment elated members of the Immigration Restriction League (IRL). For the first time since Ellis Island opened, a true restrictionist and New England patrician was now guarding the gate. Williams kept in contact with members of the IRL, telling Prescott Hall he wanted even stricter exclusionary laws. In the meantime, he would work within the law to prevent undesirable immigrants from entering the country.

Immigrants were on notice. Take the case of twelve-year-old Raffaele Borcelli, suffering from an advanced case of the scalp disease favus. When lawyers tried to intervene on behalf of the young boy, Williams bluntly informed them that America did not want "diseased people in this country and I intend that they shall not come."

Williams believed that the current laws were not going far enough. He told President Roosevelt that what was needed to "meet the real evils of the situation" was new legislation. In its absence, Williams was going to do his part to protect American civilization. In November 1902, he provided guidance for Ellis Island inspectors in interpreting the law: "Any inspector who passes an alien who may not be 'clearly beyond a doubt' entitled to land, violates his oath of office," Williams informed his subordinates. "The purpose of the statute is to exclude undesirable aliens, not to invite aliens to come here. It casts upon them the burden of proving that they are entitled to admission." He was going to take the broad and vague classifications for immigration exclusion and tighten them.

Compare Williams's 1902 edict with McSweeney's interpretation of the same immigration law from three years earlier. "I have seen cases where an immigrant would fall within the letter of the law [of exclusion] and still in the opinion of the inspectors be a desirable immigrant," McSweeney told the Industrial Commission on Immigration. Williams believed he was appointed to end just that kind of laxity. He had every reason to believe that the president who appointed him also believed in tightening the law.

In his first Annual Message to Congress, Roosevelt allotted two long paragraphs to the immigration problem, calling the present system unsatisfactory. He went on to call for adding anarchists to the list of excludable categories, as well as some type of education or literacy test. Just as important, Roosevelt felt that all immigrants who were "below a certain standard of economic fitness to enter our industrial fields as competitors with American labor" should be excluded. Immigrants had to prove they could earn a living in America and had to have enough money to make a new life here.

So it must have surprised Williams when he received two letters from Roosevelt informing him that reports had been filtering into the White House from the president's German-American and Jewish friends in New York objecting to the treatment of immigrants at Ellis Island. They criticized what they called the "star chamber" quality of the boards of special inquiry, complaining that immigrants were deported before their relatives were notified of their landing, that immigrants were no longer allowed to have counsel during their exclusion hearings, and that Williams no longer allowed the issuance of bonds for those believed likely to become a public charge.

Roosevelt warned Williams that he needed to avoid the appearance of arbitrary harshness. While the president heartily approved of the exclusion of immigrants who "would tend to the physical or moral deterioration of our people," such actions needed to be tempered with compassion. Sending an immigrant back home, Roosevelt understood, was "to inflict a punishment upon him only less severe than death itself." Roosevelt asked Williams if he could include members of German, Jewish, and Italian immigrant societies in board of inquiry hearings.

It was a mild scolding, but a scolding nonetheless. The letter opened a window into Roosevelt's conflicted view of immigration. Good immigrants were welcome; bad immigrants need not apply. Yet above all, such sifting of immigrants had to be done with the utmost sensitivity and without regard for race, religion, or ethnicity.

Williams at first responded with uncharacteristic deference. "I have carefully noted all that you say," Williams wrote. "It will be my earnest endeavor at all times to execute the immigration laws rigidly, but fairly and without unnecessary friction and I think I can satisfy any reasonable person that I have never exhibited any anti-foreign feeling." He informed the president that, as per his orders, he had invited some representatives from immigrant societies to lunch at Ellis Island.

Williams then shot back with another letter to Roosevelt in a more characteristic style. "Any reliable person asserting that the immigrant is judged by 'star chamber' methods must be densely ignorant of the facts," wrote an indignant Williams. While admitting the need to enforce the immigration laws without "friction," he added snippily, "Of course, I do not call it lack of discretion in proper cases to show up thieves, dismiss missionaries for revenue, or expose fraudulent practices of steamship agents."

Williams continued a few days later even more unapologetically. While reassuring the president that he was consulting with philanthropic groups, immigrant aid societies, and social welfare agencies, he informed Roosevelt of just what he found in these talks. "Every intelligent person with whom I converse (whether engaged in charitable work or business) is of the opinion that altogether too many low-grade aliens are entering this country," Williams wrote. He was not against immigration in general, nor did he oppose hearty immigrants ready to work, even if they were not highly intelligent. What he opposed was the minority of undesirable immigrants.

By background and temperament more of a New Englander than a New Yorker, the Connecticut-born Williams shared many of the fears of the Immigration Restriction League. In contrast, Roosevelt's background was tempered by his connections and friendships with New York City's ethnic groups. For Roosevelt, it was a constant fight between his patrician side, which looked on some newcomers with dismay, and his pluralist side, which believed that it was character, not education or race or religion, that counted the most when judging individuals.

A man like William Williams appealed to Roosevelt's patrician side.

For Williams, as for Roosevelt, the regulation of immigration was not just about preserving Anglo-Saxon culture; it was also about limiting the selfish interests of big business in the interest of the public good. With the immigration question, Williams argued, America was "confronted with problems of far greater importance than the immediate material development of the country." Just as progressives argued that unrestricted laissez-faire damaged the social fabric, Williams argued that Americans could not "sacrifice our National ideals and character for mere pecuniary gain."

For many progressives, being critical of the excesses of capitalism meant not only criticizing the selfish greed of businessmen and the unfair competition of the trusts; it also meant regulating the tide of immigrants fueling industrial America. Being progressive meant being in favor of a strong national government to rein in private interest; it also meant, for Roosevelt, Williams, and many others, that "National ideals and character" existed.

Williams ended his 1903 Annual Report by arguing that a "too rapid filling up of any country with foreign elements is sure to be at the expense of national character when such elements belong to the poorer classes in their own respective homes." Although the words were published under Williams's name and no doubt reflected his views, they were actually penned by Theodore Roosevelt, who added those words in his personal edits of the original text.

Perhaps for that reason, Williams was not deterred by Roosevelt's mild rebuke. Williams could proudly note that "the worst riff-raff of Europe" was kept out of America, yet more work was needed. Many of those who technically did not fall under the excludable categories of immigration law, in Williams's opinion, were still undesirable.

Over 857,000 immigrants arrived during Williams's first year, of whom about 60 percent were Italians, Jews, and Slavs. These new immigrants were overwhelmingly male (including 89 percent of all Croatians and 81 percent of all Italians), overwhelmingly unskilled (including 96 percent of all Ruthenians and 89 percent of all Lithuanians), and mostly between the ages of fourteen and forty-five. Anywhere from one-third to one-half of these groups were illiterate, and, on average, they came with about $9 per person. They were mostly young, unskilled, illiterate males with little money but the necessary muscle and brawn to run the country's mills, factories, and powerhouses and build the subways and skyscrapers. Hardly paupers, yet definitely not professionals, these new immigrants were raw labor pure and simple.

In a 1906 book sympathetic to the plight of immigrants, Edward Steiner, an immigrant himself, described the lumpen masses from which new Americans would be created: "It is true that many criminals come, especially from Italy. Many weak, impoverished and poorly developed creatures come from among Polish and Russian Jews, but they are only the tares in the wheat. The stock as a whole is physically sound; it is crude, common peasant stock, not the dregs of society, but its basis."

Not everyone agreed. The poet Wallace Irwin took to the pages of a New York newspaper to express his thoughts at the rising tide of immigrants washing ashore in a poem entitled "Ellis Island's Problems":

> Down the greasy gang-plank
> See the motley pack
> Nothing in the pocketbook
> Tatters on the back
>
> Pauper, cripple, criminal
> Halt and blind and slow
> Has Uncle Sammy room enough
> To give 'em all a show?
>
> Crime, disease, and wretchedness
> Of a hundred lands
> All the world's incompetence
> Dumped upon our hands.

It was a sentiment with which William Williams would have agreed. In his 1903 Annual Report, Williams boldly estimated that at least two hundred thousand of the immigrants arriving that year "will be of no benefit to the country." Had they all stayed home, he argued, nobody "would have missed them," except of course the steamship companies that made money on their passage. Most of these immigrants came from "some of the most undesirable sources of population" of Italy, Austria, and Russia.

It was a cold but not uncharacteristic sentiment, but Williams was unapologetic. He called it "sheer folly" to allow in so many immigrants when the nation was trying to fix the social and economic problems brought about by rapid industrialization and urbanization. Besides, he declared, "aliens have no inherent right whatever to come here," asserting the right of sovereignty at the heart of American immigration law. Congress, acting upon the wishes of the people, had the right to determine who could and who could not enter America. Williams was an officer of the state, faithfully administering the wishes of the people, transmitted through Congress into law and executed by the Immigration Service. In such an arrangement, sentimentality was banished from the stage.

Though immigrants did not have a legal right to come to the United States, once they entered the country, became citizens, and put down roots, they too joined the immigration debate. It is no surprise that immigrant groups, especially Jewish and German leaders, would voice their concerns. These were men of high character, as Roosevelt would say, leaders in their community striving toward social betterment. They were the right kind of immigrants. Of Leopold Deutschberger, though, Roosevelt would probably not go so far.

A reporter for the *New Yorker Staats-Zeitung*, the city's German-language newspaper and one of the most influential of its kind in the country, Deutschberger covered the Ellis Island beat. Throughout Williams's tenure at Ellis Island, the German-American journalist published numerous inflammatory articles about alleged abuses of immigrants, with titles such as "Men Weep," "Disgrace to Country," "Unrestricted Despotism," "Barbarous Treatment of Immigrants," and "No Pity."

To his supporters, such criticism was proof of Williams's success. Keeping a close eye on affairs from Boston, Prescott Hall congratulated Williams "on the great tribute which the *Staats-Zeitung* is paying your administration. . . . I have never known the *Staats-Zeitung* to abuse anything as much as it has your administration, which of itself is the highest praise." To Hall, if the paper was criticizing Williams, then he must be doing something right.

Germans were the largest ethnic group in New York City and a political power in the Midwest. The number of German immigrants, however, had fallen off substantially in recent years. In the first two years of Williams's administration, only about 93,000 German immigrants arrived through Ellis Island, less than 6 percent of all immigrants. Of those only 696 were excluded, or about 0.7 percent.

Although Williams's animus was largely directed against southern and eastern European immigrants, it was the German press and the German-American community that were most angered by his administration. When Williams banned a German missionary from Ellis Island, a woman from Washington, D.C., wrote him to complain. If he did not reconsider his actions, she warned, millions of German-Americans would condemn him. "Do look over the matter again or you have struck trouble," she warned. With self-confidence bordering on arrogance, Williams responded: "I hasten to assure you that I am utterly indifferent to any unfavorable resolutions that may be passed upon me, or to any that may come to me, as a result of doing the right thing."

Even the *American Hebrew*, a prominent voice in the Jewish community, defended Williams, writing that the agitation "is not based on firm ground, and seems to be inspired by some motive other than the unselfish one of securing justice for the immigrant." The newspaper credited Williams with creating an atmosphere at Ellis Island where immigrants were well-treated and no longer "hauled about like foreign baggage." The editors encouraged Jews not to complain and "for the sake of their own self-respect refuse to ask for special treatment."

Despite the support from some quarters, Williams's problems with the German community continued. In early September 1903, Deutschberger published another article about Ellis Island, entitled "Hell on Earth." Among its accusations was that "people on the Island were literally eaten up by vermin."

Williams may have appeared indifferent to the criticism, but when combined with his frenetic work schedule, it was all beginning to take a toll. During 1903, Williams was at his desk for all but five days of the year—including Sundays and holidays. "I have for a long time felt that you were overworked and that it was only a matter of time when you could no longer stand up under the strain," Commissioner-General Frank Sargent wrote Williams a year into his tenure. Just after the publication of the *Staats-Zeitung*'s "Hell on Earth" article, Robert Watchorn was hearing rumors that Williams had already "run his mile" at Ellis Island.

For Roosevelt, it was time for a presidential visit to Ellis Island, the first one by a sitting president. The visit was scheduled for Wednesday, September 15. Roosevelt would leave his home at Sagamore Hill, on Long Island's Oyster Bay and arrive with his party on the presidential yacht *Sylph* in time for lunch. In addition to his wife and son Kermit, Roosevelt was joined by special guests, including friends Jacob Riis and Owen Wister, as well as local politicians, journalists, and academics. The Ellis Island dining facility added oysters and champagne fritters to its usual bland fare of stewed prunes for the occasion.

Roosevelt's trip began badly. The *Sylph* left Oyster Bay a little before 10:00 A.M. Strong winds and heavy rains beat hard against the *Sylph* as it made its way from the North Shore of Long Island, southwest into the East River toward Manhattan. The winds reached near hurricane force as the *Sylph* made its way around Fort Schuyler off the coast of the Bronx. As waves continued to break over the presidential vessel, Mrs. Roosevelt and Kermit were sent below deck. Upon reaching Hell's Gate, the notorious cross-current patch of water where the tidal straight known as the East River meets the Long Island Sound just off the Upper East Side of Manhattan, the presidential party saw a tugboat that had been capsized by the winds and waves. The *Sylph*'s pilot suggested that the trip be canceled, and the group landed at the Brooklyn Navy Yard until the storm blew over.

The weather eventually improved and Roosevelt's group got back on the ship and continued its journey. It was not until sometime after 2:00 P.M. that the presidential party approached Ellis Island, where they were met by a small tugboat that transferred the passengers of the *Sylph* to the slip at Ellis Island. The president stood at the front of the tugboat, in his raincoat and slouch hat, waving to a small crowd of officials, including William Williams, waiting in the rain to welcome the tardy presidential party. After more than four hours, Roosevelt and his party had finally made it to Ellis Island.

After a quick lunch, Roosevelt began his whirlwind tour of the facilities. Over two thousand immigrants were on the island when Roosevelt arrived, and the president dove right into the process. Not content simply to watch, he joined inspectors in questioning immigrants, including a fifteen-year-old Slavic orphan named Ildra Andras. After a few questions, the president gave Andras a hearty slap on the back and the boy was off to Minnesota to be with his uncle. When Roosevelt saw Adele Walte, a young German women carrying her sleeping baby in a wicker basket, he passed Jacob Riis a five-dollar bill to hand to the woman. "It's for the baby," Riis told a startled Adele, "It's from the President of the United States."

Little escaped the president's curious mind. Roosevelt was dismayed by the eye exam performed on immigrants, complaining that doctors had dirty hands and did not clean their instruments between patients. The eye exam, designed to uncover cases of trachoma, was the most infamous test at Ellis Island. Given at this time only to those who exhibited symptoms of the disease, by 1905 every immigrant passing through Ellis Island would be subjected to it. Usually using a buttonhook, a doctor would flip back, or evert, the immigrant's eyelid to look for signs of trachoma. For some, it was a painful and traumatic experience. "The eye of the unsuspecting arrival is so brutally pulled open by the doctor," noted a German-language newspaper, "that the poor unfortunate is unable to see anything for the next two or three hours because of the pain." With some exaggeration, the paper called it "a brutality without equal." From 1904 to 1914, almost 25,000 immigrants would be debarred for trachoma, nearly two-thirds of all those excluded for loathsome or contagious diseases.

After this, Roosevelt and his party were off to a hearing room to witness the boards of special inquiry. One case dealt with a Hungarian man heading to his son-in-law in Pennsylvania with a railroad ticket and $12 in his pocket. Was he likely to become a public charge? Two members of the board voted to defer the decision for further investigation, while one member voted to allow the man to land. "Why should there be any doubt about this man," the president chimed in. Williams, an upholder of the strict interpretation of the law, tried to explain to Roosevelt that immigrants had to be beyond a doubt entitled to land. Since the old Hungarian had only $12, Williams declared him certain to become a public charge. To that, the German-born Arthur von Briesen, a member of Roosevelt's party and president of the Legal Aid Society, interjected: "Under the law, Jake Riis should have been sent back when he came over." That sealed the deal in favor of the Hungarian.

Von Briesen's presence at Ellis Island that day was important to more than just that Hungarian immigrant. Roosevelt used the trip to announce that he was appointing a commission to investigate the operations at Ellis Island. This was news to William Williams, who had not been previously informed of the decision.

Among those invited to the island that day were the five men Roosevelt had already chosen to sit on the commission, including von Briesen, who would head the commission. From the dramatic, rain-soaked arrival to the surprise announcement, it was vintage Roosevelt. Everyone assumed that Roosevelt created the commission in response to the charges from the *Staats-Zeitung*. What better way to counter the complaints that Roosevelt's immigration service was anti-immigrant than to appoint a committee composed of, in the words of a newspaper critical of the president, "two Germans, two Irishmen, and a Jew—not a single native American."

Roosevelt could not have picked a better commission from his perspective. As von Briesen wrote the president after the completion of the report, the commission was unanimous in agreeing that "desirable immigrants are men and women of good repute and good character and that undesirable immigrants are persons of bad repute and bad character." This was the Roosevelt party line on immigration, which he had earlier reiterated in a letter to another member of the commission: "My own feeling is that we cannot have too many of the right kind of immigrants and that, on the other hand, we should steadily and consistently endeavor to exclude the man who is physically, mentally or morally unfit to be a good citizen or to beget good citizens."

It was not only pure Roosevelt, but his immigration axiom neatly encapsulated the broad American consensus toward immigration. A few Americans may have supported unrestricted immigration, and a larger number may have supported a complete shutting of the nation's gates. Yet even the Immigration Restriction League did not go so far as to lobby for such extreme measures. Public opinion polling was still decades away, so it is difficult to gauge accurately what exactly the American public believed, but the consensus, as witnessed through immigration policy and elite opinion, seemed to support some kind of regulation and selection of immigrants, while upholding the nation's traditional views on the benefits of good immigrants.

The devil, of course, was in the details. How does one define good and bad immigrants? Each person who worked at Ellis Island, from commissioner to inspector to doctor, had his own interpretation of that dividing line, as did officials in Washington.

The Von Briesen Commission was the fifth investigation of Ellis Island in eleven years—and certainly not the last. It was the first to deal exclusively with the concerns of pro-immigrant groups. Williams's presence at Ellis Island satisfied that part of Roosevelt's patrician psyche worried about the wrong kind of immigrants, but the pluralist side of Roosevelt needed to be soothed as well. Appointing an ethnic commission to investigate his handpicked Ellis Island commissioner dutifully following Roosevelt's own beliefs about immigration was a masterly, yet cynical political stroke.

For native-born Americans fearful of the rapid changes of the late nineteenth and early twentieth centuries, the man at the gate at Ellis Island was a comforting idea that made mass immigration a more palatable concept. As immigration continued and first- and second-generation immigrants entered the American mainstream, they too wanted Ellis Island to reflect *their* values. As Roosevelt was well aware—and William Williams would soon discover—the growing political power of immigrant groups meant that operations at Ellis Island had to take into account the sensitivities of immigrants as well. The tension of serving as a symbol for both immigrants and restrictionists would define—and haunt—Ellis Island its entire history.

The Von Briesen Commission served as a sounding board for complaints from numerous ethnic and religious groups. The first to testify was Leopold Deutschberger and his editor at the *Staats-Zeitung*, repeating their charges of maladministration against Williams. The German newspapermen were followed by a long procession of ethnic representatives, members of the German Lutheran Society, the Irish Emigrant Society, the Austrian Hungarian Home, Our Lady of the Rosary, United Hebrew Charities, the Leo House for German Catholic girls. All of these witnesses testified on behalf of Williams. Sure, they had their quibbles. Most complained about overcrowded conditions, small waiting rooms, not enough bathrooms or benches. And then there were the steamship company representatives, who had their own complaints.

Two months after Roosevelt's trip to Ellis Island, the commission completed its report and sent it to the president. It was largely an exoneration of Williams, although it included some criticism of the sanitary conditions on the island, the money exchange, and the overcrowding. As to the criticisms brought by the German-language press, the report declared them unfounded.

Roosevelt was happy with the report, except for one detail. Though it dismissed every charge against Williams, the president regretted that it "did not in one telling sentence embody what it in effect said, and back up Williams not merely by inference but by positive aggressive statement." Despite all the criticism and despite Williams's personality quirks, Roosevelt still held him in very high regard.

The final report contained one sentence declaring that Williams was "entitled to the highest commendation for the indefatigable zeal and intelligent supervision" at Ellis Island. That was not enough for Roosevelt. After all, the point of the commission was the vindication of Williams. The president wanted the report to speak specifically of his integrity. Roosevelt was going to use the commission, entirely made up of ethnic members, to both soothe ethnic concerns and absolve the restrictionist Williams. Perhaps uncomfortable with his role in this Rooseveltian play, commission member Eugene Philbin answered the president's charges with some odd logic of his own. He believed it was "absolutely necessary that the report should avoid, as far as possible, anything like actual praise, but that it should be so worded as to have the inference irresistibly created that the administration of the island was a most commendable one."

In any case, Williams weathered the storm and continued his zealous enforcement of immigration laws. Roosevelt, up for reelection in 1904, could legitimately appeal to Americans in favor of immigration restriction by pointing to Williams, but also appeal to immigrant ethnic groups by pointing out his deep concern for conditions at Ellis Island.

Williams, however, continued to speak out against what he saw as the large numbers of undesirable immigrants streaming through Ellis Island. His writings showed the strain of dark pessimism exhibited by New England restrictionists. "It is full time, however, for us to appreciate the fact that the settlers who made the country great belonged to a totally different class of people from those described and came here with loftier views of their prospective future," Williams wrote, "and that a desire to emigrate can no longer be regarded as evidence of initiative, thrift, or courage."

As proof, Williams offered a story about a family of eight from eastern Europe. The family had little money and was heading to a tenement district in New York City. When asked how he intended to provide for his family, the father responded by saying that his family did not care for a big house and would be satisfied with one room to sleep in: "That is all we want; that is the way we did it in Russia." To some, this might be a sign of an appropriately humble immigrant who was not demanding great riches from his adoptive country. Perhaps the father thought such a modest answer would impress officials. If that was the case, he thought wrong. To Williams, the family's aspirations were too narrow, and he sent the entire clan back to Europe.

Though Roosevelt said more about immigration than any previous president, he remained remarkably quiet about the issue during the 1904 campaign. "There seems to be a good deal of uneasiness as to saying anything about immigration this year," he wrote to Lodge. "It is not believed it would help us to getting legislation. There is no question but that there will be a sharp lookout kept to see if they cannot catch us tripping on it." Roosevelt may have wanted tougher immigration laws, but he felt it was best not to make any such references in the party's platform.

Roosevelt's campaign manager heard rumors that a Democratic operative had gone to Ellis Island to investigate conditions and warned Williams that Democrats saw the potential to make Ellis Island a campaign issue. A month later, Williams complained to Roosevelt about Congressman Richard Bartholdt. Though a Republican, the German-born former newspaper editor represented a heavily immigrant district in St. Louis. "He is very hostile to the Ellis Island administration, although he has been here, seen things as they are and had ample opportunity to satisfy himself that the *Staats-Zeitung* articles are false and malicious," Williams wrote. He warned that Democrats had recently produced a campaign document that attacked Ellis Island based on the *Staats-Zeitung* articles and using Bartholdt's comments.

All of this meant little to Roosevelt's reelection bid. He handily won reelection over a lackluster Democratic candidate, winning every state north of the Mason-Dixon line. While he lost heavily immigrant and Democratic Boston and New York City, Roosevelt ran well nationally among Germans, Poles, Italians, and Jews.

Roosevelt could be all things to all people. Restrictionists were heartened by the selection of Williams to run Ellis Island and the president's words calling for continued regulation and sifting of good immigrants from bad immigrants. However, immigrants and ethnic communities could also find comfort in Roosevelt's words and deeds.

In the end, it was not the accusations of insensitivity toward immigrants that ultimately drove William Williams out of Ellis Island. It was Joe Murray. The patrician Williams simply could not stand the unsophisticated machine politician. He described Murray as lazy and dull-witted and complained that he was "unable to write any kind of a letter. He can neither write nor speak correctly." Murray arrived late to work, could not complete basic tasks given to him, and, according to Williams, failed "to show any intelligent interest in anything that was going on to give me the slightest assistance in rooting out deviltry."

It galled Williams that the easygoing Murray had been on a first-name basis with John Lederhilger, even while Williams was drumming him out of government service. An exasperated Williams could do little to spur Murray to work harder, so he finally decided to leave him alone to do as he pleased, which turned out to be spending an inordinate amount of time shooting the breeze around the Ellis Island barber shop.

As a Harvard man, Roosevelt saw the problem clearly. "The trouble with Williams," the president wrote his friend Gifford Pinchot, "has been that owing to his past associations and education he has found it difficult to get on with men of inferior education and social status." In other words, Williams was an officious snob. Yet Roosevelt could not admit that his experiment in old-time patronage, while pleasing to Murray, not only stained Roosevelt's reform image, but also made the job of reforming Ellis Island more difficult.

Apparently, Williams's problems with his subordinates went beyond just Murray. On two different occasions, the Ellis Island workforce was on the verge of going out on strike. In cleaning out incompetent and abusive workers, Williams made enemies with his uncompromising personality. "They say he has his peculiarities and I presume he has," Robert Watchorn said of Williams, but if "he hadn't he would not be of much account."

Roosevelt appeared more than willing to overlook those peculiarities, remarking that he didn't "know anyone who could have done quite the work that he did." Roosevelt lauded Williams as fearless, energetic, and pubic-spirited—all the qualities that Roosevelt so admired. At the same time, he admitted that his dear friend Murray was not exactly the most engaged employee on the federal payroll.

In December 1904, Williams's patience finally ran out and he went to the White House to tell Roosevelt he could no longer work with Murray. Williams accused him of being "ignorant, inefficient, and wholly worthless" and said that he had played absolutely no part in helping to reform Ellis Island. Because Roosevelt held Williams in such high regard, he was willing to jettison Murray and keep Williams, although he hoped to place Murray in another government job.

But Williams did not just want Murray out as his assistant. He wanted his friend Allan Robinson, a fellow New York lawyer, named as his replacement. This Roosevelt could not abide. Frank Sargent informed the president that Robinson "possessed in even accentuated degrees the failings of Williams in dealing with other men." If Williams and Robinson were both in charge, Sargent feared a full-scale mutiny among Ellis Island's employees. Failing to get the assistant he wanted, Williams resigned in January 1905 and returned to his Wall Street law practice.

The Immigration Restriction League's Robert DeC. Ward was saddened at the news of Williams's departure. "It has been a source of constant satisfaction to me to feel that the gates at Ellis Island were so well guarded," Ward wrote Williams. Madison Grant, another patrician restrictionist, also sent his regrets.

Some immigration defenders praised Williams on his departure. The Society for the Protection of Italian Immigrants passed a resolution lauding Williams. While the editors of the *Staats-Zeitung* were no doubt rejoicing at the news, the *American Hebrew* was not. "He has transformed the internal affairs at Ellis Island to such an extent that visitors today will find very few of the evils complained of before he came," the paper concluded. "His retirement will be a distinct loss to the immigrant department."

In many ways, Williams personified George Washington Plunkitt's reformer. He had made a great show of reforming Ellis Island and ferreting out corruption, but he had his difficulties managing both immigrants and employees. Williams also took Roosevelt's division of immigrants of good character and bad character to extremes. Roosevelt could temper his concern about new immigrants with a positive view of American national character, the miracles of assimilation, and the benefits of good immigration. For Williams, there was little but pessimism.

The book on Williams's government service was not yet closed. There would be a second act that would both refute and confirm Plunkitt's suspicions about reformers.

Chapter 9

The Roosevelt Straddle

We can not have too much immigration of the right kind, and we should have none at all of the wrong kind.
—Theodore Roosevelt, 1903

LEANING OVER THE SECOND-STORY RAILING IN THE MAIN hall of the reception room at Ellis Island, H. G. Wells surveyed the mazelike rails herding immigrants through the inspection line. "You don't think they'll swamp you?" a concerned Wells asked his companion, the new Ellis Island commissioner, Robert Watchorn. Wells had taken the ferry trip to the island as part of research on a book about the future of America. Wells was pessimistic about the future in general, especially regarding technology. Yet as these two Englishmen debated the effects of throngs of southern and eastern Europeans on America, Wells's question hit upon another uncertainty.

"Now look here," Watchorn gently rebuked his famous literary guest, "I'm English-born—Derbyshire. I came to America when I was a lad. I had fifteen dollars. And here I am! Well, do you expect me, now I'm here, to shut the door on any other poor chaps who want a start—a start with hope in it, in the New World?"

Wells had cemented his reputation as the premier science fiction writer a decade earlier with a string of successes, including *The Time Machine*, *The Invisible Man*, and *The War of the Worlds*. Now Robert Watchorn was hosting the famous writer at Ellis Island. Both Wells and Watchorn were sons of the British working class who had made good. After the visit, the two men continued on friendly terms. Wells entertained Watchorn on a number of occasions in England, and Watchorn proudly kept an autographed photo of Wells in his office for the rest of his professional life.

This perk of the job, rubbing elbows with the famous and powerful, appealed greatly to Watchorn, whose life story was truly one of rags to riches. It began in the English coal mines and continued through his arrival at Castle Garden in 1880, to his ascension to commissioner of Ellis Island in 1905, and would continue after his time in the immigration service.

Watchorn was the second of seven children born in Alfreton, Derbyshire, to a doting mother and an alcoholic coal-miner father. At age eleven, Watchorn himself went down into the coal pits, where he worked for the next ten years. An intelligent boy, he went to night school, and at the age of twenty-two left for America.

Once there, Watchorn ended up loading coal in the Pennsylvania mines. Soon after, he brought his family over and became involved in the local Knights of Labor chapter, where he befriended Terence V. Powderly, who would remain a lifelong friend and mentor. Watchorn then went on to become the first secretary-treasurer of the newly created United Mine Workers.

Filled with ambition and drive, Watchorn did not remain long with the union. Like another determined member of the working class, Edward McSweeney, Watchorn made the leap from labor activism to politics. The thirty-three-year-old Watchorn became the state's first chief factory inspector under Robert E. Pattison, Pennsylvania's first Democratic governor since the Civil War.

Driven to succeed as only one who had escaped the coal pits of both England and Pennsylvania could, Watchorn cleverly amassed important friends, including Powderly and the Pennsylvania senator Matthew Quay, the state's Republican boss. Politically ambidextrous, Watchorn began his political career working in a Democratic administration but later became a staunch Republican. His ties to Powderly led to a patronage post as an inspector at Ellis Island. During the controversy there with McSweeney, Watchorn became an important ally and friend to Powderly, who later plucked Watchorn from the maelstrom at Ellis Island and promoted him first to Washington and then to

Montreal, where he put Watchorn in charge of the immigration service along the Canadian border.

When Roosevelt was searching for a suitable replacement for William Williams in early 1905, he quickly settled on Watchorn, whom he remembered from his first visit to Ellis Island, when Roosevelt was police commissioner and Watchorn a mere inspector. Morally upright, Watchorn could be expected to continue the vigilance against corruption, patronage, and abuse at Ellis Island, but would accomplish it without the abrasive air of the patrician Williams. As an immigrant himself, Watchorn might enforce immigration law without Williams's restrictionist touch. Also, Watchorn needed the job—unlike the independently wealthy Williams—and might be less difficult to manage.

On the issue of Joe Murray, Roosevelt only asked that Watchorn give him a fair shake. If Watchorn decided that Murray was incompetent, Roosevelt would transfer his friend. "You will be the absolute judge of his competency or incompetency," Roosevelt wrote. Watchorn, who had not escaped a life in the coal mines by bucking authority, was not about to take the bait. "I shall respect your wishes, Mr. President, in regard to Mr. Murray, whom I know very well," Watchorn responded. Murray would end up staying at Ellis Island for the rest of the Roosevelt administration.

Watchorn assured the president that they shared a common vision of immigration. Such agreement was important because America was about to witness its biggest wave of immigration ever. For the first time, more than 1 million immigrants entered the country. Roosevelt put this in historical perspective by noting that more people entered the United States in 1905 than had arrived in the 169 years between the first landing at Jamestown and the signing of the Declaration of Independence. Despite stringent laws, Roosevelt believed that a large number of immigrants were still undesirable because they came not of their own initiative, but were instead enticed by agents from steamship companies interested only in increasing their profits.

Roosevelt was adept at finding that perfect fulcrum of American opinion on immigration, melding fears of alien newcomers with respect for the country's open-door tradition. "In dealing with this question it is unwise to depart from the old American tradition and to discriminate for or against any man who desires to come here and become a citizen, save on the ground of that man's fitness for citizenship," Roosevelt wrote. An immigrant's character, not his ethnicity or religion, should determine whether he or she be allowed into the country. To him, a Slav of good character was far more preferable than an Englishman of poor character. Of course, the status of excluded Chinese immigrants complicated the president's argument.

It was a fine statement of the assimilationist credo, but one that rested on the vigorous enforcement of immigration laws at the nation's gates, with Roosevelt calling for "an increase in the stringency of the laws to keep out insane, idiotic, epileptic, and pauper immigrants." He already had had four years to push for this, but achieved little more than banning anarchists and prostitutes. Now he wanted not just anarchists excluded, "but every man of Anarchistic tendencies, all violent and disorderly people, all people of bad character, the incompetent, the lazy, the vicious, the physically unfit, defective, or degenerate."

If Roosevelt wanted a stricter application of immigration laws, Ellis Island was in the best shape since it opened to accomplish that. And just in time. From 1905 to 1907, some 3.5 million immigrants would come to America, nearly 80 percent passing through New York's inspection station. Having visited at the beginning of this period, novelist Henry James called Ellis Island "a drama that goes on, without a pause, day by day and year by year, this visible act of ingurgitation on the part of our body politic and social, and constituting really an appeal to amazement beyond that of any sword-swallowing or fire-swallowing of the circus."

With each passing week in the spring and fall—the peak arrival seasons for immigrants—a new record would be broken. In one week during April 1906, an estimated 45,000 immigrants arrived at Ellis Island. Ships seemed to pile one on top of the other, many forced to dock for two or three days as their passengers remained on board while they awaited inspection. Bigger steamships that could carry as many as 2,300 steerage passengers, like the White Star's Celtic and Republic, brought these immigrants on a daily basis.

"Immigrant Type Low, But 1,100,735 Get In" read a Times headline about the record number of immigrants in 1906. Of that figure, Ellis Island processed roughly 880,000 immigrants, 10 percent of whom were detained for board of special inquiry hearings, and 7,877 were excluded, less than 1 percent of all those who arrived. Ellis Island witnessed 327 deaths, 18 births, 2 suicides, and 508 marriages that year.

If Americans thought 1906 was bad, the following year would be even worse. In fact, Americans would not see as many immigrants in one year as they saw in 1907 until 1990. Some days, the flood was unmanageable. On March 27, 1907, 16,000 immigrants entered New York Harbor; May 2 brought 21,755. Ellis Island had to process over a million people in 1907 alone, which came to over 2,700 per day, every day.

Robert Watchorn, who oversaw this flood, was a man apart from his predecessor. "A man of brawn, a man who knows how to use his hands in both the sporting and industrial sense of the phrase," was how the *Times* described him. He repeated Roosevelt's mantra that America could not have enough of the right kind of immigrant and too little of the wrong kind. Unlike Williams, however, Watchorn believed that America *was* largely getting the right kind of immigrants.

This was a bit of an intellectual shift for Watchorn, a man who would prove himself nothing if not flexible in his beliefs. While working under Powderly, Watchorn portrayed himself in favor of strict regulation of immigrants, especially regarding the contract-labor laws. Now, working under Roosevelt, the former United Mine Workers official changed his tune. He found himself harangued for his pro-immigration views while speaking before crowds of workers.

Watchorn told a Jewish audience on New York's Lower East Side that "the immigrant has done as much for this country as the country has done for him." While he supported a careful selection of immigrants to keep out those likely to become a public charge, he hated to order deportations. Even though the editors of the *American Hebrew* had praised William Williams, they noticed a change in tone at Ellis Island. "Since Mr. Robert Watchorn entered upon his duties as Commissioner, there is an entirely different atmosphere about the place," the paper wrote. "The immigrant is no longer looked upon as one to be kept out, if the law is strained to do so."

College professor Edward Steiner dedicated his sympathetic book about the new immigrants to Robert Watchorn.

> He does not share the feeling that the immigration of to-day
> is worse than that of the past; in fact he will say quite freely
> that it is growing better every day. He has his fears and
> forebodings; but he knows that the miracle of transformation
> wrought on us, can still be wrought on this mass of clay

in the hands of the potter, which may be moulded just as millions of us have been moulded, into the likeness of a new humanity.

Men like Steiner and Watchorn held a deep faith in the transformative power of America on European immigrants.

Watchorn had a chance to explain his views to a group of female college students visiting Ellis Island. Unanimously opposed to immigration, these well-off young women heard the case of a sixty-six-year-old Italian man heading to his son in Lynchburg, Virginia. They believed him too old and weak to be admitted, especially since the son was not there to pick up the father. In a scene out of Hollywood, the son showed up at the last moment to an emotional reunion with his father. Should the father be sent back to Italy, Watchorn now asked the young women? "No, no, no, certainly not," was the unanimous response.

Those young women discovered the difference between discussing immigration in the abstract as opposed to dealing with the concrete—and very human—reality at Ellis Island. "There are those who vehemently protest against the landing of aliens on these shores en masse," Watchorn later wrote, "so long as their protests are made in abstract form, but who, Pilate-like, say, on being brought face to face with the units of the mass, 'I find no fault with him.'"

Watchorn's tenure marked an evolution in how Roosevelt handled immigration. Practical politics played no small hand in this change. In 1906, William Randolph Hearst used his fortune to run for governor of New York as a Democrat. Roosevelt could not abide Hearst and resented his "enormous popularity among ignorant and unthinking people." Hearst used the pages of his *New York Journal* to take on the mantle of defender of immigrants. He further expanded his reach to the city's largest ethnic group by starting the German-language paper *Morgen Journal*. The populism of Hearst's papers filled the patrician Roosevelt with disgust. He had to be stopped.

Roosevelt threw himself heart and soul into helping the Republican Charles Evans Hughes defeat Hearst. Hughes was a bit of a stiff, but enough of a progressive for Roosevelt—anything to keep Hearst from defiling Roosevelt's old office. The path to stopping Hearst, Roosevelt soon realized, began with New York's ethnic communities.

When an opening appeared for secretary of Commerce and Labor, Roosevelt jumped at an opportunity to make a point. Roosevelt conferred with Jewish leaders like New York banker Jacob Schiff and named Oscar Straus to the post. Roosevelt now had a Jew and a Catholic in his cabinet. (Charles Bonaparte, the grandnephew of Napoléon, was attorney general.)

At a dinner celebrating Straus's appointment, Roosevelt explained that he had chosen Straus without regard to race, color, creed, or party. To that, an elderly and increasingly deaf Jacob Schiff nodded and said in his thick German accent: "Dot's right, Mr. President. You came to me and said, 'Chake, who is der best jew I can appoint Segretary of Commerce?'" Though probably apocryphal, the spirit of the story contains a germ of truth. Roosevelt had begun a long tradition, followed by most of his successors, of choosing cabinet members to satisfy various racial, ethnic, and religious groups.

Straus, along with Schiff, belonged to an earlier generation of German Jewish immigrants. Oscar Straus was born in Bavaria in 1850. His father, a Reform Jew and grain merchant, left for the United States in 1852, where he ran a general store in Georgia. Oscar, his brothers, and his mother followed him there two years later. The family's future was not to be in the South, but rather in New York City. There, the Straus family ran a china and glassware store and later bought out Macy's. Oscar, however, was not drawn to the world of commerce like his father and brothers. Instead, he opted for a career in law.

As part of his arrangement with Roosevelt, Straus agreed to stump for Hughes in New York, joining Schiff in blunting Hearst's appeal to the Jewish community. In the end, Hughes squeaked by Hearst with just sixty thousand votes, and Straus took up work at his new job after the election.

The Bureau of Immigration and Naturalization, as it was now called, was but one of twelve divisions of the Department of Commerce and Labor, but it was clearly the one that animated Straus the most. "Indeed, no subject in the department occupied my daily attention to the extent that immigration did," he wrote in his autobiography. Immigration was the most difficult issue because "it is the most human" and "throbs with tearful tragedies," Straus wrote.

On the morning of December 17, 1906, Straus sat at his desk in his new office and immediately threw himself into the heart-wrenching morass of appeals from immigrants waiting to be deported. He looked at some thirty cases that first day. "I was not surprised to find that most of these cases present difficult questions appealing to the humanity and judgment of the Secretary," Straus wrote in his diary. Straus believed that the letter of the law must be tempered by humanity.

Some cases were easily disposed of, but others were more difficult. The power Straus possessed was enormous and would determine the futures of many individuals. It was a grave responsibility. "I felt that there was a domestic tragedy involved in every one of these cases, and as the law placed the ultimate decision upon the Secretary," he wrote, "I decided this responsibility was one that should not be delegated; so day by day I took up these decisions myself." So engaged was Straus that he brought a number of the toughest cases home with him that first night to examine in more depth.

"I would be less than human if I failed to interpret the laws as humanely as possible," Straus wrote his brother Isidor. "I propose to remain on the side of the angels come what will, and I shall defy hostile criticisms—to do less would be cowardly." Straus was especially sensitive to the plight of Russian Jewish immigrants, thinking it the height of cruelty to send Jews back to the nightmare of czarist Russia.

Straus made his first official visit to Ellis Island in February 1907, witnessing some 2,600 immigrants passing through that day. He appeared there again two months later, examining every detail of inspection from the time immigrants got off the ferries to the time they passed inspection.

Straus also heard a number of appeals cases, including that of a Scots-Irish family of seven ordered deported because one son was certified as feebleminded. The family was faced with a decision: Should they split up, with the mother or another sibling returning to Europe with the son and the others remaining in America? The family decided that they would all stick together—either the entire family would stay or the entire family would go back home. Straus thought the family, with the exception of the twenty-year-old feebleminded son, was "an exceptionally fine lot" and decided to allow the entire family to remain in America, including the son. Upon hearing the good news, the family burst into tears of gratitude.

Straus made yet another visit to Ellis Island in June 1908, joined by the commissioner-general of immigration, Frank Sargent, and other immigration and medical officials from East Coast inspection stations. Straus convened the conference to deal with medical cases that had caused him concern. They first took up the case of a fifty-nine-year-old Russian immigrant named Chena Rog, who was headed to her five children and thirty-six grandchildren in Reading, Pennsylvania. Rog had been diagnosed with trachoma, an infectious disease of the eye. Should she be ordered deported or held in a hospital for treatment?

When Straus asked Ellis Island's chief medical officer, George Stoner, about his opinion on the case, an agitated Stoner answered: "Just what I have stated in my certificate." Stoner and his staff had recommended deportation since trachoma was a contagious disease. They felt they were now being second-guessed by Straus. Can't she be treated for the disease, Straus asked? Stoner was not optimistic, arguing that it would take an "indefinite period which must be counted by years rather than by months." Straus kept pushing to see whether there was any way to avoid deporting Rog, who had no relatives back in Russia and whose children had become successful members of their community, as was attested by the presence of their congressman at the conference. Stoner became impatient by Straus's line of questioning and argued that there was nothing in the law that said that the officials had to treat Rog or any other immigrant suffering from a loathsome or contagious disease.

Clearly, Straus wanted the woman admitted, but Watchorn and Sargent argued that any ruling allowing diseased immigrants to land would be seen by steamship companies as an invitation to relax their own standards in Europe. They also sensed that their boss had already made up his mind, so they put their concerns aside and agreed to have the woman treated at the Ellis Island hospital. Chena Rog was permitted to land for medical treatment, practically guaranteeing that she would not be deported.

Stoner was not happy with the decision and had to have the last word. "I doubt very much whether she will be in any different condition at the end of six months' treatment than she is today," the doctor said. In fact, he believed that her condition could worsen and any other diagnosis was sheer folly. Straus ignored Stoner's speech and went on to the next case.

Schimen Coblenz was a forty-two-year-old butcher from Lithuania diagnosed with psoriasis, a skin condition. The disease was not particularly attractive, but in no way contagious. However, the law stated that immigrants could be deported for a "loathsome or dangerous contagious disease." Watchorn ordered him deported because psoriasis was loathsome and would be problematic in his profession as a butcher. If Coblenz were a factory worker, Watchorn argued, the disease would not cause his exclusion.

The case hinged on whether the law required a disease to be both loathsome *and* contagious or whether an alien could be deported just for suffering from a loathsome disease. It was clear that the law read "or," instead of "and," meaning that a loathsome disease alone could certify an immigrant for exclusion. However, since loathsome was a subjective term, and not a medical one, it was at Straus's discretion to decide the fate of immigrants diagnosed with such diseases. He was willing to read the law loosely and Coblenz was admitted.

Unfortunately, Straus could only appear at Ellis Island on rare occasions. Most of his influence would have to be exerted from Washington. While working late at night on immigration appeals in the library of his enormous Italianate villa on the capital's stately 16th Street on Meridian Hill, Straus came up with an idea. While his sympathies led him to find every means to allow an immigrant to stay in the country, Straus was also bound by the law. Though faithfully executing the law, he felt pangs of guilt for his role in excluding and deporting immigrants. He knew the devastation such decisions caused. Many immigrants had sold all of their possessions to come to America. Those excluded would return home broken in spirit, as well as financially ruined.

With this in mind, Straus sent a personal check for several hundred dollars to Watchorn, instructing him to dole out the money to unfortunate immigrants excluded at Ellis Island. Watchorn was to use his judgment in disbursing the funds. The only stipulations were that he was to disburse the funds without regard to "creed, country or race," and that the source of the money should remain anonymous. The move speaks volumes of Straus's humanity, as well as the heavy weight on his conscience caused by his work.

By 1907, it was clear that a perceptible shift in immigration policy had occurred. While the law remained the same, the tone of those in charge of enforcing the law had changed dramatically. Only someone like Theodore Roosevelt could have pulled off such a transformation. The shift was also reflected in the president's own rhetoric. In most of his earlier Annual Messages to Congress, Roosevelt reiterated his support for the strict regulation of immigrants. In his December 1906 message, he abruptly changed course.

"Not only must we treat all nations fairly," Roosevelt wrote, "but we must treat with justice and good will all immigrants who come here under the law. Whether they are Catholic or Protestant, Jew or Gentile; whether they come from England or Germany, Russia, Japan, or Italy, matters nothing." It was a far cry from his first message, five years earlier, when he called for weeding out immigrants of "low moral tendency" and "unsavory reputation."

While many worried that immigrants dragged down the standards of civilization and morality, Roosevelt now saw a different threat. "It is the sure mark of a low civilization, a low morality, to abuse or discriminate against or in any way humiliate such stranger who has come here lawfully and who is conducting himself properly," he argued. There would be no more talk of immigrants of the wrong sort or preservation of America's national stock. "I grow extremely indignant at the attitude of coarse hostility to the immigrant taken by so many natives," Roosevelt wrote editor Lyman Abbott.

Throughout the first decade of the new century, a more organized, pro-immigrant voice began to be heard. Political organizing on immigration had previously been the sole preserve of the Immigration Restriction League. In 1906, the National Liberal Immigration League was formed as a counterweight, opposing any further restrictions on immigration, as well as "all unjust and un-American methods of administering these [current immigration] laws." Yet even the most liberal immigration defenders did not support a completely open-door policy. The group wanted "to preserve for our country the benefits of immigration while keeping out undesirable immigrants."

The new organization's board included luminaries such as Princeton's president, Woodrow Wilson; Andrew Carnegie; and the president of Harvard, Charles Eliot. In addition, it was strongly allied with German-American organizations and received funds from German-owned steamship companies, lending credence to the charge that the pro-immigrant movement consisted largely of businessmen concerned with profits.

The pro-immigrant group also drew support from Jewish Americans, who wanted to make it easier for their coreligionists to escape religious persecution. Back in the 1890s, the German Jewish community had looked askance at the new immigrants from eastern Europe and many had even favored a strict interpretation of immigration laws. This stemmed partly from the snobbishness of cultivated and assimilated German Jews toward their poorer and more orthodox coreligionists, but also from the fact that needy Jewish immigrants from eastern Europe might become a burden on Jewish charities. It took repeated crackdowns in czarist Russia for America's German Jews to throw themselves wholeheartedly into the battle against further restriction.

The public debate over immigration revolved around how strict the regulation of immigrants should be, not on whether there should be any regulations at all. It was hard to find someone arguing either for completely restricted immigration or for a completely open door. Oscar Straus came close when he told the National Conference on Immigration that "the right to move from one part of the earth to another is a fundamental part of personal liberty." However, he prefaced the remark by saying, "We all agree there should be some restriction of unnatural immigration."

Closer to the general consensus on immigration policy was a 1907 *New York Times* editorial.

> It is well understood and admitted by all men of enlightened
> and unprejudiced opinions that selection, not exclusion,
> should be the guiding principle in any amendments of our
> immigration laws undertaken by Congress. . . . An immigrant
> capable of adding to the productive energy of the country
> is desirable. On the other hand, immigrants who are clearly
> beyond all dispute undesirable, who would be a burden or a
> source of danger to health, morals, and the public peace, are
> already under the ban of our statutes.

As an official devoted to upholding the law against undesirables as well as staying true to his belief in the positive contributions of immigrants, Robert Watchorn had to maintain a careful balance. As his friend Edward Steiner explained, Watchorn "must be both just and kind, show no preferences and no prejudices, guard the interests of his country and yet be humane to the stranger." It was a tall task for any individual and perhaps unrealistic to expect anyone to satisfy.

Not only did Watchorn need to strike the right balance in enforcing American immigration law, but he also had to manage a difficult workforce. One who tried Watchorn's patience was Marcus Braun, the president of New York's Hungarian Republican Club who received his patronage position thanks to his friendship with Roosevelt. In fact, Braun increased his stature when Roosevelt agreed to attend a dinner in his honor put on by Braun in January 1905, which over four hundred people attended.

With pull like that, Braun was no ordinary inspector. Soon after his appointment, he was sent to Europe to investigate conditions there. He charged that officials of the Hungarian government were scouring the countryside, encouraging people to come to America and making money from steamship tickets, since the government owned the steamship company. Braun implicated high government officials, including Prime Minister Stephen Tisza.

The charges angered Hungarian authorities, who put Braun under constant surveillance. On a subsequent trip to Budapest in 1905, Braun caught a policeman opening his mail and slapped the man, leading to his arrest. After paying a fine, Braun was released and returned to the United States, where he made the episode public, turning the case into an international diplomatic incident and forcing his patron, President Roosevelt, to privately condemn him for acting with "extreme folly."

Upon returning home, Braun was given a month's leave from Ellis Island, after which time he would have to return to work. However, Braun had little desire for the mundane work of immigration inspection and instead asked for a year's leave, which was denied. Upon returning to work, Braun refused to wear his blue inspector's uniform. Instead, he resigned. "He didn't like the uniform because it was a sign of a condition against which he revolted," said a frustrated Watchorn.

Braun's situation did not elicit much sympathy. The New York Times headlined its editorial on the incident "In Mockery of Marcus." Yet his political patron, Theodore Roosevelt saved Braun. The president reinstated him to government service and transferred him to the Immigration Bureau along the Canadian border. In early 1906, Braun resigned yet again, only to be reinstated later that year. Only Roosevelt could say whether the support of the Hungarian Republican Club was worth the trouble of dealing with Marcus Braun.

Theodore Roosevelt showed more judgment when he named Philip Cowen, the editor of the American Hebrew and a second-generation

Polish-Jewish-American, as a special inspector at Ellis Island in 1905. In doing so, he bypassed civil service regulations as he had with Joe Murray. For more than twenty years, Cowen would be a presence at the immigration station. When he retired in 1927, the occasion attracted attention from as far away as Germany, where Adolf Hitler called Cowen's presence at Ellis Island proof that American immigration policy was under the control of "Pan-Jewry."

Another appointment largely went unnoticed at the time. Unlike Cowen, this new interpreter at Ellis Island got his job in 1907 through a civil service exam, earning the top score among three test takers on the Croatian language test. In addition to Croatian, this twenty-four-year-old son of Italian immigrants also spoke Italian and Yiddish. Fiorello La Guardia earned $1,200 a year at Ellis Island while attending law school at night.

La Guardia was clearly a man on the make. At Ellis Island, he was one of the many men and women who served as an important link between English-speaking inspectors and confused, non-English-speaking immigrants. When a young child named Louis Pittman was forced to stay at the Ellis Island hospital for seventeen months until his trachoma healed, he received periodic visits from a short, round-faced La Guardia bearing gifts of chocolate for Pittman and other sick children.

La Guardia found his coworkers "kindly and considerate," a big change from the earlier patronage era. His superiors found La Guardia a good worker who showed a keen interest in his job, even if he did manage to lose his official badge once, forcing Washington to send a replacement. In recommending La Guardia for a pay raise, Robert Watchorn described him as "energetic, intelligent, and familiar with a number of foreign languages." Yet he also noted that La Guardia was "inclined to be peppery." Perhaps the weight of troubles he witnessed at Ellis Island wore on La Guardia, since Watchorn noted that the young interpreter was "inclined to be argumentative" with members of the boards of special inquiry, no doubt in defense of immigrants.

An acquaintance of young Fiorello described his personality as "a magnificent unrest coupled with a desire to be a leader on his own terms." La Guardia was a child of the new America and had little sympathy with the daily rigors through which his country put newcomers. "I never managed during the years I worked there to become callous to the mental anguish, the disappointment and the despair I witnessed almost daily," he wrote years later. As a low-level bureaucrat, he chafed at his own lack of power and at an immigration system of which he was a part, but for which he had little respect. His uncompromising personality and budding social conscience, as well as his relatively low salary, made his position untenable.

After three years at Ellis Island and now armed with a law degree, La Guardia struck out on his own, hanging a proverbial shingle in a small downtown Manhattan law office. His early practice was largely made up of representing immigrants ordered deported, referred to him by his former colleagues. Though many lawyers who plied this trade took advantage of their greenhorn clients, La Guardia did not—not at $10 a case. Years later, many of his clients would pull the lever in the voting booth to make La Guardia mayor of New York City.

THE IMMIGRATION PROBLEM WAS a conflict between abstract laws and the individual tragedies those laws sometimes created. Thanks to technological improvements in photography, this human element could now be brought directly to average Americans as they sat at home reading the newspaper or one of the growing number of magazines aimed at middle-class audiences.

For Americans who did not have close contact with immigrants, their vision of these newcomers often came from cartoons drawn by unsympathetic hands. Cartoons featured negative characteristics drawn in an exaggerated manner to reinforce stereotypes: the sneering Italian with a dagger, the Jew with a hooked nose, the anarchist immigrant hiding a bomb. The immigrant's foreignness was often highlighted, as was his general undesirability.

Jacob Riis, an immigrant and close friend of Theodore Roosevelt, had already showed the power of photos when his portrayals of life in New York's tenement district were published in the 1890 book *How the Other Half Lives*. To arouse public sentiment for tenement reform or public parks, Riis portrayed the worst aspects of immigrant life—filth, overcrowding, and child exploitation.

In the early years of the twentieth century, middle-class readers began to encounter the faces of the masses that would be transformed into new American citizens. Sometimes these new immigrants would be staring straight into the camera, while others were photographed in profile. Few had smiles on their faces and many had hardened or faraway looks in their eyes. The immigrants were usually anonymous. Photo captions read simply "Russian bookbinder," "Hungarian farm laborer" or "Pollack girls." An exception was the Mittelstadt family from Germany—father Jacob, wife, daughter, and seven sons, all lined up from tallest to shortest. "Seven soldiers lost to the Kaiser," proudly read the *New York Times* caption.

These men and women may have worn elaborate and strange native costumes and some of their faces may have betrayed a hard life that aged them beyond their years, but these photos hardly portrayed the grave threat to American society that critics feared. Instead, these subjects were proud and dignified, healthy and strong. These photos spoke of the singularity and individuality of the immigrant.

Lewis Hine was one of those photographers drawn to Ellis Island. Originally from Oshkosh, Wisconsin, he came to New York with a zeal for social reform. Though he would later gain fame with his photographic exposés of child labor and his iconic images of the construction of the Empire State Building, Hine's first large-scale photographic project was Ellis Island in 1905.

It was no easy task to photograph amid the turmoil and chaos of Ellis Island. As Hine later described his difficulties:

> Now, suppose we are elbowing our way thru the mob at Ellis Island trying to stop the surge of bewildered beings oozing through the corridors, up the stairs and all over the place, eager to get it all over and be on their way. Here is a small group that seems to have possibilities so we stop 'em and explain in pantomime that it would be lovely if they would only stick around just a moment. The rest of the human tide swirls around, often not too considerate of either the camera or us. We get the focus, on ground glass of course, then hoping they will stay put, get the flash lamp ready.

Then, with his five-by-seven camera on a shaky tripod, Hine would take his photo. The explosion of the flash pan blew smoke and sparks in the air, startling all those in the area.

The intrusiveness of the early photographic process, combined with the chaotic environment of Ellis Island, makes the subtlety and intimacy of Hine's finished products even more remarkable. The photos provide visual examples of the daily experiences of immigrants: an Italian family looking for their baggage; a Slavic woman asleep on a bench, her kerchiefed head resting on her bags; children enjoying a cup of milk poured by an attendant. A "Young Russian Jewess" stares away from the camera with her big brown eyes, searching for something or perhaps thinking of what she left behind.

Hine's photos were posed, yet this did little to take away from their immediacy. One photo was entitled "Italian Madonna." An Italian woman sits on a bench, her head covered in a black shawl and her young daughter in her lap. The mother looks down at the child, while the child looks at the mother with adoring, yet somewhat fearful eyes. Hine interrupts this classical- and religious-themed photo by placing mother and daughter in front of a chain-link fence behind which a crowd of young and old immigrants is milling about slightly out of focus. In juxtaposing the idealized mother-and-child image with the reality of immigrants penned behind a fence, Hine captures the reality of Ellis Island.

More photographs made their way into newspapers and periodicals from the camera of Augustus Sherman, an amateur photographer and inspector at Ellis Island. Sherman's subjects were largely anonymous, with captions mentioning little beyond ethnicity and occupation, such as "Romanian shepherds" and "Finnish girl." Even more than Hine, Sherman was attracted to the picturesque—Albanians, Dutch, Greeks, Cossacks, all in their native dress. He also documented the exotic, almost freak-show quality of some immigrants: heavily tattooed German stowaways, a Russian giant, Burmese midgets, and microcephalic East Asians heading to the circus.

The photographs of Hine and Sherman may have helped humanize immigrants, but they did not convince all Americans. By Roosevelt's second term, the IRL realized that its earlier faith in the president was misplaced. Roosevelt showed little desire to push for a literacy test. His appointments of Watchorn and Straus meant that the guardians of the gate were more likely to swing the door wide than hold it tightly closed. Like Oscar Straus, Prescott Hall realized that those entrusted to execute immigration law possessed a great deal of influence as to how those laws were carried out.

Labor leader Samuel Gompers also joined in the call for restriction. A Jewish immigrant from England, Gompers admitted to mixed feelings, yet the complaint about low-wage immigrant labor was a natural argument. He blamed big business and "idealists and sentimentalists" for opposing restriction, but the National Liberal Immigration League was more than willing to turn that argument around. "The selfishness of their [union] efforts is perfectly plain," Harvard president Charles Eliot wrote. "As a rule they have only been a few years in this country themselves and are now trying, for their own supposed advantage, to keep other people out."

The test for both sides would come in 1907 when Congress again took up the literacy test. Henry Cabot Lodge managed to get a bill through the Senate, but it got bogged down in the House. Though there was enough support in the House, the powerful Speaker, Republican Joe Cannon, managed an end run around the literacy test.

Cannon was a laissez-faire, pro-business Republican who opposed nearly every attempt by government to regulate private business. He was also adamantly anti-union, so it was natural for Cannon to support a steady stream of low-wage workers for which his business constituents clamored. He was also a member of the National Liberal Immigration League, which, in addition to German-American, Irish-American, and American Jewish groups, came out against the bill. Ultimately it was Cannon's manipulation of the legislative process that won the day. In place of the literacy test, Cannon substituted the creation of a federal commission to investigate immigration.

The Immigration Act of 1907 was a victory for opponents of restriction in the sense that the literacy bill was defeated. In reality, the bill was much more complicated and restrictionists got far more than most people realized. The head tax on immigrants was raised to $4 per person, though some had wanted to raise it as high as $25. More importantly, Congress once again expanded the categories for exclusion. First, in addition to the insane and epileptics, feebleminded immigrants were now excludable. Second, Congress expanded the exclusion for prostitutes to include the "importation into the United States of any alien woman or girl for the purpose of prostitution, or for any other immoral purpose." Lastly, any immigrant determined by doctors to be "mentally or physically defective," and whose defect would "affect the ability of such alien to earn a living," could be excluded. Loosely worded legislation opened up new grounds for debate over policies at Ellis Island. The key question boiled down to the definition of terms like "mental defective," "immoral purpose," "feebleminded," or "ability to earn a living."

As for the commission to investigate immigration, it combined two features of twentieth-century commissions. First, it would collect data and investigate various conditions throughout the country to give lawmakers better information. Second, it would allow short-term-minded politicians to postpone any further discussion of immigration, giving them cover on an increasingly sensitive issue.

President Roosevelt, who years earlier had criticized Grover Cleveland's veto of the literacy test and who spoke earlier in his presidency in favor of one, was spared the agonizing decision of whether or not to veto such a bill. "When it came to a showdown," Alabama congressman John Burnett said of Roosevelt's behavior during the congressional fight, "the President was not to be seen, and his hand was not to be felt."

Writing to Speaker Cannon, Roosevelt saw the commission as an opportunity to achieve restriction without jeopardizing his political capital. "I would want a Commission which would enable me . . . to put before the Congress a plan which would amount to a definite solution of this immigration business," he told Cannon. He hoped this would occur after the 1908 election but before he left office. Roosevelt wanted legislation that would keep "out the unfit, physically, morally, or mentally." These were words that came easily in private, but which the president was increasingly loath to speak publicly.

It would be four more years before this new commission would make its report to Congress. In the meantime, the focus of immigration left Washington and returned to the increasingly busy island in New York Harbor.

THE DEFEAT OF THE literacy test showed the growing influence of immigration supporters, but also led to virulent attacks against Oscar Straus and Robert Watchorn. Not surprisingly, one of their sharpest critics was Prescott Hall, who complained that Straus was reversing half of the exclusion cases that reached his desk on appeal and that such behavior was demoralizing the department. The first Jewish cabinet secretary also attracted complaints that he was less sympathetic to appeals from non-Jewish immigrants. By early 1908, there was such a steady drumbeat of protest that Watchorn complained to Straus about "the growing impression among many officials—both state, county, and municipal—that your administration is not disposed to execute the expulsion feature of the immigration laws."

Hall took his case against Straus directly to President Roosevelt, who did not seem terribly disturbed by the charge. Nevertheless, he would pass along Hall's criticism to Henry Cabot Lodge for further investigation.

Lodge had been a staunch ally of the IRL and the main point man in Congress for the literacy test. After looking into the charges against Straus, however, Lodge came away unimpressed. "Hall is both honest and able but he is extreme and does not understand that it is one thing to make general charges on hearsay and another to sustain them by proof," he wrote to Roosevelt. Lodge admitted that Straus was "averse to the laws which affect the entry of poor Jews," a fact he found unfortunate. Nevertheless, he could find no proof that Straus had ordered any easing of the enforcement of the law. In fact, Lodge told Roosevelt that reversals of deportation orders on appeal to Washington had not increased under Straus's tenure.

Lodge, however, was not quite correct. In the first full year before Straus took office, almost 52 percent of immigrants who appealed their deportations to Washington lost their case. In 1908, Straus's first full year as secretary, that figure dropped to 44 percent. In 1910, the first year after Straus left office, the number of lost appeal cases jumped to over 60 percent. On the whole, however, this relatively minor dip hardly proves a lax administration of the law.

That even Henry Cabot Lodge was defending Straus must have galled Hall. He later told Roosevelt that Straus "has deceived you time and again in regard to many immigration matters . . . he is one of the most subtly insidiously unscrupulous officials that ever breathed." Such words would do little to dent Roosevelt's admiration and respect for Straus.

Hall also aimed his fire at the man he saw as the other villain in this piece. "Watchorn has been a crook ever since he immigrated to this country," Hall told Roosevelt. "His naturalization papers were fraudulent." He also accused Watchorn of stealing the addresses of union members in a political campaign in 1890. "I am absolutely sure of Watchorn's dishonesty and unscrupulousness," he raged.

Roosevelt had sought to mollify Watchorn's critics in late 1906 by asking IRL member James B. Reynolds to investigate operations at Ellis Island. When he had completed his report, Reynolds did not come up with an indictment of Watchorn's administration, but instead issued a strong condemnation of the treatment of mentally ill immigrants in detention. This is not what Prescott Hall was looking for.

Part of Hall's anger stemmed from the fact that Watchorn had tried to play both sides of the immigration debate. He accurately sensed a slippery nature to Watchorn's personality. He had already proved himself a man a little too eager to please his superiors, someone who easily switched from Democrat to Republican when it suited his career. Watchorn appeared tough on immigration early in his term, but later trimmed his sails when he began reporting to Oscar Straus.

In July 1905, Watchorn wrote to Robert DeC. Ward, explaining that he had no qualms about separating families when one member was ordered excluded and the rest admitted. "What sort of protection would be afforded the United States," wrote Watchorn, "if any such minor children, wife or parents are of the kind who are going to furnish as a legacy a progeny of the sort which you and I and all thoughtful persons must of necessity view with no little apprehension?" In words that would have shocked those who saw him as an advocate for immigrants, Watchorn told Ward that he wondered whether "misplaced sympathy is not responsible for more evils than the so-called callousness of which we are occasionally accused."

Keeping up a correspondence with the Boston restrictionists, Watchorn wrote Hall in 1906 to discuss a paper that William Williams had recently delivered. Watchorn was hurt that Hall remarked that it was a shame that Williams was no longer at Ellis Island, implying that a lax enforcement now existed there. Watchorn was eager to correct that impression, writing that he was in near complete agreement with Hall and Williams, and that it was his "unremitting endeavor to prevent the landing of any and all such persons" defined as mental or physical defectives. Hall responded by calling Watchorn an "exceptionally capable and energetic official."

That was before Oscar Straus. Now Prescott Hall was not the only one unhappy. Judson Swift of the American Tract Society wrote to Roosevelt to complain that Watchorn, supposedly under orders from Straus, was hampering the efforts of missionaries at Ellis Island.

Protestant missionaries looked upon the crush of immigrants streaming through the inspection station not so much as a fearful deluge as an evangelical opportunity. In his 1906 book entitled *Aliens or Americans?* Baptist minister Howard Grose called the new immigrants an opportunity for evangelists and asked: "Will we give the gospel to the heathen in America?" Some were truly ministering to the newcomers, while others were busy targeting Catholics and Jews with Protestant pamphlets written in their native language.

Jewish leaders complained of the situation to Watchorn, who ordered missionaries to stop proselytizing to Jewish immigrants. Rumors began to circulate among Protestant churches in New York that Watchorn had threatened to banish from Ellis Island anyone using the name of Jesus Christ. Although Swift insinuated that Straus's Judaism was the cause of Watchorn's actions, Straus himself was unaware, though not unsupportive, of what his subordinate had done.

Roosevelt had little sympathy for the criticism and dispatched his secretary, William Loeb, to deal with Swift. Speaking for the president, Loeb chastised Swift for bringing Straus's religion into the matter, calling it "unwarranted slander" to which "missionaries of the Gospel should be most averse." Loeb also noted that Watchorn, a devout Methodist himself, could hardly be antagonistic toward religion since his own brother was a Protestant minister.

At the same time that Swift was complaining about the treatment of Christian missionaries, New York police commissioner Theodore Bingham blasted Watchorn for failing to deport immigrants convicted of crimes, calling on the president to appoint a new commissioner dedicated to keeping the "bars up against the criminal class."

Watchorn noted that in the preceding year, warrants for deportation had risen almost 50 percent. Still, Bingham would not relent and repeatedly stressed the connection between immigration and criminality. He furnished Watchorn and Straus with a list of Italian immigrants in New York with criminal records, baiting officials to deport them. Straus told Watchorn he was "ready to cooperate in ridding the country of the class that can be deported under the immigration laws." Warrants soon arrived from Washington for their arrest.

Immigration restrictionists saw further proof of the nefarious influence of Oscar Straus in the case of the commissioner-general of immigration, Frank Sargent. Many people noticed a change in Sargent after he began to report to Oscar Straus. Public Health Service official Victor Safford recounted the tale of a hearing in Boston. When the doctors recommended sending home a Swedish girl with trachoma, Sargent replied: "If you exclude this alien and the case comes to Washington on appeal, backed by the political influence which the relatives evidently can command, I can assure you that your decision will be reversed and the alien admitted to the country." Safford noted that by early 1908, Sargent had become "discouraged, sick, entirely dependent upon his official salary and wondering what was to become of his family after he was gone."

Samuel Gompers, another friend of Sargent, noticed that he had become so "disappointed and crestfallen" working under Straus that he sought reelection to his old post as president of the Brotherhood of Locomotive Firemen. When he lost his bid, Sargent was faced with the realization that he had to remain in his government job. Needing money to support his family, he could not resign on principle and would have to continue upholding interpretations of the law that compromised his beliefs.

By the summer of 1908, the pressures began to get to Sargent. He struggled with severe stomach problems and would eventually suffer a stroke. After two more strokes and a serious fall, Sargent died in early September at the age of fifty-three. "If ever a man died of a broken heart it was he," wrote Gompers, "because he found himself in a position which he deemed it necessary to retain and yet was unable to carry out his ideals of public service and righteous conduct." Remarking on Sargent's death in his diary, Oscar Straus spoke well of his subordinate, calling him, with a touch of mild condescension, "a good and conscientious official and whatever defects he had were not the result of lack of human sympathy, but education."

Straus's views on immigration also had an effect on another old labor restrictionist. Terence V. Powderly had been out of steady work for over three years. By 1906, Roosevelt had made amends with him and sent him on a fact-finding mission to Europe to investigate the causes of European immigration. After Powderly submitted his report, Roosevelt named him to a new position. The old union leader needed a steady government paycheck, but the man who once led Washington's immigration office now had to take a subordinate position in the agency he once ran.

Powderly was now in charge of the new Division of Information. Its goal was to "promote a beneficial distribution of aliens admitted into the United States." This was a reform supported by both sides of the immigration debate. In fact, the motto of the National Immigration Restriction League was "Distribution and Education Rather than Restriction." What Powderly's new organization did was more prosaic. It collected information on wages and employment throughout the country, put the data together, and got the information into the hands of immigrants at stations like Ellis Island.

It was a rather naïve view of how immigrants behaved. When most immigrants arrived in America, they usually stayed with friends and relatives from their homeland in immigrant ghettos. No matter how overcrowded and bleak these neighborhoods might seem to the outsider, they served as a safety blanket that provided the greenhorn with a foot into America's golden door. The air of the familiar—language, newspapers, food, music—was more enticing than job opportunities elsewhere. The Lower East Side of Manhattan or the West Side of Chicago were more attractive than the steel mills of Alabama or the farms of Texas.

It is no surprise that Powderly's efforts were relatively unsuccessful. Between 1908 and 1913 only 23,000 immigrants made use of Powderly's information. Despite this seeming failure, labor leaders pounced on the new agency. Gompers, who never had much respect for Powderly, called the Division of Information "a strike-breaking agency." The head of the Brotherhood of Locomotive Firemen told Powderly that his division would only be a success if it could "convince the people of Europe to stay at home." Gompers's deputy, John Mitchell, told Powderly he wanted him to distribute unemployment statistics to immigrants to discourage them from coming.

The in-house journal of the Knights of Labor remarked that its former leader had once been known as a restrictionist until he started working for Oscar Straus. Powderly, the journal mused, "must feel greatly embarrassed, when to keep a job, he manufactures new speeches and opinions at variance with those of only yesterday." It may have been a change of heart brought about by age, but the reality was that Powderly now reported to Oscar Straus. Desperate to keep his government paycheck and avoid another embarrassing dismissal, he quietly modified his views.

AMERICANS TRIED TO BALANCE concerns about newcomers with the country's traditional role in welcoming immigrants. Allan McLaughlin, a doctor with the U.S. Marine-Hospital Service, was one of those who framed the immigration debate within the boundaries of the political center. The complete exclusion of immigrants, he argued, was "illogical, bigoted, and un-American," while a completely open door was "an act of lunacy" and a "crime against the body politic."

Instead, McLaughlin called for the strict enforcement of the present law. That was also Frank Sargent's position. He made clear he had no desire to see a closed-door policy and believed that America had need of "a high class of aliens who are healthy and will become self-supporting." The real question was how to divide desirable immigrants from the undesirable.

"The advocates of absolutely unrestricted immigration are too few to be taken into account," noted *The Outlook*. Prescott Hall could only count about a handful of people who believed in a completely open-door policy, the most prominent being William Lloyd Garrison Jr., the son of the famed abolitionist. When pro-immigration lawyer Max Kohler debated restrictionist academic Jeremiah Jenks in 1911, he applauded the fact that twenty-four thousand immigrants had been rejected in the previous year, thereby proving the effectiveness of the law. He also wanted no restrictions on "healthy, willing, industrious immigrants, whom this country needs as much as they need this country."

With remarkable flexibility, Theodore Roosevelt found himself operating within that debate. When immigration supporters complained about the restrictionist leanings of William Williams, Roosevelt named an ethnically diverse panel to investigate him. Later, when restrictionists complained about the lax enforcement of laws under Robert Watchorn, the president named an IRL member to investigate. Only Roosevelt could have pulled it off.

For all of his early bluster about immigration, Roosevelt was surprisingly mute about the issue in his final years in the White House. The young patrician who had once supported the literacy test and corrected a New York newspaper for calling him an opponent of restriction, was replaced by an older, more politically savvy man. Roosevelt began his presidency bemoaning the deficiencies of immigration law and calling for more categories of exclusion. In his last Annual Message to Congress, Roosevelt never once mentioned immigration.

Dr. Victor Safford struck at the heart of Roosevelt's conflicted mind. A close friend of Edward McSweeney, the doctor believed that Roosevelt had discovered "that while it was good politics to have stringent immigration laws to point to, it was poor practical politics to enforce them impartially."

This was how Roosevelt straddled the immigration question. In his openness to ethnic and religious groups, he satisfied immigrants and their defenders. In his rhetorical concerns about the quality of new immigrants, he satisfied restrictionists, but at the end of the day all of his talk about restriction was little more than bluster. On immigration, the straight-talking reformer blurs into an amorphous, but highly successful politician.

It was the kind of ideological flexibility and pragmatism that would have pleased George Washington Plunkitt.

Chapter 10

Likely to Become
a Public Charge

*In many important respects, indeed, the foreign immigrant is
the very anti-type of the pauper. . . . Their very presence here
shows the desire for bettering one's condition and the energy
to set about it that is so characteristically a lack in the true
pauper.*
—Kate Holladay Claghorn, 1904

*It is therefore high time that aliens of poor physique should be
debarred from our shores. When we raise horse or cattle or dogs
or sheep, we select good, strong healthy stock. If we have any
concern for the physical development of our race, we should
certainly be no less careful in the selection of our human stock.*
—Robert DeC. Ward, 1905

BY FEBRUARY 1910, THEODORE ROOSEVELT HAD RETURNED
from his post-presidential big-game hunting trip to Africa. A private
citizen now ensconced in the Manhattan offices of his new employer,
The Outlook magazine, the fifty-one-year-old Roosevelt was not a man
accustomed to retirement. His discomfort was made even worse by his
increasing annoyance with his handpicked successor, William Howard
Taft.

One day, the former president received some friends, including
Robert Watchorn, at his new office. Pulling him aside by the arm and

leading him to a quiet corner away from the others, Roosevelt asked: "Tell me, Mr. Watchorn, why did you leave your post at Ellis Island?"

"Because you left the White House," Watchorn responded. "Or in other words because you are a friend of mine," the former president said with a mischievous grin, receiving the confirmation he had sought. Watchorn's answer reinforced Roosevelt's belief that Taft had been forcing out Roosevelt loyalists. Yet Roosevelt was also suffering from a bit of selective memory. As president, he had brushed aside complaints about Watchorn's supposed lax enforcement of immigration laws, but not the charges of corruption that had surfaced by 1908.

While he had expressed a desire to renew Watchorn's appointment, Roosevelt also asked Herbert Knox Smith, solicitor of the Department of Commerce and Labor, to look into the accusations. The charges were relatively minor, dealing with accusations that Watchorn had forced the owner of the food contract at Ellis Island to cater private parties for him without charge. Watchorn denied the accusation, claiming he reimbursed the company except where "extravagant and extortionate charges" were made. Roosevelt seemed uneasy about the arrangement, but did not ask that formal charges be brought and renominated Watchorn. The request died in the Senate before Taft's inauguration.

Prescott Hall had been filling Roosevelt's ear with negative stories about Oscar Straus and Watchorn. Now he turned his attention to the incoming president, telling Taft that Roosevelt had been deceived in appointing Straus to his cabinet, calling him a man who "has done all in his power to interpret and apply the existing laws in such a way as to practically nullify some sections entirely and to weaken and demoralize the whole service." As for Watchorn, Hall doubted his sincerity in enforcing immigration laws.

Despite the criticism, many people entreated Taft to renominate Watchorn. That was not to be. Noting that those around Taft were "not only not friendly to me but were distinctly unfriendly," Watchorn knew his days were numbered. Taft had already eased out Straus as secretary of Commerce and Labor by naming him ambassador to Turkey and replacing him with Charles Nagel, a German-American lawyer from St. Louis. With Roosevelt and Straus gone, Watchorn lost his strongest defenders. When it became clear that Taft would not renominate him, he resigned. The official White House statement noted that Taft had found Watchorn's administration "unsatisfactory."

The personal attacks took their toll on Watchorn. When a reporter asked him after his resignation to comment on affairs at Ellis Island, Watchorn bluntly declined: "When I left the island I cut all connections. There has always been trouble down at that place and always will be. I'm out of it." As late as 1913, Watchorn was still seething over his dismissal, complaining to Taft's chief aide that he had been "very shabbily treated in the manner of my elimination from the service."

If Watchorn was upset with Taft, the president complained that conditions at Ellis Island were "not what they ought to be" and sought out a man who could put them "on a proper basis." Much as Roosevelt had been disgusted with the Powderly-McSweeney imbroglio when he took office, Taft seemed unhappy with the controversy that surrounded Watchorn.

Just as Roosevelt had done seven years earlier, Taft turned to William Williams to put things in order at Ellis Island. Williams, four years younger than the new president and a fellow graduate of Yale, was reluctant to reenter public service, but eventually agreed to return. One of Williams's first orders of business upon coming back to his old job at the end of May 1909 was to declare that Assistant Commissioner Joe Murray, his former nemesis, would be leaving. Replacing him would be Byron Uhl, who had been working at Ellis Island since its opening in 1892.

Now that Williams was returning to Ellis Island, he was eager to get to work on what he felt was the most pressing concern: tightening the sieve that would strain out larger numbers of undesirable immigrants.

Even after leaving Ellis Island in early 1905, Williams continued to be an outspoken advocate for greater restriction. Writing in the *Journal of Social Science*, he charged that the immigration law was good as far is it went, but failed to sift out "a certain minority of immigrants who are generally undesirable because unintelligent, of low vitality, almost, though not quite, poverty-stricken." He argued that roughly 25 percent of immigrants currently admitted were "not wanted." He called them "the *undesirable minority* of immigrants." Williams carefully pointed out that he was not against all immigration. "I will say that I have as little sympathy with those who would curtail all immigration as I have with those who would admit all intending immigrants, good, bad, or indifferent," he wrote.

Part of his concern with this "undesirable minority" was with what he called its "racial effects." Immigrants flocking to the United States in the early twentieth century were different from earlier settlers and immigrants. "We owe our present civilization and standing amongst nations chiefly to people of a type widely different from that of those now coming here in such numbers," Williams wrote. Those older immigrant groups largely hailed from northern Europe and consisted "mainly of the rugged types that were kindred to the native stock." These groups were "as good as the new immigrants are bad," he told the *New York Times* after returning to Ellis Island.

Although he questioned whether newer immigrants could ever be assimilated into American society, Williams found it "impracticable to legislate directly or discriminate against any race or locality of Europe as we have done in the case of the Chinese." Even he understood that any exclusion of Europeans based on nationality or ethnicity, as had occurred with the Chinese, would violate America's basic understanding of immigration. However, Williams's emphasis on this difference between new and old immigrants foreshadowed drastic changes to come.

The letter of the law, which Williams deeply respected, forced him to accept even those immigrants he believed would bring down the nation's standard of living and weaken American democracy. Working within these legal and ideological confines, Williams set out to do what he could to protect the Republic. Seven days after taking over, he distributed the following note to his staff:

> It is necessary that the standard of inspection at Ellis Island be raised. Notice hereof is given publicly in order that intending immigrants may be advised before embarkation that our immigration laws will be strictly enforced, and that those who are unable to measure up to its requirements may not waste their time or money in coming here, only to encounter the hardships of deportation.

Williams had now put everyone—from inspectors to politicians to immigrants—on notice. America was receiving too many "low-grade immigrants" and "riffraff," he insisted. The country's present laws, in his opinion, kept out "only what may be termed 'scum,' or the very worst elements that seek to come here." William Williams was determined to do something about it.

If Taft thought his appointment of William Williams would quiet the storms at Ellis Island, he was sorely mistaken.

HERSCH SKURATOWSKI ARRIVED AT Ellis Island in late June 1909 with $2.75 in his pocket. The twenty-nine-year-old Russian Jewish butcher appeared to be a desirable immigrant in every way. He was in good health, literate, intelligent, and neither a criminal, a polygamist, or an anarchist. Yet Skuratowski was ordered excluded by officials at Ellis Island because he was deemed "likely to become a public charge."

This little phrase became a stumbling block for many individuals coming to America. Between 1900 and 1907, 63 percent of all immigrants barred from the country were kept out because officials deemed them likely to become public charges.

The public charge clause had been a feature of American immigration law since 1882, although the law originally barred those *"unable* to take care of himself or herself without becoming a public charge." In 1891, that was changed to *"likely* to become a public charge." With this new phrasing, the government could bar paupers, who were already dependent on public funds for support, as well as those whom immigration officials suspected *might* end up as public charges *in the future.*

The clause embodied a basic American belief: immigrants should be able to take care of themselves. Although this was an era before the federal welfare state, persons were considered a public charge if they were being taken care of by either private charities or local government institutions such as poorhouses or asylums.

It also possessed another characteristic of American immigration law: it was vaguely defined. As one legal scholar would write in the 1930s: "Likely to become a public charge is used as a kind of miscellaneous file into which are placed cases where the officers think the alien ought not to enter, but the facts do not come within any specific requirements of the statutes." It was the responsibility of officials at Ellis Island to decide which immigrants were likely to become public charges.

Realizing this, and wanting to tighten inspection standards as he stated in his first notice to Ellis Island employees, Williams issued a second one at the end of June.

> Certain steamship companies are bringing to this port
> many immigrants whose funds are manifestly inadequate
> for their proper support until such time as they are likely
> to obtain profitable employment. . . . In the absence of a
> statutory provision, no hard and fast rule can be laid down as
> to the amount of money an immigrant must bring with him,
> but in most cases it will be unsafe for immigrants to arrive
> with less than twenty-five dollars ($25) besides railroad ticket
> to destination, while in many instances they should have
> more. They must in addition, of course, satisfy the authorities
> that they will not become charges either on public or private
> charity.

The money test had occasionally reared its head in the past. In his first Annual Message to Congress, Theodore Roosevelt called for immigrants to show "proper proof of personal capacity to earn an American living and enough money to insure a decent start under American conditions." Williams had informally tried such a money test during his first administration, but Watchorn disavowed it when he took over.

Now Williams was reinstating the test. Recognizing that he was entering murky legal territory, he said that the $25 requirement was not a fixed rule, but instead "a humane notice to intending immigrants" that they should have a certain amount of money on them when they landed. As for Hersch Skuratowski, he had arrived on June 22, six days *before* Williams issued his new rule. Further stretching the law, officials kept Skuratowski in detention until his board of special inquiry hearing, which was conveniently held on the same day that Williams made his $25 edict.

Twenty-five dollars was a significant amount of money in 1909. In 2007 dollars, it would equal roughly $570. Add to that steerage tickets that cost between $30 and $40, and the cost of coming to America would now become an onerous financial burden.

Williams's edict had an immediate effect. On its first day of enforcement, 215 of the 301 passengers on Holland-Amerika's *Ryndam* liner were detained for possessing less than $25. Most would not be sent back, but the burden of proof would now fall on immigrants to convince authorities they would not become public charges.

Conditions worsened as more immigrants piled up because of the new rule. "Trouble Feared from the Excluded," read a *New York Times* headline, "The 800 Immigrants Held on Ellis Island Not Taking Deportation Easily." One of the detainees was a twenty-three-year-old Russian medical student named Alexander Rudniew, who was ordered deported as likely to become a public charge because he arrived with less than $25. At one point, a frustrated Rudniew lashed out in Yiddish at Ellis Island officials, who feared that the doctor might stir the detainees to take over the station. A night watchman pulled a gun on Rudniew, which seemed to settle down the crowd. Rudniew would eventually be admitted into the country.

On July 4, Rudniew was one of a hundred detained Russian Jews, ranging in age from eight to fifty-eight, to sign a letter to the *Forward*, New York's Yiddish-language newspaper, complaining of crowded conditions at Ellis Island. The editors printed the letter on page one. "Everyone goes around dejected and cries and wails," the letter read. Many of the detainees had deserted from the Russian army and feared deportation. They called Williams's $25 rule an "outrage" and "nonsense" and hoped to alert fellow Jews as to "how we suffer here." The *American Hebrew* sent a correspondent to Ellis Island and found that none were sick, although most were pale and flustered from their ordeal.

Williams was unmoved by the protests and thoroughly unapologetic. "I have enforced the laws," he told a reporter. "Why shouldn't I? That's what I am here for."

Many Americans were glad that Williams was there. Russell Bellamy, a member of the Immigration Restriction League, told him his "appointment is most agreeable because we know you will enforce the laws your predecessor and his Chief so shamefully ignored." Chiming in from Boston, the gloomy Prescott Hall was cheered by Williams's appointment: "Nothing has made me as happy for a long time as feeling that you are there and seeing, as far as I do from the papers, how you are cleaning things up."

Eighty-two-year-old Orville Victor, a leading editor in the world of dime-novel publishing, was less genteel. Calling himself an American "of early colonial ancestry," Victor congratulated Williams on his appointment and wrote: "What a stench in the nostrils of true Americans are the dirty Jew lawyers who rush to the 'defense' of their kin whom you would exclude. . . . More power to you, and success to your efforts to keep out the dirty scum of European fields, bogs and warrens." William Patterson, who described himself simply as an "obscure

American," wrote Williams that "God only knows what havoc is going to be brought upon the United States by the influx of Europe's scum. . . . You cannot render the country a greater service than by restricting the inflow of worn-out, decadent, and impoverished Europeans."

Not all of Williams's correspondents were as sympathetic. An anonymous student from PS 62 on Manhattan's Lower East Side complained to Williams in ungrammatical, yet heartfelt, prose:

> You don't realize what you are doing. You kill people without a knife. Does money make you a person? A person who has a mind and hands and has not $25 cash is not a person? Has he to be killed? Here is the free America. People how much do they suffer until they come here. [sic] If you would have conscious [sic] in you would not do such things. You think that they are not people but animals. . . . I do not see what do foreigners do harm [sic].

PS 62 had opened its doors a few years earlier to deal with the massive influx of mostly Jewish immigrants on the Lower East Side, a neighborhood bursting with overcrowded tenements. Though Williams's edict may have looked like a patriotic gesture in the Upper East Side, it had a very different effect on those living on the Lower East Side.

The child who wrote that letter was not a lonely voice, as groups such as the Hebrew Immigrant Aid Society (HIAS) worked to assist Jewish immigrants. Irving Lipsitch, the organization's New York representative, worked closely with Ellis Island officials, while Simon Wolf lobbied Washington.

For Jews who fled anti-Semitic oppression in Europe, protest in America had to be carefully calibrated so as not to stir up ancient hatreds here. However, there was a change in the way American Jewish groups approached immigration. In the 1890s, they had been largely deferential to authorities during the typhus fever and cholera crises. During Williams's first term at Ellis Island, Jewish groups were cautiously supportive, but not without concerns. By 1909, however, some took a more oppositional approach. To Max Kohler, a lawyer working for the American Jewish Committee, the courts seemed a more appropriate place to challenge immigration law.

The HIAS took on the cases of Hersch Skuratowski and fourteen other Russian Jews detained for possessing less than $25. The organization was unable to prevent the deportation of eleven of them, but did convince Kohler and another lawyer named Abram Elkus to file habeas corpus petitions with U.S. District Court Judge Learned Hand for Skuratowski and the other three—Nechemie Beitz, Meyer Gelvot, and Gershon Farber—who arrived together on the steamship *Raglan Castle* from Rotterdam.

They argued that the $25 rule had created an extra-legal means of exclusion. "The retroactive character of these regulations makes them all the more unjust and oppressive," according to the brief. They countered the idea that these men were likely to become public charges. Skuratowki was a butcher by trade, literate, and had a cousin and other family in the country. He left behind his wife and two children in Russia, where he owned his home, as well as a cow and some farming equipment. This was hardly the profile of a pauper. Similar cases were made for the other three men, two tailors and a baker.

The lawyers were not just arguing the merits of these four immigrants; they also set out to attack much of the administrative and legal apparatus for making decisions at Ellis Island. They claimed that the men did not receive due process and were not allowed the benefit of counsel during their hearings. The brief also charged that members of the board of special inquiry were not "free agents," because they were also subordinates of Williams and felt pressured to carry out his orders. "The officers here are afraid to decide cases on their merits," Kohler said. It was as if assistant district attorneys were sitting in judgment of cases brought to court by their boss, the district attorney.

There was something else that bothered Kohler and Elkus. Toward the end of their petition, the lawyers included a lengthy section decrying the "unconstitutional classification and discrimination . . . as to his Russian nationality and his Hebrew religion." Official government documents had referred to Skuratowski as a "Russian Hebrew." Kohler and Elkus contended that Ellis Island officials "illegally and without authority took into consideration the fact which they spread upon their minutes that applicant is a Russian Hebrew." They did not accuse Williams and his staff of overt anti-Semitism, but argued that "things foreign to our own conception produce at least a subconscious feeling, and that we may entertain prejudices of which we have no distinct consciousness." Religious classification by government authorities was unconstitutional and un-American, the lawyers claimed, a situation only made worse by the fact that Judaism was the only religion to be so defined by immigration officials.

The controversy over the classification of immigrants had begun with an 1898 report by then commissioner general of immigration Terence Powderly, Ellis Island assistant commissioner Edward McSweeney, and Victor Safford, a doctor at Ellis Island. Officials had been unhappy that immigrants were being classified solely by country of origin, which meant large multi-ethnic political divisions such as Russia or the Austro-Hungarian Empire. Lost within those groupings were myriad ethnic identities. People coming from the Austro-Hungarian Empire, for instance, could have been German, Jewish, Polish, Magyar (Hungarian), Bohemian (Czech), or Croatian.

In its place, the commission recommended that all immigrants be classified by their nationality and their race. That is why officials marked Hersch Skuratowski as a Russian Hebrew. As Powderly explained, "an Englishman does not lose his race characteristics by coming from South Africa, a German his by coming from France, or a Hebrew his, though he come from any country on the globe."

Powderly made clear that officials were using race in the "popular rather than in its strict ethnological sense." Basically, they meant to use the word as modern Americans would use the term "ethnicity," while they used the term "color" for what is now called race. However, that qualification did little to settle the controversy.

Powderly and his colleagues also made clear that their endeavor had nothing to do with targeting undesirable immigrants based on their race. Instead, they hoped that better classifications of ethnic backgrounds would assist officials in understanding the nature of immigration, especially as it related to labor issues. "It is not intended as a history of an immigrant's antecedents but as a clew [sic] to what will be his immediate future after he has landed," the report concluded.

As Safford noted, Russian Jews had been coming to the United States in larger numbers in the 1890s. "They have for the most part entered well defined fields of labor here and have given rise to special labor problems," he wrote to Powderly in 1898. "The Immigration Bureau fails to give a clew [sic] to the size of this movement. They are lumped up with Poles, people of a distinct race and of different capacities and who have gone into entirely different fields of industry." Officials sought better information about who was coming to America, what kinds of work they did, and where they were heading.

For Jews, this new classification was a double-edged sword. Over his many years with HIAS, Simon Wolf protested the Hebrew classification to government officials, arguing that Jews were not a distinct race. However, when he sought to compile the opinions of leading Jewish authorities on the matter, he found that his views were not universally shared. Many Jews, especially Zionists, did consider themselves a "race or people" and had no objection to the government's classification scheme.

Decades later, even Max Kohler had a change of heart. Such a system, he explained, enabled "the Government to furnish Yiddish-speaking interpreters quickly in the majority of Jewish cases pending before the immigration officials," he wrote, and "it enabled the Jewish immigrant aid societies quickly to identify their prospective protégés."

Now it was William Williams's turn to offer a point-by-point rebuttal of the lawyers' brief. He defended not only his $25 rule but also the entire structure of administrative law at Ellis Island. He admitted to "certain shortcomings" on the part of the boards of special inquiry, but argued that better training of those who sat on the boards was the answer, not legal challenges. Not only were the lawyers wrong about the law, Williams believed, but their real goal was not the improvement of immigration regulation, but rather "to facilitate *the admission of one particular class* of immigrants," no doubt referring to Jewish immigrants.

Williams was not happy about having to defend himself in court. He resented that the court case was brought by "four ignorant aliens who do not know a word of English" and was indignant at the "objectionable manner" in which Kohler and Elkus put forward their critique of his administration. Williams claimed that even before he received the petition, he had already delayed the deportation of these four men in order to rehear their cases. Now that the case was going to court, he told a representative of HIAS that if he was "expecting to compel the granting of a rehearing through the threat of habeas corpus proceedings, I assure you that you will not succeed."

Williams assumed that the petitions would be quickly dismissed. After he had won in court, he would grant a rehearing to the four men and most likely allow them to land. "Any other course," Williams wrote, "might place the immigration authorities in the attitude of wish-

202 / AMERICAN PASSAGE

ing to be vindictive." However, government officials were receiving some disturbing news. Simon Wolf told the assistant commissioner-general of immigration, Frank Larned, that he hoped that the immigration service "would act in such a way as not to embarrass itself." He gave confidential information that led Larned to believe that if the government did not relent on these cases, "the Government's exclusive control of these matters might be put in hazard to a certain extent." Perhaps Judge Hand was not going to dismiss this case.

A court decision could have upended the entire inspection and exclusion apparatus at Ellis Island and opened up every decision to a judicial appeal. Not wanting to risk this, Williams immediately ordered a new hearing for the four immigrants before Judge Hand could make a ruling. Witnesses appeared vowing employment for the four men, money was deposited in their name by Jewish organizations, and Williams declared himself satisfied that, especially in light of the publicity surrounding the case, the four would not become public charges. All were allowed to land and their petitions withdrawn. "We have now shown them that the immigration authorities can do full justice without the necessity of the intervention of the courts," Williams wrote somewhat disingenuously. "What has now happened is exactly what would have occurred sooner had they not rushed to the court."

After the resolution of the case, Secretary Charles Nagel came to Ellis Island with Elkus and Simon Wolf. In public, Nagel was supportive of his commissioner, but the visit signaled that he took the charges against Williams seriously. In private he made his displeasure known, albeit in a cautious and respectful manner. Nagel told Williams that his $25 rule—which Nagel had just a few weeks earlier approved—had already served its purpose to warn immigrants and steamship companies that the supposedly lax policies of the Watchorn years were over. He believed the rule was now "of no value, but on the contrary is calculated to give you and the Bureau and the Department trouble."

"There is no more need for making suggestions as to the amount of money than there is for saying how short a leg must be to constitute lameness," Nagel wrote after his visit. He was also concerned that the $25 rule was operable at Ellis Island but not at other inspection stations and would create confusion if each station created its own rules. In conclusion, he reminded Williams that "we can well afford even to err on the side of fairness and toleration." The $25 rule appeared to be history. Though a personal victory for Hersch Skuratowski and his friends, it was no victory for Kohler and others who sought to liberalize the process at Ellis Island.

The following year, the U.S. District Court heard another habeas corpus petition challenging the detention and deportation of Vincenzo Canfora. The sixty-year-old Italian bookbinder had lived in America with his wife and six children since 1895, but he got sick and had his leg amputated below the knee. He then returned to Italy for a brief visit with his mother. Before his arrival at Ellis Island, a letter arrived from one Joseph Ruggio alerting officials to Canfora's arrival and alleging that he had been a public charge when recovering from his amputation at Bellevue Hospital, where doctors performed the surgery for free. Upon his arrival, Canfora was ordered excluded as suffering from a physical defect that would likely make him a public charge, despite his skill as a bookbinder, his $200 in savings, and the presence in America of his family, including self-supporting children.

The judge called the deportation order against Canfora "an act of cruel injustice," yet he ruled that he was "compelled to dismiss this writ," since immigration laws "confer exclusive power upon the immigration officials to determine such questions" as to the admissibility of immigrants. As long as officials were following the law and their own procedures, the judge had "no jurisdiction to interfere" with the decision to deport Canfora. Having said that, the judge expressed his personal hope that officials would reconsider Canfora's deportation. Officials did just that and Vincenzo Canfora was allowed to rejoin his family.

These cases show that while a noncitizen living in the United States would be covered by constitutional protections, noncitizens stopped at the gates of Ellis Island were not. This was upheld in a 1905 Supreme Court decision dealing with the due process rights of a Chinese-American named Ju Toy. Writing for the majority, Justice Oliver Wendell Holmes declared that at stations like Ellis Island, an immigrant, "although physically within our boundaries, is to be regarded as if he had been stopped at the limit of our jurisdiction and kept there while his right to enter was under debate."

In essence, the Court created a legal fiction that Ellis Island was not part of the United States. Immigrants arriving at Ellis Island may have thought they were on American soil, but by law they had not technically crossed the border until they were officially declared "free to land" by officials. Ellis Island had become the nation's premier border; few immigrants standing in the Great Hall would have realized that, in the eyes of the courts, they were still on the wrong side of that border.

This peculiar legal situation brought up another issue. Is a child born to an immigrant woman detained at the hospital at Ellis Island and not yet legally admitted to the country, an American citizen? According to the Fourteenth Amendment, it would appear that the child would be. In granting citizenship to freed slaves, that amendment defined citizens as "persons born or naturalized in the United States, and subject to the jurisdiction thereof." Hence the idea of birthright citizenship, that birth on American soil automatically conferred U.S. citizenship.

However, the Department of Commerce and Labor issued a legal memorandum stating that such a child would not receive automatic citizenship solely by being born at Ellis Island or any other inspection facility, if the mother had not yet been legally admitted to the country. Focusing on the words "subject to the jurisdiction," officials argued that although the mother had offered her allegiance to the United States by attempting to enter the country, "her offer has been refused and she does not acquire even a momentary residence."

Such rulings, combined with Supreme Court precedents, would create a legal twilight zone around Ellis Island where immigrants had the potential for being trapped in limbo, having forsaken their native country and been rejected by their desired adoptive country. The creation of this legal fiction would pose challenges to American law, national security, and concepts of human rights for decades to come.

DESPITE THE SETBACKS, WILLIAMS would not completely give up on his monetary test. In March 1910, he was still announcing that "immigrants will not be allowed to land without funds adequate for their support until such time as they are likely to find employment." He did not mention any specific dollar amount, but referred people to his earlier memo laying out the $25 rule. Williams was nothing if not stubborn. However, he needed to find other tools with which to weed out undesirable immigrants. Now officials began to focus more closely on immigrants who possessed "poor physiques" or were suffering from "low vitality."

These supposedly weak and listless new immigrants would never make it in industrial America. Their lack of strength would mean unemployment and poverty. Some Americans believed that the poor physiques and low vitality of immigrants indicated a genetic disposition, not caused by environment or circumstances. Those genes would be passed down to their children and grandchildren, lowering the overall vitality and strength of the American people for generations to come.

In 1902, commissioner-general of immigration Frank Sargent warned William Williams that an immigrant should be excluded unless it "was positive from appearance and the physical condition of the aliens that they could immediately obtain employment, with good wages, whereby they could support themselves and not become public charges." Hinting at a monetary test, Sargent wrote that, "sturdy Scotchmen, Irishmen, or Germans who land at Ellis Island with but a few dollars can enter immediately and find employment." Those of other nationalities with little money, however, "should not be permitted to enter unless they produce satisfactory proof of their ability to work and support themselves."

Medical officials initially classified immigrants with "poor physique" as those suffering from what was called "chicken breast" or displaying symptoms of pulmonary tuberculosis, but not necessarily the disease itself. Immigration officials, however, wanted to stretch out the term to encompass a wider range of alleged physical defects.

Sargent defined a "poor physique" as those who were "undersized, poorly developed, with feeble heart action, arteries below standard size . . . physically degenerate." An immigrant with a poor physique was not just more likely to become a public charge; he would also "transmit his undesirable qualities to his offspring."

William Williams agreed that a broader interpretation of the term "poor physique" might achieve greater restriction within the law. "I am glad that you approve of my remarks as to the low vitality of many immigrants," Williams wrote Prescott Hall. "I would like to see some steps taken to keep out immigrants who do not come up to some proper physical standard."

He had been quite taken with an article written by Allan McLaughlin, a doctor at Ellis Island who noted that "thousands of immigrants of poor physique are recorded as such by the medical inspectors at Ellis Island." The problem, as McLaughlin understood it, was that nothing in the law mandated that immigrants of poor physiques be excluded, therefore "this mere note of physical defect carries little significance under the present law, and the vast majority of them are admitted by the immigration authorities, because it does not appear that the physical defect noted will make the immigrant a public charge."

Doctors with the Public Health and Marine Hospital Service were conflicted about the term "poor physique." Dr. George Stoner, who was stationed at Ellis Island, listed a number of physical maladies that might constitute a poor physique, including respiratory problems, "deficient muscular development," poor circulation, and an inadequate proportion between height and weight. However, he was concerned that the term itself "does not imply a clinical or pathological entity." Surgeon General Walter Wyman agreed. "Poor physique is not a diagnosis," he said.

The Immigration Act of 1907 would provide restrictionists with a way to shoehorn "poor physique" into immigration inspection. Many immigration defenders thought the bill a great success for what was not included: a literacy test. However, the new law expanded classifications for excluding aliens, including one that allowed doctors to certify immigrants with a mental or physical defect that might affect their ability to earn a living. In effect, the new law would allow immigrants to be excluded from entering the country if their physical appearance was poor, to the extent that officials felt they would not be able to survive in America.

This did not mean loathsome diseases or dangerous and contagious diseases (such as trachoma), which were already excludable under the law. Medical officers certified immigrants with these diseases, as well as those suffering from insanity, epilepsy, and low intelligence, as Class A, which meant they were automatically excluded by law. Immigrants with poor physiques and other physical deficiencies were certified as Class B, which meant that their exclusion was at the discretion of immigration officials.

In the first full year that the new law was in effect, 870 immigrants were barred from the country under this new clause. By 1912, that number would grow to over 4,200. However, this did not satisfy Prescott Hall. He wrote to Robert Watchorn, commissioner of Ellis Island at the time, that only 34 percent of immigrants initially classified as having a poor physique were deported, while the rest were allowed to enter on bond or on appeal. Watchorn responded that most of those admitted were older parents coming to stay with their adult children already in the United States, and therefore "not prospective progenitors, for the most part." There is, of course, something comical about the gaunt and sickly Prescott Hall complaining about the poor physiques of incoming immigrants.

When Williams took over in 1909, he focused more closely on the new law. He made a long list of ailments that might qualify an immigrant for exclusion. It included ankylosis (stiffness) of the joints; arteriosclerosis; chronic inflammation of lymph glands; hernia; goiter; lupus; and varicose veins. Even otherwise productive and healthy immigrants who were deaf or mute might be excluded under the new rules. All immigrants "not clearly and beyond a doubt entitled to land" would be pulled aside for a hearing, but only those with ailments "in aggravated form," to the extent they would affect their ability to earn a living, would find themselves excluded.

In his first annual report since returning to Ellis Island, Williams noted that many of the previous year's detentions were due to "serious physical defects discovered by our surgeons," placing these immigrants under what Williams termed the "excellent provision of the law of 1907." Williams did not believe that these ailments were randomly assigned across the racial and ethnic spectrum. "Relatively few immigrants from Northern Europe are so held," he wrote. "It is those coming from the other parts of Europe (particularly the southern and south-eastern parts) that constitute the great majority of the doubtful cases."

The amount of work being done was astounding. In 1911, there were 70,829 board of special inquiry hearings. That came to almost two hundred hearings a day, seven days a week, twelve months a year. Ellis Island possessed a staff, not including medical officials who did not sit on the boards, of 523 workers, although that included many, such as watchmen and maintenance staff, who did not perform inspection duties.

Williams worried about the ability of his staff to carry out that work. "Some of these men will never understand the meaning of the phrase 'likely to become a public charge' or how to apply it," he wrote to Daniel Keefe, the new commissioner-general of immigration. "The fact is we are executing here some of the most difficult laws in the world with much green material." Ellis Island was running as many as eight board hearings at any one time, necessitating more than thirty officials to sit on those boards. "We have not 32 men here who are qualified to do good Board work," Williams lamented.

To the thousands of immigrants who passed in front of those boards, like Wolf Konig, Williams's strict enforcement of the law had real consequences. The seventeen-year-old arrived at Ellis Island alone and penniless in June 1912; doctors certified him as "afflicted with lack of physical and sexual development for age claimed which affects ability to earn a living." Wolf was headed to his uncle, Nathan Waxman, in Chicago, who owned a stationary store and property worth $3,800. Nathan signed an affidavit that he would support Wolf so that he would not become a public charge.

Irving Lipsitch of HIAS took up the case. He argued that Wolf was sixteen years old and therefore not underdeveloped for his age. "We believe that he can improve himself and his development with the assistance of his relatives who are prepared to help him to get better nourishment and exercise," Lipsitch told officials. "Being a young boy, and not accustomed to travel, it is quite natural that this first long journey should cause him to become fatigued and to look poorly developed when examined." Nathan Waxman enlisted the help of two Chicago-area congressmen to write to Secretary Nagel about their interest in Wolf's fate. Ultimately, though, Williams stood his ground. "This boy is frail and obviously weakling [sic]," he concluded. Unfortunately for Wolf, Nagel was away from his office when his appeal reached Washington. In his place, Solicitor General Charles Earl agreed with Williams and ordered the boy back to Galicia.

Michele Sica was also a victim of the new regime at Ellis Island. He was a "bird of passage," an immigrant who would come to America for a number of years to work and then return to his wife and children back home in Italy with the money he had saved. This would have been Sica's fourth time in the country, having first arrived in 1901. He had lived in America for seven out of the last ten years.

On his fourth visit, in June 1911, Sica ran into problems. Though he arrived with $21, had a brother-in-law and friends in New York, and had resided in the country for a number of years over the previous decade, Sica was declared likely to become a charge. He was now forty-five years old and diagnosed with a hernia. "Although there are some favorable features in the case, he is certified to be physically defective," wrote Assistant Commissioner Byron Uhl, "his general appearance is not good, he is considerably older than when previously in this country and there is great doubt as to his ability to earn a livelihood, afflicted as he is, as a laborer."

Sica was ordered deported, but his expulsion was postponed until September. That meant Sica would spend the entire summer cooped up at Ellis Island, where temperatures would often rise to over 100 degrees in the poorly ventilated dorm rooms. In the meantime, Sica hired Fiorello La Guardia to argue his case. One year out from his work as an interpreter at Ellis Island, La Guardia now had his own small practice dealing with cases like Sica's. If Williams believed that too many undesirable immigrants were getting into the country, La Guardia thought that too many desirable immigrants were being kept out.

La Guardia appealed Sica's case to Washington. During Sica's last stay in New York, he had worked for a Manhattan lumber company for more than three years and would be rehired if admitted. La Guardia could not claim that Sica already had a job lined up with the firm, since that would mean he was in violation of the contract-labor law. "Considering all these facts it is clear that the medical certificate cannot even incidentally be the cause of this alien's becoming a public charge," La Guardia concluded. "He is now in good physical condition and well able to secure and keep profitable employment." However, La Guardia's efforts were for naught. After three months in detention, Sica was shipped back to Italy.

Although much younger than Sica, sixteen-year-old Bartolomeo Stallone also faced exclusion. Arriving from Italy in September 1911, Stallone was headed for his brother's home in St. Louis, where he would work as a barber. At Ellis Island, Dr. E. H. Mullan certified the young man as "afflicted with flat deformed chest, lack of muscular development (poor physique), which affects ability to earn a living." Stallone appealed his case, but Augustus Sherman, acting in place of William Williams, reaffirmed the deportation order, noting that Stallone was "quite frail in appearance."

When Stallone's case landed on the desk of Secretary Nagel, he ordered that the immigrant be admitted on a $500 bond, most likely posted by his brother. After three weeks in detention at Ellis Island, Stallone was released. Two years later, Bartolomeo requested that the bond be canceled. He had to report to officials in St. Louis, who found that the young man was making $12 a week as a barber. Though he had no savings, he told officials: "I live well, dress well, and send money home to my father and mother in Italy, so I haven't anything saved up." Impressed by the now-eighteen-year-old, officials canceled the bond and declared him: "Physically fit for admission and that there is little or no likelihood that he will become a public charge."

Williams himself was not completely hard-hearted in his application of the law. Jacob Duck, a twenty-one-year-old Turkish Jew, arrived in March 1910 headed to a cousin who owned a wholesale lace business on New York's Lower East Side. Doctors certified Duck as lacking in physical development, and Williams agreed that "he does not present a very robust appearance." On the other hand, Duck arrived with $47. Williams personally interviewed him and found him intelligent and unlikely to become a public charge. The commissioner agreed to admit Duck, even though he was "the type of immigrant that I care not to see come into this country." It was a startling admission, both that Williams would put his personal prejudice against Jewish immigrants on an official document and that he still felt compelled to follow the law despite that prejudice.

Though Williams may have occasionally shown leniency, he found that his superiors in Washington often had a different interpretation of the law. In January 1912, Chaie Kaganowitz arrived at Ellis Island with her nine children, ranging in age from three to twenty. Williams declared that the forty-two-year-old Russian Jewish widow with poor eyesight and her youngest children were likely to become public charges. The older children were also ordered excluded for poor physical development. Williams was sitting in his office with Commissioner-General Keefe when he asked to see the family. Both Keefe and Williams agreed with the decision owing to "their extremely poor appearance." The two older sons were carpenters, but Williams found them to be "frail appearing" and not very "robust." Only the oldest daughter, a seamstress, made a favorable impression upon Williams and Keefe.

The family appealed its exclusion to Secretary Nagel. As confident as Williams and Keefe were that the Kaganowitzes were undesirable, Nagel thought them to be admissible. Apart from the mother's poor eyesight, there was not a single medical certificate against any family member that would have classified any of them as excludable. All that officials had stated was that the Kaganowitz family looked poor and weak. Nagel was impressed that "every member of this family who is old enough to work does work." What more was needed to prove that this family was self-supporting? he argued. "This evidence of a willingness and capacity to work is worth more than all the ordinary money tests that may be applied," he concluded, in a direct slap at Williams's beloved monetary test. After almost a month in detention at Ellis Island, the family was admitted, although the six youngest children were released on bond.

Meier Salamy Yacoub, a thirty-seven-year-old Syrian Jew, arrived at Ellis Island three days after the Kaganowitz family. Williams found him to be "an undersized man and his appearance is not good." He ordered him excluded as likely to become a public charge. "The indications are that he has come here with the expectation of entering that non-producing class of peddlers of which there are now so many," Williams wrote. For him, the pushcart peddlers who crowded the streets of the Lower East Side and other immigrant ghettos were a nuisance and the nation did not need any more of them. While Keefe agreed, Assistant Secretary Benjamin Cable, acting in place of Nagel for the day, overturned the decision and allowed Yacoub to land. "I do not see how this man is likely to become a public charge," Cable wrote. Yacoub was allowed to leave Ellis Island after only five days there, leaving behind the Kaganowitzes as they awaited word on their fate.

Jewish groups were sensitive to charges that Jewish immigrants were being certified as having poor physiques, especially peddlers like Yacoub. Simon Wolf defended peddlers, calling them "the pioneer merchants of our country at one time," adding that "there is no telling what a peddler might become in the course of time, or at least his children, as evidence of what a rail-splitter and a tailor had accomplished when they finally were located in the White House." Wolf continued with his critique before the National Jewish Immigration Council. "If the immigration officials at the ports of entry and some of those in the Bureau had a little more imagination and a little more red blood percolating for the human species," Wolf said, "there would not be so many likely to become public charges . . . which provoke a smile and not a tear—that is hernia and double hernia, diseases that never prevent a man from labor if properly cared for."

A few months after Williams took office, the New York Times featured a story about complaints that the immigration law was being directed at Jews. "I found that Jews were being singled out among immigrants of other nationalities and the rule of physical development applied to them rigorously," a reporter for a local Jewish paper told the *Times*. Assistant Commissioner Uhl admitted that for "some unknown reason there has lately been an unusual number of young men of the Jewish faith who were unable to come up to the physical requirement," but denied that it was due to any discrimination.

Whether Uhl was being honest or not, it was not unusual to find arguments about the alleged physical weakness of Jews. "On the physical side the Hebrew are the polar opposites of our pioneer breed," claimed social scientist Edward Ross. "Not only are they undersized and weak-muscled, but they shun bodily activity and are exceedingly sensitive to pain." Besides fighting against these stereotypes, Jewish groups also argued that since many Jews were not headed for coal mines or steel mills, great physical strength was not always a necessity. Some, like Meier Yacoub, would become peddlers. Still others, like Solomon Meter, were tailors, a job that required delicate skill rather than physical brawn.

Meter was detained at Ellis Island when doctors certified him as suffering from "atrophy, partial paralysis, club foot, shortening and lameness of right lower extremity which affects ability to earn a living," and therefore likely to become a public charge and ordered excluded.

Irving Lipsitch tried to intervene on behalf of Meter. He admitted that the diagnosis at first appeared damning, but went on to note that "if reduced to plain language," it "simply means that the immigrant is slightly lame." Lipsitch claimed that Meter was a good tailor and his was an occupation "which does not require him to make much use of his lower extremities, nor does it mean that he has to stand on his feet for any length of time." Officials did not buy Litsitch's argument and Meter was deported.

While HIAS continued to appeal immigrant cases, Max Kohler stepped up his criticism of immigration officials in harsh tones. In a well-publicized speech, he declared that America was "in the midst of a new 'Know-Nothing Era' and only a campaign of education can safeguard the best interests of the country and maintain the 'open door' to continued national prosperity."

He noted that the exclusion rates of Jewish immigrants were increasing, although still less than 2 percent of all arrivals. More than two-thirds of Jewish exclusions were deemed likely to become a public charge. That number was roughly in keeping with overall rates for all groups, but to Kohler it was occurring because "of ever newer misconstructions of the law, furtively forced upon inspectors at Ellis Island, day by day, breaking down their judicial attitude and creating an atmosphere of uncertainty and anarchy and cowed timidity." Kohler was not going to let Williams off the hook.

Not all Jewish leaders followed the adversarial lead of Kohler. Responding to Kohler's speech, an editorial in the *Times* stated that it would have "been more effective if he had adopted a somewhat less controversial tone." Others went even further. Nissim Behar, one of the leaders of the National Liberal Immigration League, defended Williams. "Nothing is gained by making absurd charges," Behar warned. "No man could please everybody and do his full duty."

Simon Wolf also defended Williams, saying he placed "no stock" in the accusations against him. He instead urged Jewish leaders to work more closely with immigration officials, rather than antagonize them. Irving Lipsitch, who had to deal on a daily basis with Williams, counseled against appealing every case of Jewish immigrants ordered deported. He worried that such an aggressive move might backfire. "I believe that if that were done," he wrote, "we would lose the privilege."

HIAS President Leon Sanders expressed the same fear, telling Kohler there was "much discontent" among officials in Washington with Jewish immigrant aid societies. "It has been hinted also that Jewish societies are making themselves obnoxious by calling upon the Department frequently and repeatedly for trivial matters," wrote Sanders.

Secretary Nagel, who showed himself sympathetic to immigrant appeals, urged immigrant aid societies to work with, not against, government officials. "Your societies are, of course, carrying out your individual views," he said in a speech before a Jewish group, "but you cannot expect me in my official capacity to accept anything that they say." He pointed to Simon Wolf as a model for cooperation between government and immigrant advocates. "The way Mr. Wolf approaches us is calculated to get best results because he comes to us fairly, good-naturedly, and when he is defeated he recognizes our point of view," he lectured the audience. "This is the spirit in which you ought to come."

He also joked about Wolf's continual presence in Nagel's office. "If we ever miss him," Nagel said, "we think the world is going to stop."

Jewish groups attempted a kind of détente with William Williams when they invited him to address the HIAS annual meeting in January 1910. Jacob Schiff led the applause for Williams, setting "the example of paying due respect to a government official by rising and the audience followed him," the *American Hebrew* reported. In turn, Williams extended an olive branch to the crowd. He repeated his basic philosophy of immigration: whereas some Americans believed all immigrants should be let in and others believed few should be allowed to enter, "I agree with neither." In a surprising turn, the blue-blood Williams told the crowd that he particularly disagreed "with the latter, especially when I see what promising citizens the Jewish immigrants make."

The warm feelings did not last long. Following Williams to the podium was a rabbi who addressed the crowd in Yiddish and criticized the debarment of immigrants with poor physiques. "The strong man by his very strength may be a menace to the peace of the country," the rabbi said, "but the man physically weak may be mentally strong and able to help build up the nation." Williams "seemed to be under the impression that he was being criticized, which was not exactly the case," according to the *American Hebrew*, and grew visibly angry. This cultural and linguistic misunderstanding seemed to have negated any of the initial good will.

After that, Williams's relationship with the Jewish community continued to deteriorate. In his 1911 annual report, Williams spoke dismissively of new immigrants, singling out the crowded Italian and Jewish ghettos of lower Manhattan.

> The new immigrants, unlike that of the earlier years, proceed in part from the poorer elements of the countries of southern and eastern Europe and from backward races with customs and institutions widely different from ours and without the capacity of assimilating with our people as did the early immigrants. Many of those coming from these sources have very low standards of living, possess filthy habits, and are of an ignorance which passes belief. Types of the classes referred to representing various alien races and nationalities may be observed in some of the tenement districts of Elizabeth, Orchard, Rivington, and East Houston Streets, New York City.

In response, members of the "Citizens Committee of Orchard, Rivington, and East Houston Streets" fired off a letter to President Taft. They called Williams's remarks "false," "libelous," and a "gratuitous insult" and maintained that "no public official should be permitted with impunity to malign a large and populous section of this great city." Williams denied that he was targeting Jews, but was only stating "economic, industrial, and sociological facts which are open to the observation of anyone." However, in a 1912 letter to Theodore Roosevelt, Williams complained that many Jews put "the interests of their race before those of their country."

HIAS officials continued to lobby the government on behalf of Jewish immigrants. At Ellis Island, immigrants were too often reduced to words on a sheet of paper: transcripts of hearings, summaries of fact by officials, and medical inspection records. Immigrant aid societies were able to add the human element to this often two-dimensional bureaucratic story. Though immigrants were barred from having lawyers represent them at board of special inquiry hearings, men like Irving Lipsitch served as combination defense attorney and lobbyist.

While it was the job of William Williams and his inspectors to execute the law faithfully, immigrant aid societies became the immigrants' advocates, tilting the scale in the immigrants' favor when no one else would.

WILLIAM WILLIAMS SET OUT to rigorously enforce the law against those he considered undesirable, especially those deemed likely to become public charges. Rather than focusing on markers of personal character to determine desirability, as Theodore Roosevelt had encouraged, Williams increasingly linked undesirability to southern and eastern Europeans. As the enforcement of the law at Ellis Island became tighter and the rhetoric of the commissioner more pointed, opposition to Williams was building. More and more people came to believe that something had to be done to stop "Czar Williams."

Chapter 11

"Czar Williams"

———————

The more humanely the immigrant is treated at Ellis Island,
the more humanely he will deal with us when he becomes the
master of our national destiny.
　　　　　　　—Edward Steiner, 1906

A saint from heaven actuated by all his saintliness would fail
to give satisfaction at this place.
　　　　　　　—Robert Watchorn, 1907

GEORGE THORNTON HAD THE GOOD FORTUNE TO ARRIVE at Ellis Island in October 1910. The Welsh miner and widower was accompanied by his seven children, ranging in age from two to nineteen. The family had over $100 with them and was headed to George's sister in Pittsburgh. However, George was missing fingers on one of his hands, suffered from a hernia, and was therefore certified as likely to become a public charge. He and his family were ordered excluded.

It was Thornton's luck that when William Williams heard the family's appeal, sitting in the commissioner's office was all three hundred and twenty pounds of the president of the United States. Theodore Roosevelt had handpicked William Howard Taft to be his successor and continue his policies, so it is no surprise that Taft emulated his predecessor and paid a presidential visit to Ellis Island. If Roosevelt braved torrential rains and near-hurricane-force winds to arrive at Ellis Island, Taft had to make his way by ferry across New York Harbor through dense fog. Once there, Taft threw himself into the visit, spending almost five hours examining the entire process.

Taft listened to a number of appeals that day and took a special interest in the nicely dressed Thornton family. He proceeded to question the elder Thornton, who was unaware of the identity of his new interrogator inquiring about the singing abilities of the Thornton children. Taft then asked George if he knew who the head of the U.S. government was. "The President," replied George. Did he know his name? "Mr. William H. Taft," responded George. The scene must have given the president a good laugh, as he then revealed his identity to the shocked Thornton. "It appears to me that this respectable-looking family . . . will all grow up to be good, self-supporting citizens of the country," Taft concluded. The family was allowed to land.

The poignant story of the Thornton family barely saved from deportation by the intervention of the president of the United States was enough of a public-interest story to make the newspapers. However, some people in Wales heard the story and wrote to Williams stating that George Thornton had left the country without paying his debts. When Williams contacted George two months after his arrival, he admitted that he had not been able to secure work and his sister was unable to support the family. So George asked to be deported back to Wales, a wish Williams was no doubt happy to fulfill.

But President Taft's personal judgment was on the line, having publicly vouched for the promising character of the family. Therefore, the secretary of Commerce and Labor, Charles Nagel, who had accompanied Taft to Ellis Island on that foggy October day and also strongly urged that the family be allowed to land, intervened to help Thornton find work. The results of Nagel's efforts were disappointing. "In the Thornton case I have ignominiously surrendered," Nagel wrote Taft only a few weeks later. "I find that he does not feel able to do work and that the doctors at Ellis Island evidently knew more about the case than we did."

These were hard words for Taft to hear. Members of the American Association of Foreign Language Newspapers visited Taft at the White House in January 1911 to voice their concerns about the treatment of immigrants at Ellis Island. In response, the president told the group about his visit there a few months earlier. "I have since followed those cases in which I influenced him [Williams] against his better judgment," he told the group, "and I am obliged to make the humiliating confession to you that the outcome vindicated him and showed that my judgment was at fault for lack of experience.

"There are certain parts of this Government that I understand very well, but immigration is new to me," Taft further admitted, "and it is a subject to which I must give as much study as I can, being dependent, however, on the men whom I have selected to administer the law." Such humility clearly marked Taft a different political animal than Theodore Roosevelt. It also led Taft to place even more faith in William Williams.

For the remainder of his term, no matter how heated the criticism got, Taft always stood behind his fellow Yale man. "In selecting Mr. Williams, I have selected a man whom I thought to be a very just and kindly man, and that is what you need there," Taft told the foreign-born newspapermen. Moreover, Taft offered a mild criticism of the group, noting that when one is "continually pulling a man's coattail when he is making a speech you can't expect anything but a poor speech, and so it is with reference to the administration of the Federal law." As for the Thornton family, Nagel wrote to Taft shortly after this meeting at the White House to inform the president that he had just "reluctantly signed the warrant for his deportation."

Never again would Taft meddle in another immigrant case. However, immigrants at Ellis Island did not lack for vocal defenders. During Williams's second tour of duty, the more he tried to tighten the enforcement of the law, the louder the roar from his critics. In his own mind, William Williams was a fearless upholder of the law who ran Ellis Island as a bulwark against undesirable immigrants. The foreign-language press had other ideas. To them, he was a dictator ruling over his fiefdom with an iron fist, enforcing his will upon powerless immigrants and servile employees. He was Czar Williams.

"AWAY WITH CZARISM AT Ellis Island," screamed an editorial from the German-language newspaper *Morgen Journal*. "Bestiality Rampant in the Name of the Law," cried another. The English-language *Evening Journal* chimed in with an editorial castigating "Brutality at Ellis Island." Both papers were owned by William Randolph Hearst and were part of a relentless drumbeat of criticism that Williams would face during his second term at Ellis Island.

The *Morgen Journal* listed almost two dozen German-language papers from Baltimore to Cincinnati, from Buffalo to Denver, from Davenport, Iowa to Sandusky, Ohio, that ran editorials condemning the Ellis Island administration. The *Chicago Abendpost* complained that the members of the boards of special inquiry were "mostly ossified and grouchy bureaucrats of the first order to whom the dead letter of the law is more precious than sound common sense." The protests against Williams's rule went beyond the German-American community. A Hungarian paper in Cleveland, the *Szabadsag*, described "The Terrors of Hell's Island: The Calvary of an Old Hungarian Couple."

O. J. Miller of the German Liberal Immigration Bureau sent out a mass mailing to "Citizens of German Blood" calling attention to the "bias and prejudices of ignorant government hirelings" and the "tyranny" they practiced at Ellis Island. Noting that Jews had "organized a powerful system for the shielding of immigrants of their race from political ruffianism and from the chicane and bias of the immigration officials," Miller called for German-Americans to do the same. He called for every German organization in the country to demand the resignation of William Williams.

Groups such as the Alliance of German Societies of the State of Indiana, the Deutsch-Amerikanischer National Bund of East St. Louis, Illinois, and the German-American Alliance of Hartford, Connecticut, all joined the calls for Williams's resignation. The Brooklyn League of the National German-American Alliance (NGAA) pronounced "the tyrannical and inhuman practices of Commissioner Williams and his staff of inspectors a blot upon civilization."

Likening Williams to a "czar" or "pasha" turned the Ellis Island commissioner into a brutal authoritarian who used his power to suppress helpless immigrants. It was imagery designed to raise the hackles of those who had escaped czarist Russia or other monarchical regimes. The use of terms such as "inquisitors," "star chamber," and "catacombs" were also meant to hit the raw historical nerves of foreign-born Americans.

At first, Williams was surprised by all the heat he was taking from German groups. "If this hostility were confined to papers representing south Europeans I could at least understand the philosophy of it all," he wrote to Charles Nagel. "But we are so fond of Germans, so anxious to have them come here, and we send back and detain such a negligible quantity of those who arrive, that we must look for this hostility elsewhere than in the application of the immigration law to Germans."

Nor could Charles Nagel understand it. The overall rate of rejection of immigrants was "smaller than the general public is prepared to hear," Nagel told President Taft's secretary. He believed that Germans and Jews, the two ethnic groups complaining the loudest about Williams, "have fared if anything better than any other race."

German immigration had slowed. Between 1900 and 1913, nearly 1 million Germans entered the country, but that was only 7.7 percent of all immigrants. In the great divide between old and new immigrants, Germans fell on the right side of the equation. By the early twentieth century, most Americans saw Germans as hearty pioneers who were easy to assimilate, especially when compared to Italians, Greeks, or Russian Jews. Teutonic blood was seen as relatively compatible with that of Anglo-Saxons, as people like Henry Cabot Lodge remembered the origins of their beloved Saxons.

German immigrants had a slim chance of being excluded and were kept out at a rate lower than the average. Between 1904 and 1912, less than 1 percent of all German immigrants were excluded. German-Americans would have noticed that the percentage of exclusions was increasing, although that began before William Williams returned to Ellis Island. Still, this was hardly a crusade against German immigrants. There had to be some other reason for these ferocious attacks on Ellis Island.

Harper's Weekly asked: "Who Is Stirring Up the Germans?" William Williams and the magazine both agreed that the answer could only be explained by the influence of German-owned steamship companies. As Williams stepped up deportations, each one cost the steamship companies $100 in fines, plus the cost of shipping the excluded immigrant back home. Williams may have contributed to the heartache of immigrants concerned about passing through the inspection process, but he was also making a dent in the finances of the steamship companies.

The stricter enforcement of immigration law may not have seriously affected German immigrants, but there was no denying that Williams was now turning away more immigrants at Ellis Island. He believed that

Robert Watchorn, with the approval and oversight of Secretary Oscar Straus, had kept the gates at Ellis Island wide open.

Between 1907 and 1909, less than 1 percent of all immigrants arriving at Ellis Island were rejected. Williams had set out to rectify that situation, and the numbers demonstrate his success. In 1910, Williams's first full year back at Ellis Island, the rate of exclusions doubled to 1.8 percent of all arrivals. That would decrease over the next three years but never dip below 1 percent, as it had under Watchorn. Immigrants faced tougher scrutiny at Ellis Island than they would at any other major inspection station in the country, with the exception of those along the Mexican and Canadian borders.

Nor was it just a question of immigrants having a tougher time getting through inspection at Ellis Island. Those already landed could be deported within three years of their arrival if found to be public charges, prostitutes, criminals, anarchists, feebleminded, or any one of a number of categories that would have labeled them as undesirable under the law. Such deportations were steadily increasing over the years and continued under Williams. During Williams's second tenure at Ellis Island, over 6,000 immigrants found themselves returned to Ellis Island and deported back to their homelands.

Even with the stricter enforcement of the law and increasing number of deportations and in spite of Williams's rhetoric about undesirable immigrants, over 98 percent of all who arrived at Ellis Island were eventually admitted. This speaks to the powerful legal, political, social, economic, and ideological consensus that allowed America to accept millions of new immigrants despite the grumbling of those made uneasy by the disruptions that this human wave brought. Every exclusion was a personal tragedy; in 1910 there were over 14,000 such tragedies at Ellis Island. However, when compared to the hundreds of thousands who easily passed through, it is hard to describe Ellis Island as a restrictionist nightmare.

What is not fully known is how many potential immigrants were stopped at European ports from emigrating in the first place. Steamship companies set up their own inspection process there to weed out individuals they felt were not qualified to land according to American immigration law. If someone did not pass that inspection, he or she could not purchase a ticket. It was simple economics for the steamship companies, who did not want to incur fines and the added expense of transporting rejected immigrants back to Europe. In many ways, that inspection was much tougher and more intrusive than the one immigrants experienced at Ellis Island.

It is hard to come by official figures on the number of people rejected by steamship officials at European ports. Journalist Broughton Brandenburg investigated the conditions of immigrants on both sides of the Atlantic and found that at the ports of Hamburg, Bremen, Liverpool, Naples, and Fiume, from which most American immigrants sailed, some 68,000 people were refused steamship tickets during 1906. At Naples, roughly 6 percent of immigrants seeking passage to America were turned away in 1906. The following year, Robert Watchorn estimated that a total of 65,000 immigrants were barred at all European ports.

For some immigrants, their obstacle course to the New World began even earlier. Russians had to first make their way to German ports like Hamburg or Bremen. Since most of these Russians were Jews, German officials were not happy about having them tramp through their lands, although they were more than willing to have German steamship lines take their passage money. Therefore, Russians could not enter Germany unless they had a ticket to America and a sufficient amount of money on their persons. To enforce the law, Germany erected a series of fourteen border stations in the east. According to one estimate, German border guards turned away some 12,000 Russians in 1907.

An American congressional committee toured these border stations and found that things were even worse for Russian Jews who were deported. Since Germany did not want them in their country, the law demanded they be returned to their villages in Russia. Agents from steamship companies met these unwanted individuals at the Russian-German border because, according to the congressional report, "if emigrants so rejected were turned over to the Russian frontier guards they would be severely treated and subjected to great hardships." For these Russian Jews, the tragedy of rejection at Ellis Island was just the beginning of their hardships, which is why organizations like the Hebrew Immigrant Aid Society fought so hard against many of the deportation orders.

WILLIAM WILLIAMS WAS NOT the only one to feel the sting of criticism from ethnic groups. Secretary Nagel felt the barbs more keenly because he was the son of German immigrants and a member of St. Louis's large German-American community. His continued support for Williams made him a villain among his own landsmen. He told Williams that he was "sick and tired of being accused of prejudice against people when my position is such that suspicion, if any, might well come from the other side."

Nagel won no friends among restrictionists since his natural inclination, like that of his predecessor Oscar Straus, was to side with immigrants in appeals cases. "I am frank to say that my sympathy is all for the human side," Nagel admitted. "I have sometimes felt that I forgot my own country and the law of my country in my desire to help out and to relieve the hardships of individual cases." He was sensitive to the power he possessed in controlling the fate of tens of thousands of individuals. "I can send back anybody," he remarked. "It is an awful power, but I try to use it to the best of my ability."

This was not the kind of introspection that occupied the mind of William Williams. Yet Nagel still maintained good relations with Williams and continued to defend his work. As a sympathetic contemporary noted, Nagel was "never liberal enough to suit the one group, although he became almost a law-breaker in the eyes of others."

The agitation among German-Americans led New York congressman William Sulzer to offer a resolution in the House of Representatives to investigate the affairs at Ellis Island. Before allowing a vote on the resolution to come to the House floor, the Rules Committee began hearings on the matter in late May 1911. Sulzer began the hearing by noting the "deplorable condition" of the immigration service and calling attention to the "atrocities, cruelties, and inhumanities practiced at Ellis Island."

The committee then heard from a procession of German-Americans who had been vocal critics of Williams. Gustave Schweppendick, a journalist for the *Morgen Journal*, admitted that while Ellis Island officials were not specifically targeting German immigrants, he and his colleagues felt the need to stick up for other immigrant groups. Ernest Stahl of the National German-American Alliance described his opinion of an immigrant's Ellis Island experience. "He goes through hell," he told the committee, "that is the only expression that I know of." He called the inspection process "barbarous." Alphonse Koelble, of the United German Societies, complained both about the odor that pervaded Ellis Island and the increasing percentage of exclusions under Williams.

The ubiquitous Marcus Braun also testified, calling Williams "one of the ablest and most honorable men in the service," even though he disagreed with the commissioner about immigration. "The great trouble with Mr. Williams is that he is too strict," Braun said, "not only with the enforcement of the law, but also too strict with his subordinates." Perceptively, Braun noted that Robert Watchorn had "played to the galleries and Mr. Williams does not."

Before the hearings had taken place, Williams tried to play down the criticism from the German press, calling them "so silly and extravagant as to make it seem beneath one's dignity to notice them." He told Prescott Hall that while the charges might poison the minds of "ignorant persons" concerning the operations at Ellis Island, "on the whole I have paid little or no attention to this matter."

Williams was much more thin-skinned than he let on. He obsessively kept detailed records of the attacks against him by the German press, having each article translated into English. He answered nearly every allegation of abuse against an immigrant at Ellis Island, usually in letters or memos to his superiors in Washington.

Still, while the earlier criticism had been merely irritating, the charges made against him before a congressional committee caused Williams to fume. He had not been present at the May hearing, but had received a transcript of the hearing from Senator Henry Cabot Lodge. "I may feel differently tomorrow but just now I am outraged at the falsehoods told about my administration," he wrote to Charles Nagel. "These criticisms pass the bounds of decency and in some manner the persons making them ought to be told what decent people think about them." He would not appear before the committee until it reconvened in July, but promised not to let the accusations go unanswered.

"The law is a difficult one to administer," Williams claimed in his written opening statement to the committee, "particularly in regard to the determination of who is 'likely to become a public charge.'" While recounting some of the accusations against him, Williams said it was all part of the uncertainty and vagueness of immigration law that bedeviled those sworn to execute it. However, he resented "as wholly untruthful the use of such words as those quoted to characterize the work at Ellis Island." He produced details of specific cases of deported immigrants that had appeared in the German press and contradicted the claims of administrative misconduct, all while defending the necessity to tighten the inspection process in the name of enforcing the law.

Williams again repeated his standard line regarding his personal views of immigration. "I will say that I have as little sympathy with those who would curtail all immigration as I have with those who would admit all intending immigrants—good, bad, or indifferent," he said. Daniel Keefe and Charles Nagel both appeared and defended Williams. Unlike Williams, Nagel was torn by the human tragedies that daily crossed his desk. "It is well enough to make a rule under lamplight," Nagel told the committee, "but it is very hard to enforce that rule when you see a pair of eyes looking at you."

In Williams's previous two years at Ellis Island, most of the criticism came from German- and Jewish-American groups. One of the bitterest witnesses against him at the hearing, however, was one of those few Anglo-Saxons who sometimes passed through Ellis Island. The Reverend Sydney Herbert Bass, a minister from England, told the committee about what he was forced to endure while temporarily detained at Ellis Island. "I, too, have photographs," he testified, "but mine are engraved on my mind and heart and burned into my soul as by a red-hot iron."

He had arrived in January 1911, headed for a new congregation in Pennsylvania. Bass traveled in steerage—"for purposes and reasons of my own"—on the White Star's *Adriatic*. The steerage passengers were marched single file into the main building of Ellis Island. As they headed up the stairs towards the Great Hall, Bass remembered that an inspector yelled at them: "Anyone who comes steerage is cattle, you will soon have a nice little pen."

Bass was then marked in chalk with "2 hieroglyphics" on his overcoat that designated him for further inspection. After a quick examination, a doctor found that Bass suffered from "atrophy and partial paralysis of right leg; deformity of right foot; shortening of right leg and lameness due to old poliomyelitis," defects that would affect his ability to earn a living. To this diagnosis, Bass responded, "If that were so it was fortunate that my brains were the other end and I earned my living with them as I did not preach and lecture with my feet."

He would remain at Ellis Island for almost thirty hours, an experience that enraged and deeply scarred the British preacher. He was placed in a holding room with some six hundred immigrants from various nationalities. Although freezing outside, the overcrowded room was steaming hot. Bass took off his overcoat, placed it on the floor, and sat on it. Only when he arose later did he notice that the coat was now stained with "a portion of Italian phlegm, as large as a silver dollar piece."

"The noise alone was a diabolical experience to sensitive people," Bass later remembered, "and I shall never doubt again the literal truth of Scriptures especially with reference to the Tower of Babel." No matter what he did, Bass could not escape the rabble with whom he now found himself detained. "I was standing hemmed in on all four sides by Italian immigrants very taller than I," he told the committee. "They were eating garlic and you can imagine how offensive it was. . . . It made it difficult for me to breathe. The smell was worse than I ever smelled before."

Forced to put up with such conditions, Bass complained to officials, "as any self-respecting Englishman, or American, or those self-respecting Germans . . . would do under similar conditions." That injury to pride and racial superiority, more than the loss of a day at Ellis Island, seemed to drive Bass's anger. What particularly galled him was the treatment of those "delicately nurtured English ladies of much culture and refinement" placed with the rest of the rabble in detention. The whole place was so shocking that it reminded him of Dante's Inferno and the Black Hole of Calcutta.

Bass would get a chance to explain himself before a board of special inquiry the following day. Meanwhile, he was forced to spend the night at Ellis Island. Bass successfully appealed to an official to have all of the detained Englishmen—and one "perfect French gentleman"—put together in the same room, while four English women, a French woman, and a Swedish woman were kept together in another room.

The men slept on canvas hammocks, suspended from the ceiling, numbering three from top to bottom and nine rows for a total of twenty-seven "beds." The canvas mats were damp, and the men were left without blankets for hours. To add insult to injury, these English and French gentlemen were not alone in their dormitory room. "The insects were fearful, and I think I can safely say the English at any rate were all a mass of bug bites," Bass said about the bedbugs.

He was released the following day. Hearst's muscularly populist

New York Evening Journal, always happy to give Ellis Island officials a black eye, ran Bass's story with the headline: "Pastor Calls Ellis Island Hell on Earth." The publicity brought Bass's plight to the attention of Washington, as he complained to the British consulate.

Williams explained his decision to Charles Nagel, calling Bass "an undersized, badly crippled man." He noted that Bass was showed special consideration at Ellis Island, considering the fact that he was a steerage passenger. Williams concluded that he thought "that this badly crippled alien was fortunate in securing admission," a view seconded by Nagel, who told Taft's personal secretary that Bass "was lucky to get in, or rather that we were unlucky to get him in." Now, seven months after his ordeal, Bass was telling his story to a congressional committee and demanding a full investigation of Ellis Island.

At the end of the second hearing, Sulzer testified that Ellis Island could be improved for the benefit of immigrants and that its problems were not the fault of Secretary Nagel or Commissioner Williams. The fault was with the government for not appropriating more money to expand the facilities and hire more inspectors. Despite the hearings, the House Rules Committee never acted on Sulzer's resolution and there was no full-scale congressional investigation of Williams and Ellis Island.

With the failure of Congress to do more than give lip service to their complaints, German groups did not give up the fight. At its annual convention in 1911, just a few months after the end of the congressional hearing, the National German-American Alliance lashed out against William Williams. Henry Weisman, president of the Brooklyn branch of the organization, called Williams's interpretation of the law "arbitrary" and claimed that he excluded many desirable immigrants. The NGAA called for the removal of Williams. Weisman, a lifelong Republican, declared that if Taft did not remove Williams from office, he would never again vote for another Republican for president. A few months later, the *Morgen Journal* demanded: "Williams Must Go." When that did not happen, the newspaper followed up with another editorial asking: "How Much Longer, Mr. Taft?"

Williams had his defenders. In the midst of the congressional hearing, *Harper's Weekly* called Williams "a resolute, upright person, a terror to all scamps who try to plunder the immigrants, and a considerable terror to the steamship companies, who know him as a man not to be trifled with." The editorial concluded that the "suggestion that he is brutal does not match with anything in his record or with his known character."

Arthur von Briesen, president of the Legal Aid Society and chair of the committee that had looked into conditions at Ellis Island during Williams's first term, wrote President Taft about his organization's recent investigation. A member of the Legal Aid Society was sent to Ellis Island in 1911 to see if there had been any changes there since von Briesen's 1903 report. He told Taft that the investigators "were filled with admiration at the manner in which the business was being conducted and the manner in which the immigrants were treated." The facilities at Ellis Island were still too small and the detention quarters too cramped, causing great discomfort for detainees. However, von Briesen's investigators absolved Williams of blame.

Williams's most steadfast ally and friend turned out to be President Taft. "I want you to know that every day, as I think over the Government, I rejoice that I have a commissioner like you in the place you fill," Taft wrote to Williams in November 1911. He then set out to dispense some advice to his fellow Yalie. "Now, brace up!" he wrote. "Life is not so infernally serious that we can not take an interval at time for enjoyment." Taft thought his friend "too darn conscientious" and in working so hard to save the Republic from the evil influences of undesirable immigrants, "you are neglecting your own health, thus defeating the very objects you have in view by curtailing your usefulness in a short time through a break-down."

In his own way, Taft was both bucking up the spirits of a friend and telling him to ease off a bit. Taft saw Williams as faithfully executing the nation's immigration law, but he did not share his overall view of the world. Taft loved his country no less than Williams, but did not find that the procession of aliens streaming into the country marked the downfall of the Republic. "Don't let each trouble weigh on you with its intrinsic weight," advised the weighty president. For Williams there would be no letup. It was not in his makeup. He was, as Taft would later write jokingly, "a severe old bachelor."

Williams continued with his work despite the criticism. While he was making headlines—and enemies—with his strict policies, he also displayed a more typical bureaucratic mentality. Williams wanted a bigger budget from Congress. "I have repeatedly asked for more money and Congress usually has given me only from one-third to one-half of what I asked," he complained. This plaint became a staple of his yearly annual reports, with Williams ever concerned about Washington's "false economy."

The Immigration Service was supposed to be self-supporting, since it received a head tax of $4 for every immigrant, paid by steamship companies but passed along in their ticket prices. During 1910, the United States welcomed over 1 million immigrants, which meant over $4.1 million for the federal government. However, the head tax receipts simply went into the federal government's general operating fund. In fact, in 1910 Congress only appropriated a fraction of that money— $2.6 million—for the operations of the immigration service. Washington was making a profit from immigration.

The economic effects of immigration went beyond the head tax. Immigrants brought more than $46 million with them to the United States in 1910 and sent back roughly $154 million to their relatives in Europe. From 1890 to 1922, GNP increased by nearly 400 percent, as millions of immigrants lent their labor to the factories, mines, and construction crews that built industrial America and created the near unprecedented wealth upon which the American Century was built.

When the U.S. Commission on Immigration, chaired by Vermont senator William Dillingham, finally released its report in 1911, it concluded that immigration was largely an economic issue. It found an oversupply of unskilled labor that lowered the standard of living for American wage earners. Newer immigrant groups, the commission concluded, no longer came over for the idealized reasons that supposedly drove previous immigrant groups. Instead, complained economist Henry Parker Willis, who served as an adviser to the commission, many new immigrants came only "to temporarily take advantage of the greater wages paid for industrial labor in this country."

"Voluminous." "Encyclopedic." "Multitudinous." These were some of the adjectives used to describe the forty-one volumes of the final report of the Dillingham Commission. Clocking in at just under 29,000 pages, it still stands as one of the most impressive reports ever conducted by the U.S. government. How many Greek bakers came to America in 1907? Seventy-three. How many Polish Jewish boys were in the fifth grade of Chicago's public schools? One hundred and thirty-two. The Dillingham Commission had the answers for these questions and many, many more.

The commission's findings were hardly the stuff of hard-core restrictionists, and its data debunked many myths about immigration. It ultimately recommended that immigrants convicted of a crime within five years of entering the country be deported and that immigrant banks and employment agencies be more strictly regulated. It also considered barring unskilled immigrants who arrived without a wife or family, as well as limiting the number of immigrants per year by "race." However, the commission's favored method of restriction, after almost 30,000 pages of data, three years of research, and $1 million in expenses was . . . the literacy test.

Theodore Roosevelt had been a staunch supporter of the literacy test in his younger days, but had done little to secure its passage in his seven years as president. In 1912, Roosevelt was again running for president as the leader of the newly formed Progressive Party. Despite Roosevelt's long record on immigration, there would be no talk of literacy tests, undesirable immigrants, or any kind of immigration regulation during his campaign.

His new party's platform contained one section on "The Immigrant" that concerned itself only with dealing with the problems immigrants faced once here. It promised to secure greater opportunities for immigrants; denounced the "fatal policy of indifference and neglect" that left immigrants prey to abuse; recommended a policy of distributing immigrants away from overcrowded urban ghettos; and called for promoting assimilation. Considering that it was Roosevelt who first brought William Williams to Ellis Island, it can only be described as pure political chutzpah when the *Progressive Bulletin*, the mouthpiece for Roosevelt's new party, denounced Taft's appointment of Williams and his "reign of terror at Ellis Island."

The candidate who was forced to confront immigration most directly during the campaign was the Democratic candidate. Governor Woodrow Wilson of New Jersey, a former political science professor and president of Princeton University, had published a five-volume history of the United States in 1901. In the final volume, the professor delved into immigration. Fitting with the tenor of the times, he condemned the "alteration of stock" brought about by the "multitudes of men of the lowest class from the south of Italy and men of the meaner sort out of Hungary and Poland, men out of the ranks where there was neither skill nor energy nor any initiative of quick intelligence," thereby lowering the American standard of living. Wilson contrasted these "sordid and hapless" individuals with Chinese immigrants who, despite possessing "many an unsavory habit," were at least more intelligent, harder working, and driven to success.

Thanks to newspaper mogul William Randolph Hearst, who despised Wilson, these long-forgotten words now became front-page news across the country. Wilson was soon put on the defensive and tried to explain away his words. He wrote letters of apology to Italian, Polish, and Hungarian groups. "America has always been proud to open her gates to everyone who loved liberty and sought opportunity," Wilson declared in one of these letters, "and she will never seek another course under the guidance of the Democratic Party." He pointed to his membership in the National Liberal Immigration League and began to speak positively about the contribution of immigrants on the campaign stump. "I should be an ignorant man, indeed," Wilson said, "if I did not realize that America has been built up by the blood and the sinews and the brains of those born in the Old World who recognized an opportunity for freedom denied them there."

Despite the controversy, Wilson won by a plurality of votes. Taft came in third, and the lame-duck president had one more issue to deal with after his defeat. The Dillingham Commission provided the momentum for another attempt by Congress to pass a literacy test, which it did in early 1913. It was now up to Taft to decide the bill's fate. The president, whose earnest and guileless temperament was better suited to the judicial bench than the White House, was conflicted about the literacy test. Two years earlier, Taft told Harvard president A. Lawrence Lowell that while he had once been in favor of a literacy test, "I am not quite so clear in my mind now."

For two decades, those who wanted to restrict immigration looked to the literacy test to achieve their goals. However, 122,735 immigrants would have been excluded if the law had been in effect in 1911. Though a large number, it represented only 14 percent of all immigrants that year. Over 90 percent of those who would have been barred as illiterates in 1911 came from eastern and southern Europe. Even still, the proportion of immigrants from these areas would have only dropped from 68 percent to 63 percent. The literacy test would have done little to stem Jewish immigrants; in 1911, only 6,400 who entered the United States were illiterate.

The numbers support the contention of opponents who argued that it was a poor judge of the worth of incoming aliens. "The literacy test is an admirable test of a man's ability to read, and it tests nothing else," said noted rabbi Stephen Wise. Others observed that the law would exclude many a "hard-working industrious man who can add to the country's wealth by his labor," yet "admit many a shifty, adroit, and conscienceless scamp who will add merely to our sufficient supply of gamblers, grafters, and thieves." Nor would it keep out educated anarchists and radicals.

With only a few weeks left to his presidency, Taft finally announced his veto with "great reluctance." In defense of his decision, he appended a long memo from Secretary Charles Nagel, a longtime opponent of the measure, laying out the weaknesses of the literacy test. "To Hell with Jews, Jesuits, and steamships," a depressed Prescott Hall wrote to himself in the wake of Taft's veto.

A few weeks before Taft's veto of the literacy test, Woodrow Wilson paid a visit to Ellis Island. Accompanied by his wife, two of his daughters, and a number of friends from New Jersey, the president-elect was ushered around the inspection station by William Williams. Whereas Roosevelt and Taft had thrown themselves into every aspect of the process on their visits to the island, injecting themselves into the decision-making process and even interrogating immigrants, Wilson was remarkably passive during his trip. "If Mr. Wilson was impressed or otherwise moved by what he saw he did not show it," one newspaper reported. The president-elect asked few questions and his responses to Commissioner Williams were monosyllabic. Upon leaving Ellis Island, Wilson declined to comment about what he had seen. It was his "day off," he told the reporter and he was there "for information and not for thought."

Wilson was far more interested in issues like the tariff, antitrust regulation, and reforming the nation's banking system than he was in immigration, especially after the drubbing he took on the issue during the campaign. It would have made sense for him to ask immediately for the resignation of Commissioner Williams, which would have been the kind of sweeping statement that would have reassured foreign-born Americans concerned about the new president. Wilson chose not to take that politically expedient path and allowed Williams to remain in office.

Located on the Battery in lower Manhattan, Castle Garden was America's primary inspection station from 1855 to 1890, during which time over 8 million immigrants passed through it.

The original inspection station at Ellis Island, which opened on New Year's Day of 1892, was constructed from Georgia pine. After an 1897 fire destroyed all the buildings on the island, one magazine called them "wretched barns and architectural rubbish heaps."

Ellis Island reopened in 1900. Now immigrants arriving to America would be welcomed by an imposing and elegant main building built with a steel-frame and covered with red brick.

"The Stranger at Our Gate": This 1896 cartoon captures anti-immigrant sentiment at the time. The immigrant is carrying a lot of "baggage," such as poverty and disease. Guarding the gate to America, Uncle Sam contemptuously holds his nose in the presence of this newcomer.

These immigrants are heading for Ellis Island's main building, where they will begin the inspection process that will determine whether they will be allowed to enter the country.

A row of young women passes through the line inspection under the watchful eyes of officials. Single young women were not allowed to leave Ellis Island unless accompanied by a male relative or under the care of an immigrant aid society. Officials feared that single women might fall prey to con artists or pimps.

These immigrants are passing in front of a medical inspector who is on the lookout for possible physical or mental problems. The doctor has only a few seconds on the line inspection to make his verdict. If he spots a possible problem, he will mark the immigrant with a piece of chalk, indicating his diagnosis.

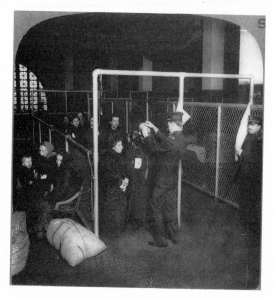

The most memorable and anxiety-provoking part of the inspection process was the eye exam. Here an Ellis Island doctor is flipping back the eyelid of a man to check for signs of trachoma.

This is the Great Hall, or registry room, at Ellis Island. These immigrants have already passed the line inspection and are waiting for an interview with a registry clerk who will make the final decision on whether they can enter the country.

Registry clerks are interviewing immigrants based upon the information found in the ship's manifests (the large white sheets on the desks). If the answers match those found on the manifest, then the individual is declared free to land.

This wonderful drawing shows an immigrant, hat in hand, waiting as an official at the registry desk ponders whether to allow the man and his family to enter the country.

Interpreters played an important role in helping immigrants through the inspection process. Here an interpreter *(middle)* speaks with an immigrant in his native tongue while an official checks the information on the ship's manifest.

Two doctors question an immigrant *(right)* in a private room to determine his intellectual capacity. Dr. Howard Knox *(second from right, in hat)* was one of the Ellis Island doctors most involved in intelligence testing. An interpreter sits at the left.

This 1908 photograph of the immigration staff at Ellis Island includes a young interpreter named Fiorello La Guardia *(top left)* who worked there for three years while attending law school at night. Twenty-five years later, he would be elected mayor of New York.

Boards of Special Inquiry, composed of three or four inspectors, were a constant feature at Ellis Island. Boards heard the cases of immigrants "not clearly and beyond a doubt entitled to land" and determined their admissibility.

This drawing shows an immigrant before a Board of Special Inquiry making his case for why he should be allowed to enter the country.

Although most immigrants spent little more than a few hours on Ellis Island, detention was a reality for those who were forced to wait while their cases were being sorted out or appealed. Some of these detentions could last weeks or even months.

"Uncle Sam, Host": Detained immigrants had to be fed. Among the items on the menu were stewed prunes, oatmeal, beef stew, baked beans, boiled potatoes, coffee, and milk. There was also a kosher kitchen for Jewish immigrants.

The Mittlestadt family arrived at Ellis Island in 1905. They were headed for North Dakota. "Seven soldiers lost to the Kaiser," read the photo caption in the *New York Times*.

"Italian Madonna": This carefully composed 1905 photo by Lewis Hine relies on iconic religious imagery to humanize this Italian woman and her child. However, the individuals standing behind the chain-link fence remind the viewer that detentions and deportations were also a part of life at Ellis Island.

Prescott Hall cofounded the Immigration Restriction League in 1894. For the next quarter century, he relentlessly fought for greater restrictions on immigration. He was dismayed by the growing numbers of eastern and southern European immigrants, and feared the decline of Anglo-Saxon America.

A descendant of New England Puritans, William Williams served as commissioner of Ellis Island from 1902 to 1905 and 1909 to 1913. Though he professionalized the operations there, he also tried to tighten the inspection process to weed out those he termed "undesirable" immigrants.

Theodore Roosevelt, wearing a slouch hat, arrives in the pouring rain in 1903 for the first presidential visit to Ellis Island. Throughout his career, Roosevelt had taken a great interest in immigration.

Following the lead of Roosevelt, President William Howard Taft paid a visit to Ellis Island in 1910. *From left to right:* Secretary of Commerce and Labor Charles Nagel; Taft; Ellis Island Commissioner William Williams; and Commissioner-General of Immigration Daniel Keefe.

From left to right: Commissioner-General of Immigration Anthony Caminetti; Secretary of Labor William B. Wilson; and Terence V. Powderly, head of the Immigration Service's Division of Information.

Edward F. McSweeney was assistant commissioner at Ellis Island from 1893 to 1902. Plagued by scandal and political feuds, McSweeney was eventually fired by Theodore Roosevelt in 1902. This photo dates to the 1920s, when McSweeney was a respected journalist and political figure in Boston.

Appointed secretary of Commerce and Labor in 1906 by Theodore Roosevelt, Oscar Straus was the first Jewish-American to serve in the cabinet. The son of German immigrants, Straus oversaw the enforcement of immigration law and sought to temper its harsher features.

Reformer Frederic C. Howe served as commissioner of Ellis Island from 1914 to 1919. He sought to humanize the inspection process but instead found himself overseeing hundreds of detainees trapped on the island during World War One, as well as German enemy aliens and political radicals rounded up for deportation after the war.

Cipriano Castro, the former ruler of Venezuela, was twice detained at Ellis Island, in 1912 and 1916. He was held on charges of moral turpitude stemming from his alleged involvement in the murder of a political opponent in Venezuela. Castro was eventually allowed to land on both occasions.

Hoping to get her play *Ashes of Love* produced in America, Vera Cathcart (*seated*) arrived in 1926. Officials barred her from the country because of a sexual affair she once had while married to the Earl of Cathcart. Her detention at Ellis Island created an international stir, and Vera was eventually allowed to land. This photo is from a rehearsal for the play taken after her release from Ellis Island.

Emma Goldman *(center, shielding her face)* arriving at Ellis Island in December 1919. Government officials ordered her deported because of her anarchist beliefs. She would remain in detention at Ellis Island for more than two weeks before being deported to Russia with 248 other immigrants.

"A Man Without a Country." In the 1950s, Ignatz Mezei spent almost three years detained at Ellis Island. Accused of belonging to a Communist-backed organization, Mezei was ordered excluded from the country. Due to confusion over his actual citizenship, no other nation would take him. Mezei remained stuck in legal limbo at Ellis Island until the U.S. government finally allowed him to land in 1954.

Ellis Island was closed in 1954. For the next thirty-plus years it sat abandoned and neglected. This photo from 1980 shows the deteriorated condition of the Great Hall before the renovations undertaken after the massive fundraising campaign by the Statue of Liberty–Ellis Island Foundation.

The renovated main building at Ellis Island, which today houses the immigration museum, sits on the north side of the island. The buildings on the south side once served as the inspection station's medical facilities.

In his first month in office, Wilson split the Commerce and Labor Department into two separate cabinet posts. The Bureau of Immigration and Naturalization would reside in the newly created Labor Department. Woodrow Wilson chose the former United Mine Workers official and Scottish immigrant, William B. Wilson, as the first secretary of the new department.

Ethnic groups were heartened by the choice of Secretary Wilson, but concerned that after a few months in office, in the words of the *New Yorker Staats-Zeitung*, he had "not moved a finger in order that the brutal government at Ellis Island may come to an end." Keeping Williams in office would "brand President Wilson as an enemy of immigration of the same type as Williams is in his heart." Yet nothing had been done to end Williams's "reign of terror," which had turned Ellis Island into an "Island of Horrors." Therefore, the paper called for Wilson to clean out the immigration service "In the Name of Humanity."

Throughout the 1912 campaign season and into 1913, the foreign-language press kept hammering away at Williams, who continued to catalog and counter every charge. One representative article appeared in the Yiddish paper *Warheit*: "A Victim of the Murderous Acts on Ellis Island: A Detained Child Held Up for Stammering Dies in Williams' Catacombs."

The *Deutsches Journal* told a tale of "Son Torn from His Mother's Arms: Was 'Too Educated' and 'Not Muscular' Enough." Aron Mosberg had left his home in Galicia to come to America to join his sixty-year-old mother. The twenty-six-year-old bookkeeper was listed at four feet, eight inches tall. Though he had enough money to travel second-class instead of steerage, his allegedly "poor physique," not his financial situation would lead to his deportation. "Not muscular enough" and "marked curvature of spine and deformity of chest" read Mosberg's medical certificate. The newspaper admitted that Mosberg had no "Jack Johnson shoulders," but lamented that it now appeared that "in America men are measured by their bone-structure."

"Sir, You are the murderer of my child, Emilia." Those were the only words contained in the letter that William Williams received in early May 1913 from a Chicago man named John Czurylo. The man's wife and two young children had arrived at Ellis Island on March 29. With both eight-year-old Stanislaw and six-year-old Emilia suffering from chronic inflammation of the lymph nodes, all three were forced

to remain at Ellis Island. One month later, John, still in Chicago awaiting his family, received a telegram announcing the death of Emilia.
Filled with grief, he fired off the letter to Williams.

Williams took the time to reply to Czurylo and explain the circumstances of Emilia's death. He believed that the father had fallen victim
to the misrepresentations of the foreign-language press eager to use
such tragedies to attack the government. John wrote back to apologize
to Williams for "writing you such nonsensical trash." Touchingly, he
ends his letter by asking Williams to have his wife, still in detention,
send him a letter since he had not heard from her in a while. Whatever the merits of the policies enforced at Ellis Island, stories like the
Czurylos' were heartbreaking.

Williams, though, was not a sentimentalist. He displayed little outward angst about the fates of people like Mosberg or the Czurylos. He
would probably answer that his first duty was to execute the immigration laws fairly and without bias, which meant that Mosberg had to be
deported and the Czurylo family had to remain at Ellis Island until the
children had been healed. Williams thought sentiment got in the way
of public duty. If exceptions were made, as the foreign-language press
continually demanded, then the law would become meaningless.

In an April 1913 letter to Washington, Williams said that the personal attacks from the German press did not bother him much. "I
attach no importance to them," he wrote somewhat unconvincingly.
Rather, it showed the effects of "foreign influences at work in our
midst." In a few short years, more Americans would join Williams in
his concern about "foreign influences" on American politics, especially
that of Germans.

Despite all his outward stoicism both to the pain of families such
as the Czurylos and to the constant barrage of criticism against him,
Williams had had enough of Ellis Island. He tendered his resignation
in June 1913 to President Wilson, who accepted it and expressed his appreciation for Williams's "peculiarly intelligent service." Williams had
served a total of six and a half years under three presidents. Wilson
named no immediate replacement, so Williams's deputy, Byron Uhl,
took over as acting commissioner, a move that promised no immediate
change in the execution of immigration law at Ellis Island.

The uncertainty as to who would succeed Williams highlighted the
fact that in almost twenty years of agitation in favor of greater restric-

tions on immigrants, Prescott Hall and his colleagues had precious little to show for their efforts. The Immigration Restriction League's silver bullet, the literacy test, had twice failed to become law over presidential vetoes. The one sliver of hope for Hall and his comrades had been the work of William Williams. "In a world which does not suit me in many ways," the melancholic Hall wrote to Williams, "your work at Ellis Island is a bright spot."

Others remembered Williams not for his restrictionist views but rather for his efforts to improve life at Ellis Island. A letter signed by representatives of twenty-four missionary organizations noted their "high esteem" for Williams, calling him "just always" and "charitable when necessary." Their letter noted: "Even the most casual observer must be conscious of the great improvement in Ellis Island under your guidance, both physically and officially. . . . We believe that those who have attacked your administration have done so either in ignorance or malice."

Others took issue with this sentiment. Unsurprisingly, the *Morgen Journal* shed no tears at the resignation of the man they dubbed the "Czar of the Isle of Tears" who made immigrants "dance to his whip."

Although no one would have known it at the time, the incessant attacks against Williams represented the high point of German-American ethnic identity. Though they did not succeed in removing Williams from office, German-Americans were the leading voice for opposition to immigration restriction. Both the German-language press—shrill and exaggerated—and William Williams—crabbed and snobbish—kept each other in check as the nation navigated this unsettling era of mass immigration. Within a short time, that delicate balance would be destroyed and immigration policy would never fully recover. Nor would the German-American community.

William Williams, however, would move on to other things. In February 1914, the new reform mayor of New York City, John Purroy Mitchell, named Williams commissioner of Water Supply, Gas, and Electricity. After this stint in city government, he returned to the military at the age of fifty-five as a lieutenant colonel during World War I, stationed with the army's procurement division in Washington.

After the war, Williams returned to his law practice in lower Manhattan, where he regularly went to work right up until his death in February 1947 at the age of eighty-four. He made few public remarks on immigration in the decades that preceded his death. We will never know if he lived long enough to temper his views about some of the "scum" who arrived at Ellis Island during his tenure.

TWO MONTHS BEFORE HIS 1910 visit to Ellis Island where he learned from the Thornton family a lesson in the perils of meddling in immigration cases, President Taft found himself dragged into the case of the Pocziwa family. Benjamin Pocziwa lived in Passaic, New Jersey, where he owned his own store. Earning $20 a week and having saved some $500, Benjamin was now able to bring over his wife, Mine; his six-year-old daughter, Anna; and his nine-year-old son, Lipe. All three arrived at Ellis Island in July 1910.

"This child is an imbecile and it is obvious to the layman that he is one," declared William Williams, and young Lipe was ordered excluded by law. His mother, Anna, was also ordered excluded so as to accompany her son back to Russia. Officials with HIAS requested that the deportation be delayed so that the mother could find someone else to escort Lipe back.

Benjamin sought the legal help of Leonard Spitz, who also lived in Passaic and practiced law in Manhattan. Spitz filed a habeas corpus petition on behalf of Lipe. He admitted that little Lipe was "not everything that one of his age should be; his appearance is dull," but explained that the boy had been pampered by his mother and "not allowed by her to run around like the ordinary children of his age, she considering him very precious and always having fear for his welfare."

Local newspapers took up the case of this shy and sheltered country boy frightened by his arrival in America. Spitz spoke about the case with Victor Mason, a businessman who had an office in the same building as Spitz. Mason happened to be a friend of Taft and would be visiting the president at his summer home in Beverly, Massachusetts, in early August. There, Mason explained the case to the president, who ordered that the child not be deported until Secretary Nagel returned to Washington from his extended vacation.

A few days later, Taft reversed himself. "The President has decided not to interfere in the matter of the deportation of Lipe Pocziwa," read the telegram from Taft's secretary, Charles Norton, to the Department of Commerce and Labor. Victor Mason again wrote the president asking him to reconsider his decision, and Taft dutifully changed his mind yet again. "Would the Department be embarrassed in any way if the request were sent down to hold up the deportation of Lipe Pocziwa and his sister and mother until Secretary Nagel's return," Norton again telegrammed Washington.

The Pocziwa family was neither rich nor famous, nor infamous, yet the president of the United States had become involved in their case. For immigration officials, however, Taft's interference and vacillation must have been irritating. Acting secretary of Commerce and Labor Benjamin Cable wrote back to Norton that he would again stay the deportation until Nagel's return, but warned that his boss would not return until sometime in mid-September; this would mean that the family would have to remain in detention at Ellis Island during the dog days of August.

When Nagel returned in September, he ordered that the mother and daughter be allowed to enter the country and rejoin Benjamin; however, young Lipe would have to be sent back to Europe with a suitable attendant. His decision was based on the medical certificate that Lipe was an imbecile, and not a shy and frightened child, and therefore excluded. The law was the law, and it said that no one medically certified as an imbecile could enter the country under any circumstance.

William Williams called on government "to make far greater efforts than it does to prevent the landing of feeble-minded immigrants," since mental deficiencies were "becoming more and more important in civilized countries and the nature and bearings of this taint are being carefully studied by scientists." A feebleminded immigrant would not just become a public charge, he feared, but "may leave feeble-minded descendants and so start a vicious strain that will lead to misery and loss in future generations."

Ellis Island officials would increasingly find themselves drawn into the uncharted territory of using science to determine the mental capacity of those who knocked at America's gate. The Pocziwa family was on the receiving end of those efforts. Even the president of the United States could do nothing about it.

Chapter 12

Intelligence

———— ❖ ————

It is of vast import that the feeble-minded be detected, not
alone because they are predisposed to become public charges,
but because they and their offspring contribute so largely to the
criminal element. All grades of moral, physical, and social
degeneracy appear in their descendants.
—Dr. Alfred C. Reed, Ellis Island, 1912

DURING THE DEPTHS OF THE GREAT DEPRESSION IN 1933,
a Youngstown, Ohio, steelworker named Salvatore Zitello sat down to
compose a letter to President Franklin D. Roosevelt. The new president
had been in office for less than a year, but already Americans felt comfortable enough to write to him by the thousands describing their woes
and asking for help. Salvatore Zitello was not complaining about losing
his job or his house or any other financial problem. Instead, he was
writing about his thirty-six-year-old daughter Gemma.

Salvatore's problems began in February 1916, when his wife Anna
and five children arrived at Ellis Island. (Salvatore had arrived a few
years earlier.) Gemma was the oldest at nineteen and four-year-old Dionisis, the only son, was the youngest. Having sold everything to come
to America, the Zitellos now found themselves stranded at Ellis Island.
Doctors declared Gemma an imbecile and ordered her excluded. To
make matters worse, the two youngest Zitello children, Dionisis and
nine-year-old Alessandra, were sick—one with meningitis and the other
with diphtheria—and confined to the hospital.

Three days after the family's arrival, Salvatore received a telegram from Ellis Island. In cold, blunt language, it read: "Doctors find Gemma Zitella [sic] an imbecile. If you are citizen of United States submit papers at once. Also send affidavit showing your ability and willingness to receive remainder of family."

A week later, Salvatore managed to take time off from his $3-a-day job and make his way to New York to plead for his family. Two days after Salvatore's arrival, Ellis Island commissioner Frederic Howe reiterated the view that Gemma was an imbecile, a condition he thought "obvious even to a layman." Because officials had suspended deportations to Mediterranean ports because of the war in Europe, the family was ordered to remain in detention.

Salvatore was not without help. The Reverend Stefano Testa, a minister with the Italian Mission of the Central Presbyterian Church in Brooklyn, took an interest in the case because his mother had been friendly with Anna Zitello back in Italy. He later accompanied Salvatore to Washington, where they hoped to meet with the secretary of Labor, but instead met with the commissioner-general of immigration, Anthony Caminetti. Rev. Testa asked that the family be released from Ellis Island and that Gemma be paroled into his care, but Caminetti refused. He feared that if the nineteen-year-old girl were released, she would get married, have children, and produce more imbeciles.

Unable to free his family from Ellis Island, a dejected Salvatore returned to his job in Ohio only to find more tragedy. One month after the family's arrival, four-year-old Dionisis died at the Ellis Island hospital. The emotional toll of that loss came on top of the possibility that Salvatore's family might be permanently banished from the country.

After nearly two months in detention, the Zitellos received some good news. Officials would allow the entire family to enter America, except Gemma, who was still ordered deported. On April 21, 1916, Anna and three of her daughters left Gemma at Ellis Island and took the train west to Ohio to reunite with Salvatore.

For the next year, Gemma remained at Ellis Island, excluded from entering the country because of her condition but unable to return to Italy because of the war. Her family was in Ohio, but Rev. Testa visited her often and claimed to have witnessed great improvements since her arrival. Why, Testa asked Caminetti, could she not be released on bond to her parents? Salvatore's hometown congressman also wrote to Washington on behalf of Gemma. The government's answer was always the same: Gemma was an imbecile who was "mandatorily excluded from admission into the United States."

Once America entered the war, Ellis Island was needed to house German enemy aliens, and Gemma was soon transferred to a smaller immigration center in Gloucester City, New Jersey. Her chances of joining her family looked hopeless.

More than two years after his family's arrival, Salvatore wrote directly to President Woodrow Wilson. He explained his family's sad story and complained that because his daughter could not count backwards from twenty, doctors ordered her detained. Since her transfer to Gloucester City, Gemma would write to her father often, complaining that she did not have proper clothing or shoes. She cried every day for her parents.

"I spent the last cent I earned for her and I couldn't do anything," the grieving father wrote to President Wilson. He emphasized his patriotism and boasted that he had bought Liberty Bonds to contribute to the war effort, "I do good right along," Salvatore wrote. Couldn't the president release his daughter, Salvatore wondered?

His response came a month later from Commissioner Caminetti. In coldly bureaucratic words that Salvatore had no doubt become accustomed to, Caminetti wrote: "You are, of course, aware that your daughter Gemma is mandatorily excluded from the United States, and there is no other course that can be pursued except to return her to Italy when it becomes possible to do so."

The war officially ended on November 11, 1918, and the only rationale for keeping Gemma detained had now vanished. The government wasted little time, and on November 20, Gemma Zitello was sent back to Italy. Since she had few decent clothes, authorities had to furnish her with a shirt, pants, undervest, and hose before her journey.

Salvatore, his wife, and three surviving children continued their lives in Youngstown without Gemma. Salvatore and Anna even managed to conceive another child, a boy named Anthony, who was born around the time of Gemma's deportation.

Yet Salvatore never completely gave up hope that he would be reunited with his daughter. That is why the foreign-born steelworker wrote his second letter to an American president in 1933. "I, a citizen of the U.S. and a resident of Youngstown, Ohio, am appealing to you for help as only you can under the circumstance," Salvatore began his letter to Franklin D. Roosevelt. He explained that his daughter had been deported back to Italy and for the past fifteen years he had tried many times to bring her to America. Now thirty-six, Gemma was living in Campobasso, Italy. Salvatore had received word that the people with whom she was living were tired of having to care for her and were mistreating her.

For seventeen years, the Zitello family found themselves staring at the concrete wall of American immigration law. And no letter seemed to make that wall move.

BEGINNING IN 1882, CONGRESS enshrined the word "idiot" into law. As harsh as it sounds, all of those deemed "idiots" and "lunatics" were barred from entering the country. While in most cases it was relatively easy to determine physical defects among immigrants, a bigger problem was how to probe the minds of those who knocked at America's gates. By law, immigrants had to prove not just their sanity but also their intelligence.

At first, doctors were more concerned with weeding out immigrants thought to be suffering from mental illness. Between 1892 and 1903, only thirty-six people were barred from the country for being "idiots," or in other words suffering from low intelligence. During that same period, almost five times as many people were barred for insanity.

When Dr. Thomas Salmon arrived at Ellis Island in 1904, he had no formal training in psychiatry, having begun his career as a country doctor in upstate New York and made his mark investigating an outbreak of diphtheria. At Ellis Island he was one of three doctors tasked with weeding out mentally deficient immigrants.

Salmon saw the chance to filter out immigrants with mental and emotional problems as a great professional opportunity. However, he also understood the limits. He lacked the proper equipment, possessing only, in his words, "a little knowledge of psychiatry in my head, a little piece of chalk in my hand and four seconds of time." With that chalk and the little knowledge of psychiatry, Salmon had mere moments to make a decision on the mental state of an immigrant. If someone on the inspection line struck Salmon as being mentally defective, the doctor would make an X on the individual's coat, selecting that person for further examination.

Salmon was on the lookout for what he called the "well-marked stigmata of degeneration," such as immigrants who seemed "unduly animated, apathetic, supercilious, or apprehensive" or whose facial expression was "vacant or abstracted." A tremor of the lips during the eversion of the eye for the trachoma test or an "oddity of dress," unequally sized pupils, a "hint of negativism," or any "unusual decoration worn on the clothing" could mean further examination and detention.

The results of Salmon's work were stark. In 1906 alone, 92 immigrants were certified as idiots and 139 were certified as insane. All were deported. However, a dispute with Commissioner Robert Watchorn led Salmon to be suspended from his duties. He was eventually transferred to the U.S. Marine Hospital in Chelsea, Massachusetts.

Just as Salmon was leaving Ellis Island, Congress was further expanding its categories of restriction. The Immigration Act of 1907 added two more terms—"feebleminded" and "imbeciles"—to the excluded list. In addition, immigrants deemed mentally defective to the extent that it prevented their earning a living could also be excluded. The new law shifted the focus away from those with mental illnesses and focused greater attention on measuring the intelligence of new immigrants.

As Congress expanded the list of undesirables, Ellis Island found itself testing that most difficult of concepts: human intelligence. What was the difference between an idiot, an imbecile, and someone defined as feebleminded? The Public Health Service informed its doctors that feeblemindedness was defined by a "demonstrated defective mentality" relative to the immigrant's age, but this was of little help. That is where Dr. Henry H. Goddard came in.

About a hundred miles south of Ellis Island, in the southern New Jersey town of Vineland, Goddard was working on measuring, classifying, and treating the feebleminded. Armed with a PhD in psychology, Goddard was the director of the Vineland Training School for Feeble Minded Girls and Boys. His great success was in translating and popularizing a series of tests to measure intelligence created by French psychologist Alfred Binet.

At the time, intelligence tests were a step forward compared to what had preceded them. Craniometry, the measurements of skull sizes, had been the main tool used to measure intelligence in the late 1800s. Unhappy with this crude measure, Binet created a series of tests that would measure the reasoning and comprehension skills of its subjects. Those subjects were largely French schoolchildren. Schools used the tests to help target children in need of special instruction. The tasks were classified by the age at which the subject should be able to complete them. Children who then completed the tests were assigned a mental age, as opposed to their actual age.

Intelligence tests satisfied the needs of early twentieth-century scientists for greater precision and empiricism. However, the question of whether humans possessed a single, fixed, and discrete entity called intelligence that could be accurately measured would continue to be a highly controversial idea for decades to come.

Goddard set out to define the terms "idiot" and "imbecile." An idiot was an individual with a mental age below three years, while an imbecile scored between the ages of three and seven years. These were people who suffered from obvious and severe mental retardation. What about those who scored at levels equivalent to a mental age of between eight and twelve? Their supposed infirmity was not readily apparent to the casual observer, but Goddard felt that intelligence tests could weed out these individuals.

There was also the problem of what to call such individuals. Though they were often called feebleminded, this caused confusion, since it was common to refer to all those with below-average intelligence as feebleminded. So Goddard invented the term "moron" to classify individuals with a mental age between eight and twelve years old. Goddard took the term from the Greek word for foolish. The word has so completely seeped into the English vocabulary that it is hard to believe its origin dates only to the first decade of the twentieth century.

If there was some innate quality called intelligence, Goddard believed, then it was to be found on a human gene that could be passed down through generations. If intelligence was a hereditary trait, then society should make sure that mental defectives did not reproduce. Eugenics, a term coined in the mid-1800s and derived from the Greek meaning "well born," gradually seeped into the public consciousness. In 1910, a biologist named Charles Davenport formed the Eugenics Record Office at the Cold Spring Harbor Laboratory on Long Island to encourage so-called heartier stock to reproduce and discourage the weak from having children. He was already serving as secretary of the

American Breeders Association and that same year he came out with *Eugenics: The Science of Human Improvement by Better Breeding.*

Though some advocated forced sterilization, Goddard preferred the establishment of institutions like Vineland to care for the feebleminded while making sure they did not reproduce. Though Goddard's famous study of the hereditary effects of feeblemindedness centered on a native-born, old-stock family pseudonymously named the Kallikaks, it was no surprise that advocates of eugenics would turn their attention to immigrants. "The idea of a 'melting pot' belongs to a pre-Mendelian age," Davenport noted. "Now we recognize that characters are inherited as units and do not readily break up."

In 1911, Davenport recommended the formation of a committee to study "the hereditary traits that immigrants were bringing into the country." Later that year, Davenport's Immigration Committee of the Eugenics Section of the American Breeders Association found that the unfit were not being properly excluded because of inadequate inspection, poor facilities, and too few medical inspectors.

Two members of the committee were Immigration Restriction League (IRL) officials Prescott Hall and Robert DeC. Ward. For the past two decades, these men had tried to convince their fellow Americans of the threat that immigrants posed. Although they never advocated closing the nation's gates, they continually lobbied for tougher inspection of immigrants and the exclusion of those they deemed undesirable. They had hoped the literacy test would be the vehicle that would keep out many undesirable immigrants, but they had been thwarted in their attempts for twenty years.

Now many IRL members took up the banner of eugenics. Ward hoped that immigration officials could practice "eugenic principles in the selection of the fathers and mothers of future American children." It was feebleminded immigrants more than the insane, Ward believed, who posed the greatest threat to the Republic. "The latter are to a considerable extent segregated and thus prevented from breeding," he wrote, "but the former are far oftener at liberty, and are thus usually free to breed as they will."

For Prescott Hall, the ability to sort populations by their genetic stock was a beneficial result of the spread of science. The rise of science and the decline of religion, Hall noted approvingly, "turned men's gaze in large part from the next world to this." With a heady mixture of Darwin, Theodore Roosevelt, and Nietzsche, Hall spoke of the new "Christ ideal" rooted not in religious faith but in "human perfection." He praised the "superman, working in a strenuous life to produce a better world here and now."

One answer for Hall was birth control. Both the restriction of immigration and the use of birth control should, in his words, be applied both to "defective and delinquent stocks of all races," as well as "less desirable races." Why, he asked, was science so devoted to using its new knowledge to breeding animals and plants, but not humans?

As to whether humans were affected more by their environment than by their genes, Hall sided with nature. "You cannot make bad stock into good by changing its meridian, any more than you can turn a cart horse into a hunter by putting it into a fine stable, or make a mongrel into a fine dog by teaching it tricks," he argued. Hall held out little hope that life in America would have any effect on the intelligence of immigrants. He approvingly quoted eugenicist Karl Pearson, that one "cannot change the leopard's spots and you cannot change bad stock to good; you may dilute it, possibly spread it over a large area, spoiling good stock, but until it ceases to multiply it will not cease to be."

At the intersection of eugenics and immigration restriction was the dark pessimism of native-born Anglo-Saxons that their culture would be washed away in a tide of southern and eastern Europeans. Some asked whether the Anglo-Saxon would go the way of the American Indian and the buffalo: to extinction.

Progressive sociologist Edward A. Ross was one of those asking that question. In 1913, Ross gave a lecture on immigration in which he prophesized that when "the blood of the old pioneering breed has faded out of the motley, polyglot, polychrome, caste-riven population that will crowd this Continent to a Chinese density, let there be reared a commemorative monument bearing these words: 'To the American Pioneering Breed, The Victim of too much Humanitarianism and too little Common Sense.'"

One late afternoon, Ross planted himself in New York's Union Square as garment workers left their jobs and headed back to their tenement homes. At six feet, four inches tall, the patrician academic from Wisconsin must have towered over the diverse, multi-ethnic crowd milling about Union Square. Ross took a quick scan of 368 individuals as they passed him and reported that only 38 "had the type of face one would find at a county fair in the West or the South."

Ross proudly noted that a trained eye could see that the physiognomy of many ethnic groups painted them as decidedly inferior. So just what kind of faces did Ross see in Union Square and in immigrant enclaves across the country? One was what he called the "Caliban type," defined by men who were "hirsute, low-browed, big-faced persons of obviously low mentality" and who "clearly belong in skins, in wattled huts at the close of the Great Ice Age." These were men, Ross confidently proclaimed, who were the "descendants of those *who always stayed behind*."

Whatever Ross's descriptions lacked in historical or scientific accuracy, they were not lacking in vivid language. When he saw foreign-born men, Ross was struck by their "narrow and sloping foreheads" and asymmetrical faces. The women were no better. He found them largely unattractive, with every face betraying some fatal flaw—"lips thick, mouth coarse, upper lip too long, cheekbones too high, chin poorly formed, the bridge of the nose hollowed, the base of the nose tilted or else the whole face prognathous." It seemed that almost every foreign face Ross encountered betrayed a deep inferiority that bordered on the subhuman. "There were so many sugar-loaf heads, moon-faces, slit mouths, lantern jaws, and goose-bill noses that one might imagine a malicious jinn had amused himself by casting human beings in a set of skew-molds discarded by the Creator," he wrote. That these men and women were contributing their genes to the American melting pot was enough to make men like Ross despondent.

A leading academic, Ross was also a Progressive, yet so many of his observations seemed rooted more in prejudice than in social science. To Ross, Jews were small, weak, and "exceedingly sensitive to pain." Slavs were "immune to certain kinds of dirt," while Mediterranean types were skilled at "nimble lying."

Ross predicted that these new immigrants would cause "a mysterious slackening in social progress" and an overall decline in national intelligence. All of this inferior genetic material floating in the American gene pool would create an increasingly sluggish people, in contrast to the hearty and independent Anglo-Saxon settlers. Crime, drunkenness, sexual immorality, and venereal disease would rise, while "intelligence, self-restraint, refinement, orderliness and efficiency" would decline.

These descriptions placed immigrants on the evolutionary scale far behind the vigorous Anglo-Saxons who settled America. Such stereotypes could take a tragicomic twist, as when a member of the Ellis Island medical staff, Howard Knox, told a meeting of the Eugenics Research Association at Columbia University that a recently deported thirty-nine-year-old Finnish immigrant closely resembled the "missing link" that scientists have searched for to explain the evolutionary gap between apes and humans.

To Knox, this immigrant resembled a man from the early Stone Age, with a low, receding forehead, long, shaggy eyebrows, thick, protruding lips, a massive jaw, long arms, teeth angled outward, and each finger resembling a thumb. The man's profession—a linesman for the telephone company—seemed to prove Knox's thesis, "since he may have inherited the characteristics of his ancestors who perhaps often found it necessary to climb to the tree tops to escape some giant animal of their time." He further explained that while he had never found a man with a tail, he held out hope that he would find such a creature at Ellis Island.

Amidst such pressing concerns for the future of American genetic stock, Henry Goddard offered his services to officials at Ellis Island, where he found a willing ally in William Williams. During his second term as commissioner, Williams was even more convinced that too many undesirable immigrants were entering the country. He was concerned that mentally defective immigrants would "start vicious strains which lead to misery and loss in future generations and influence unfavorably the character and lives of hundreds of persons." Robert DeC. Ward praised Williams for doing "more than anyone else to keep the blood of our race pure."

Williams complained to his superiors in Washington that under the current law "many families of poor stock are admissible who practically never rise out of a certain narrow border-land between independence and dependence." As part of his work, he sent an inspector to report on some three dozen Italian and Jewish children in New York City deemed feebleminded by local schools and hospitals. The longer the families had been in America, Williams argued, the worse off they were. These families, he wrote, came from classes that "have been going down hill for some time" due to "bad living conditions, in-breeding, over-breeding, the strain of persecution."

Neither Congress nor President William Howard Taft seemed willing to secure extra funding to weed out mentally defective immigrants, so Williams was forced to look in another direction, and Goddard offered a scientific method that would aid doctors in doing so. In 1910, Goddard and his colleague Edward Johnstone visited Ellis Island. The two men came away disappointed, discouraged, and overwhelmed, both by the enormity of the daily immigrant tide they saw on that one day—some five thousand immigrants—as well as the lack of proper facilities. Goddard felt there was little he could contribute to the effort to weed out mental defectives in such an environment.

So discouraged was Goddard that he did not return again to Ellis Island until the spring of 1912, when Williams invited him back to perform some experiments. Goddard came on a Saturday when no immigrants arrived, but there were a few still on the island preparing to leave for the mainland. Goddard picked out one young man and gave him the Binet test. He tested at a mental age of eight years old, an obvious defective to Goddard.

Williams seemed pleased enough with the results to invite the psychologist back the following week. This time Goddard brought two female assistants with him and set out to construct an experiment. One woman would stand on the inspection line and pick out immigrants for further testing, while the second assistant would sit in a room and administer the Binet tests to those selected. Based solely on observation, Goddard's assistant picked out nine individuals who appeared to be mentally deficient, as well as three more who appeared normal. The twelve were then tested, and Goddard reported that all nine suspected of being mentally deficient had tested so, while the three in the control group had tested normal.

Believing this was proof of the scientific validity of intelligence testing, Goddard requested a return engagement in the fall of 1912. For one week, Goddard and his female assistants administered Binet tests. In one experiment, Goddard's assistants selected eleven immigrants whom they believed were mentally defective, while Public Health Service doctors pulled out thirty-three. All were given the Binet tests. Goddard reported that fewer than half of those chosen by the medical staff qualified as mentally defective, while his own assistants proved correct in nine out of the eleven cases.

Confident of its ability to pick out mentally defective immigrants, Goddard's team moved to another experiment. Working with Ellis

Island medical officials, both groups stood in an inspection line of some 1,260 incoming immigrants. Goddard's assistants picked out 83 suspected mental defectives, while the medical inspectors picked out only 18. Extrapolating from his earlier experiment, Goddard argued that his assistants would have excluded some 72 immigrants, while the medical inspectors would only have caught 8. Goddard believed he had now scientifically proved what William Williams, Prescott Hall, and others believed—that mentally inferior immigrants were slipping past inadequate inspection at Ellis Island.

Goddard magnanimously said that he did not mean to disparage the quality and professionalism of the Ellis Island medical staff. They simply were not specialists, he argued, and his staff showed just what experts in psychology could provide. All that was needed was better training of the medical staff at immigration stations, something on the order of a year or two medical residency at an institution like the Vineland School. With such training, he wrote, officials could then "pick out with marvelous accuracy every case of mental defect in all those who are above the infant age." Women, he said, were best fit for the job because they possessed a keener sense of observation.

Goddard's test did not go completely without a hitch. He was concerned that most of the immigrants did not speak English, forcing his assistants to rely on interpreters to administer the tests. How could you be sure, Goddard worried, that the interpreters were correctly translating both the questions and the immigrants' responses? However, he did not ask whether cultural biases could subvert the results of the tests. Were intelligence tests conceived for use with French schoolchildren suitable instruments to measure the intelligence of peasants from southern and eastern Europe?

Nevertheless, Goddard carried on with his experiments, raising more funds to send another group of testers to Ellis Island in the spring of 1913 for two and a half months. What came from this round of testing was one of the most infamous and misunderstood psychological studies of the twentieth century.

Goddard's staff chose a total of 191 immigrants—Jews, Italians, Russians, and Hungarians—for a battery of five intelligence tests. To arrive at this group, Goddard first weeded out those of obvious low intelligence, as well as those who clearly appeared intellectually suitable for admission. What was left was a group that Goddard defined as borderline feebleminded, who may or may not be qualified for admission.

Although Goddard's staff conducted the test in 1913, the results were not presented publicly until a 1916 conference and not published until 1917. Why did Goddard, whose professional goal was to get intelligence tests accepted by the general public, take so long to report his results?

One reason is that the results shocked even Goddard. They showed that 83 percent of Jews, 80 percent of Hungarians, and 79 percent of Italians tested were clearly feebleminded. Worse still, Goddard's team could only pinpoint six individuals whose measured level of intelligence was without a doubt acceptable for admission. The remaining subjects possessed a level of intelligence that would make their legal admission to the United States unlikely.

The results, wrote Goddard, "are so surprising and difficult of acceptance that they can hardly stand by themselves as valid." Unlike Edward Ross, Goddard did not set out to prove the inferiority of immigrants. He wondered whether the tests were too hard and began omitting certain questions from the test. After rejiggering the results, Goddard lowered his estimate of those clearly feebleminded to almost 40 percent.

When Goddard finally published these results in 1917, his paper displayed less of the confidence of a modern scientist than the confused and self-contradictory response of a man working his way around complex sociological and psychological problems. Within the same article, Goddard repeatedly contradicts himself as he tries to explain the data.

How did Goddard determine the intelligence of immigrants? When asked, through a translator, to give the definition of common terms such as "table" or "horse," the feebleminded immigrant would respond only with that object's most common use. A table is "something to eat on" and a horse "is to ride." These answers showed Goddard a lack of imagination or creativity. In a similar vein, many immigrants had trouble taking three words and creating a sentence from them; nor could most dissect sentences, produce rhymes, or draw a design of an object from memory. Just as disconcerting, Goddard found that most of these immigrants did not know the current date.

Goddard asked whether these supposed failures were due to hereditary defects, as many eugenicists believed, or whether they were affected by environmental factors. To test this question, he set out to track those same immigrants to see whether their lives in the United States confirmed the original diagnosis of feeblemindedness. (Goddard's tests were not legally binding on the admissibility of the immigrants.)

Two years after these tests were conducted, Goddard's staff attempted to track the addresses of as many of their subjects as possible, traveling as far as St. Louis. Much to Goddard's chagrin, few of the immigrants could be found. His staff encountered numerous problems, from incorrect addresses, immigrants who had moved, and uncertainty about the spelling of names. Tenement dwellers were often unwilling to help Goddard's dutiful and earnest young female staffers.

The wild goose chase probably helped cause the delay in reporting Goddard's results. So did the gnawing uncertainty Goddard felt about his study. He asked in his 1917 paper: "Are these results reasonable?" Goddard had already answered that question by cutting his initial estimate of feebleminded immigrants in half.

As for whether intelligence was inherited, Goddard repeated the mantra, "Morons beget morons." Yet he also wrote in the same article that it seemed more likely that the poor showing of immigrants on these intelligence tests was due to environmental causes rather than genetic defects. Unlike the work of Edward Ross, Goddard avoided linking new immigrants to the weakening of America's genetic stock. In fact, he mused with little evidence that "a very large percentage of these immigrants make good after a fashion." On top of that, he said, these feebleminded immigrants did the work that Americans would not do. There was plenty of drudge work that needed to be done that required minimal intelligence.

Even a nonscientist can quickly point out the shoddy methodology of Goddard's Ellis Island research. His own writings betrayed his second thoughts about his scientific discoveries. Goddard was attempting to make science useful to mankind to help create a more rational and healthier society. He also sought to establish psychology as a respected and useful part of the medical profession. Yet his science too often fell victim to the popular biases of the time.

The Survey, the nation's leading periodical for social workers, helped publicize Goddard's study. "Two Immigrants Out of Five Feebleminded" ran a headline in the magazine's editorial on the subject. "If you had gone over to Ellis Island shortly before the war began and placed your hand at random on one of the aliens waiting to be examined by government inspectors, you would very likely have found that your choice was feebleminded," the editorial announced. Though the journal used the less inflammatory numbers from Goddard's study, it still treated his work as scientific proof of large-scale immigrant deficiency. The editorial failed to inform its readers that Goddard's tests were given to less than two hundred individuals who were not chosen from a representative sample.

Yet when Goddard began his work, he was agnostic on the relationship between immigrants and feeblemindedness. Before leaving for Ellis Island, Goddard had set out to test the opinion that many residents of American mental asylums and institutions were foreign-born. Looking at sixteen such institutions across the nation, he found less than 5 percent of the more than eleven thousand inmates were foreign-born. The fear that mentally ill immigrants were swamping the nation's hospitals, schools, and institutions, Goddard wrote, was "grossly overestimated."

For all the attention that Goddard received for his studies at Ellis Island, it was only a small part of intelligence testing taking place there. Not surprisingly, medical officers who sorted through thousands of immigrants each day resented Goddard and his team, who swooped into Ellis Island with great fanfare and then quickly left, leaving the heavy lifting of the daily inspection and testing to the doctors of the Public Health Service, whom Goddard implied were untrained for weeding out mental defectives and had let far too many immigrants of low intelligence slip through.

Goddard had been particularly critical of the powers of observation of Ellis Island doctors, yet their writings show that these officials also put a great deal of faith in initial observations of immigrants. Dr. C. P. Knight described in detail the easily detectable warning signs of a possible idiot, ranging from "low receding forehead" to the size of a face out of proportion the size of the head, to deformed or twisted ears, to excessively deep eye sockets created by a protruding brow. Idiots drooled, and were often apathetic or overly excited. "The expression is stupid, the eyes dull, the speech defective, the tongue swollen and protruding, while the limbs are short and bent and the skin is thick, sallow and greasy," Knight wrote.

For immigrants suffering from "dementia, mental deficiency, or epilepsy," doctors were on the lookout for "stupidity, confusion, inattention, lack of comprehension, facial expressions of earnestness or preoccupation . . . general untidiness . . . talking to one's self, incoherent talk . . . evidence of negativism, silly laughing, hallucinating, awkward manner, biting nail." In a sample of about 30,000 steerage passengers inspected at Ellis Island in the summer of 1916, some 3,000 received a chalk mark X, although after the battery of tests were completed, only 108 were certified as feebleminded.

Ellis Island doctors also paid attention to ethnic characteristics when assessing mental capacity. While it was perfectly normal for an Italian to show emotion "on the slightest provocation," if an Italian showed the "solidity and indifference" of a Pole or a Russian, that would signal a need for further testing. Similarly, English and Germans should answer questions in a straightforward manner, but if they became "evasive as do the Hebrews, we would be inclined to question their sanity." If an Englishman behaved like an Irishman, Dr. E. H. Mullan argued, inspectors would suspect him of mental problems. If an Italian behaved like a Finn, depression might be suspected.

Howard Knox was one of the leading experts on mental testing there. The twenty-seven-year-old Knox arrived at Ellis Island in the spring of 1912, around the same time as Henry Goddard's second visit. He had spent less than three years as a doctor in the Army Medical Corps before resigning in April 1911. The Dartmouth-educated doctor, whose round, fleshy face bore a resemblance to Babe Ruth, had been married three times in as many years. (When he left Ellis Island in 1916, he would be on marriage number four.) Knox then applied for a position in the Public Health and Marine-Hospital Service and was assigned to Ellis Island. Like Thomas Salmon, he had not been trained as a psychologist.

Knox shared many of the prejudices and biases of the time. He believed mentally defective immigrants were like drops of ink in a barrel of water, polluting the nation's bloodstream. If the feebleminded were not caught at Ellis Island, Knox argued, they would "start a line of defectives whose progeny, like the brook, will go on forever, branching off here in an imbecile and there in an epileptic."

Knox was also sensitive to the flaws in intelligence tests and recognized that many immigrants did poorly not because of innate inferiority but because of a lack of formal education. He warned that intelligence tests like the ones Goddard used would make nearly all immigrants from peasant backgrounds appear to be mentally defective. Another Ellis Island doctor, E. K. Sprague, argued that using Binet tests originally designed for French schoolchildren on poor, uneducated immigrants "is as sensible as to claim that with a single instrument any operation in surgery can be successfully performed."

"After studying carefully the methods used at the various schools for the feebleminded," Knox wrote, "the medical officers at Ellis Island were obliged to discard the great majority of them as unsuitable for their work and unfair to the immigrant." Knox claimed that one of Goddard's female assistants had pulled out and tested thirty-six immigrants as mentally deficient. When she turned them over to be certified by Knox and his colleagues, they refused. Using their own methods, they found that in each case the immigrant was either of normal intelligence or suffered from poor vision.

Their day-to-day familiarity with immigrants caused Ellis Island's doctors to reject the overly deterministic testing conducted by Goddard's team, and they were not shy about airing their criticisms in print. Knox repeatedly criticized the methods of Goddard and his staff, calling them "lay-workers with no knowledge of medicine, psychiatry, or neurology." He complained that they often confused temporary psychological disorders, brought about by environmental conditions, with a mental defect and "call such a patient 'stupid' or rate him as 'seven years old on the Binet.'"

Knox noted one case of an immigrant selected by the Goddard team as feebleminded because of a head shape that Knox classified as "simian reversion type with stigmata including malformation of helix." To Goddard's team, the shape of the man's head placed him lower on the evolutionary scale and signified low intelligence. When Knox's colleagues tested the man, they found that he had above average intelligence and spoke three languages fluently. He was admitted.

Another Ellis Island doctor, Bernard Glueck, told the story of a thirty-five-year-old southern Italian man. Based on intelligence tests similar to those used by Goddard, the immigrant was classified with a mental age of between eight and ten, a certifiable moron. Yet Glueck discovered that the man had been in the country before, working as a laborer for two years, during which time he sent back to his family in Italy some $400. He was married with two children, owned property in

Italy that he had bought with money earned in the United States, and was returning to earn still more money. "I have no doubt that he will succeed in doing this," recalled Glueck, who saw the story as a refutation of the Binet test's ability to measure intelligence. "I am inclined to assume in this case the existence of strongly presumptive evidence that this particular individual is not feebleminded," concluded Glueck.

Ignoring Goddard's work, Ellis Island doctors created their own system of testing the mental capacity of immigrants. Knox began with the realization that the conditions under which immigrants arrived at Ellis Island were less than ideal. "After ten days of sea-sickness, fatigue, and excitement," Knox wrote, such an individual "could not be expected to do himself justice." Therefore, immigrants should have a solid meal, bath, and good night's sleep before taking any mental tests.

The testing room should be no warmer than 70 degrees, well ventilated, and quiet, and there should be no more than three people in the room. Those administering the test should "have a pleasant and kindly manner." To ease the mind of the person being tested, Knox argued that the room should not have "an official air," but instead resemble a den in someone's home. If possible, tests should be conducted over two days. Doctors should make allowances for the "fear and mental stress under which the subject may be laboring." While these precautions may have been cold comfort for dazed and confused immigrants, they at least show that doctors were aware of the pitfalls of their assignment.

Once the conditions had been established, doctors began with a battery of questions. What day of the week is it? What is the date? Where is the immigrant? Next came questions that dealt with common knowledge, such as the number of hours in a day, months in a year, and names of flowers and animals. Immigrants were asked questions about their homeland, such as the capital of their native country and the name of their currency. Other questions were more culturally subjective, such as the significance of Easter. In a random survey of fifty uneducated Polish immigrants, Glueck found that while 98 percent knew the number of months of the year, only 66 percent knew the significance of Easter. Glueck admitted that these questions were relatively useless in judging intelligence among uneducated immigrants.

Other questions would test mathematical ability with simple addition problems. Immigrants would next be asked to repeat back a series of four to seven numbers given to them by their examiner and were then asked to count to twenty, sometimes by twos, and then count backwards from twenty. They were tested on their ability to gain new knowledge, so they were asked the name of the steamship they arrived on, what port they left from, and how the ships were powered.

This battery of questions confused Codger Nutt, a boy actor and mascot of the Drury Lane Theatre in London, who was coming to New York to appear in a play. The diminutive thespian could neither read nor write, spoke with a strong Cockney accent, and seemed lost at Ellis Island. Doctors suspected him of being feebleminded, so they asked him whether he knew the difference between a horse and a cow. "I told 'em that an 'orse could be driven but ye couldn't drive a cow," Nutt replied. Then they asked him what he would do if he saw someone in the road "cut up into a 'undred pieces," to which he responded that he would report it to the police. Officials were not convinced by the diminutive actor's answers, but Secretary Charles Nagel allowed him to enter the country and join the rest of his acting troupe, as long as he left the country after a year.

The questions that Codger Nutt and others faced were only the beginning of the testing. Doctors went beyond testing math or memory skills and tried to measure the creative powers and imagination of immigrants. "Some of us having gazed into the smoke of a choice cigar or into an open fireplace," wrote Knox, "may have seen, perhaps, the sweetheart of other days, or the vision of a farmhouse away off in some old country town." With that in mind, Knox set out to use inkblots of various shapes. Each figure vaguely resembled some object, such as a house, a strawberry, a snake, a leaf.

Knox conducted a small study using these inkblots among twenty-five Italian immigrants deemed normal and twenty-five deemed mentally defective. The answers from the mental defectives were often accompanied by a "negative tongue noise" or "I don't know." Knox also recorded his impression of each individual, which ranged from "stupid and indifferent" to "stupid, emotional, high tempered, and willful." He concluded that "there are no Jules Vernes" among the group. The reaction time for those deemed mentally defective was nearly twice as slow as the normal group and the mental defectives possessed more asymmetrical heads and faces, harkening back to Goddard's belief that observation alone could weed out mental defectives.

Immigrants were also given pictures to describe. One of them, entitled "Last Honors to Bunny," depicted three young children mourning their dead pet rabbit. Immigrants were asked six questions, including what was going on, what the boy and girl were doing, and why one of the boys was digging a hole.

Ellis Island doctor E. H. Mullan found that most of the immigrants poorly described the picture, but that should have come as no surprise. Hard as it may be to believe, some immigrants had little familiarity with pictures. More importantly, many immigrants were puzzled by what they saw in the drawing. They had rarely seen pets treated well and were not used to seeing rabbits as pets. Some were unfamiliar with the custom of placing flowers on graves. Mullan concluded that pictures were unhelpful in judging the mental capacity of immigrants unless they depicted scenes easily recognizable to European peasants.

Ellis Island doctors were increasingly bothered by the subjectivity of their intelligence tests. One manual admitted that testing the knowledge and intelligence of immigrants was a difficult, perhaps impossible, task. "What are likely to be considered matters of universal knowledge may be absolutely unknown to them on account of the extreme limitations of their surroundings," it stated. The average American, these doctors were informed, could not grasp how narrow were the lives of most European peasants arriving at Ellis Island. These men and women lived lives of "sordidness and hard-working monotony almost beyond belief, resulting in a mental equipment which is correspondingly limited and stunted."

With this in mind, Ellis Island doctors made use of nonverbal performance tests, many of which they created themselves. Most were little more than glorified jigsaw puzzles. Wooden boards had shapes of different sizes cut out, and immigrants had to put the pieces back in their proper place. Some of the figures were abstract, while others portrayed a face in profile or a horse.

Howard Knox created another test, referred to as the Knox Imitation Cube Test. It consisted of four one-inch cubes placed four inches apart. The doctor then took a smaller cube and, facing the immigrant, proceeded to touch the blocks in a set pattern in a slow and methodical manner. The immigrant would have to repeat the pattern. There were five levels of difficulty, beginning with four moves that touched each cube in order and proceeding to more difficult moves requiring up to six moves that touched the cubes out of sequence. There were five different sets of movements, and success at each level was tied to various levels of intelligence, from idiot to imbecile to moron to normal to highly intelligent.

These tests were about more than just the subject's ability to accomplish the task successfully. Immigrants were constantly being watched, observed, and judged. The inspecting doctor was not just concerned about whether the immigrant could accomplish the task. He was interested in how fast it was accomplished, the immigrant's facial expression while completing the task, his muscle control, the speed of his movement, his mental state, and attention span.

From the moment immigrants set foot on Ellis Island, they were under observation At least a dozen pairs of eyes were on them constantly. It is hard to imagine that immigrants could not feel the penetrating gaze of doctors and inspectors bearing down on them, judging them in a calculating, yet not totally dispassionate, manner. Ellis Island doctors were aware of the need to provide a proper environment, but the observational effect upon the immigrants must have caused a great deal of nervousness, performance anxiety, and even belligerence.

It is not surprising that officials began to uncover more mentally defective immigrants through the years. From 1908 to 1912, the total number of idiots, imbeciles, and feebleminded diagnosed remained relatively constant at between 160 and 190 per year. Yet 1913 proved to be a crucial year. That year, the *New York Times* warned that "15,000 Defectives Menace New York," Goddard was conducting his tests at Ellis Island, and Howard Knox first began publishing articles outlining the methods used by the Public Health Service doctors.

In 1913, the number of mental defectives detected rose to 555, and then almost 1,000 in the following year. The dramatic increase came almost exclusively from the category of feebleminded—those who did not appear at first glance to be mentally defective. From 1908 to 1912, the number of feebleminded immigrants was around 120 per year; by 1913 it had risen to 483, and in 1914 it reached 890. The reliance on intelligence testing increased the number of immigrants deemed to have below-average intelligence. Restrictionists believed that science was finally allowing the proper sifting of undesirable immigrants.

Knox sometimes shared the concerns of restrictionists and eugenicists, but he and his colleagues also stressed common sense. Immigrants would be tested on at least three separate occasions before being classified as mentally defective. No single test would seal the diagnosis, and immigrants were never deported for failing just one test. Instead, doctors looked at the entirety of the results on common knowledge, memory, reasoning, learning capacity, and performance tests. Still, mental testing at Ellis Island was fraught with cultural biases, as well as the unstated assumption that something called intelligence could be tested.

Like others involved in the immigration debate, Knox was a complex man. In June 1913, he could tell a scientific conference he was confident he would find the missing link among immigrants at Ellis Island, implying that some he saw there were subhuman. A few months earlier, though, he could warn readers of a medical journal that they "should have infinite compassion and pity for those whom the French have feelingly called *les enfants du bon Dieu* and the Scotch the *daft bairns*, and the innocents, for a soul is a soul regardless of what functional tests may show of the intellect." Such compassion would have been cold comfort to the Zitello family. For all the supposedly dispassionate science, intelligence tests were not conducted in a vacuum.

A few weeks after writing his letter to Franklin D. Roosevelt in 1933, Salvatore Zitello received his response. It came not from the president, but from the office of the immigration commissioner. They were words he had heard before. Gemma, the letter stated, "was excluded because she could not qualify under the mental requirements of the law. I am sorry to be obliged to advise you that she comes within a class manditorily excluded." There would be no leniency for Gemma Zitello. She would never be reunited with her family in America.

Chapter 13

Moral Turpitude

———•◦•———

Poor little me, why did they consider me a dangerous woman?
—Vera, Countess of Cathcart, 1926

DRESSED IN A LARGE GREEN FELT HAT WITH A MATCHING
coat trimmed with brown fox fur, flesh-colored silk stockings, and black
velvet slippers, Vera, Countess of Cathcart, was ready to take on New
York. The attractive and petite thirty-something member of England's
fashionable set had arrived in New York in February 1926 armed with a
copy of her play *Ashes of Love* and dreams of Broadway fame.

Instead of becoming a star or literary sensation, the countess ended
up a different kind of celebrity, an international cause célèbre who introduced the concept of moral turpitude to people on both sides of
the Atlantic.

Vera's problems began when immigration officials boarded her
ship as it entered New York Harbor. In a routine check of first-class
passengers, the inspectors discovered that, five years earlier, the countess's marriage to her second husband, the Earl of Cathcart, had ended
in divorce. Another member of the British aristocracy, the Earl of
Craven, was named as the cause of the divorce. Vera had left her husband—some thirty years her senior—and their three children to run
off to South Africa with the married Earl of Craven. Their positions
among England's minor nobility added to the tabloid quality of the
scandal.

By marking herself as divorced on her papers, Cathcart attracted extra scrutiny from officials. It is unclear how they managed to go from Vera's divorced status to her adulterous affair with the Earl of Craven. Maybe someone remembered the scandal, or perhaps, as Vera suggested, she had an enemy in New York who alerted authorities to both her arrival and her scandalous background.

Immigration officials declared that since Vera was an adulterer, she was guilty of a crime of moral turpitude and excludable under law. Most Americans had little idea what that peculiar phrase meant. *Black's Law Dictionary* defines moral turpitude as "general, shameful wickedness—so extreme a departure from ordinary standards of honest, good morals, justice, or ethics as to be shocking to the moral sense of the community . . . an act of baseness, vileness, or the depravity in private and social duties which one person owes to another, or to society in general, contrary to the accepted and customary rule of right and duty between people."

The term entered American immigration law as one of the excludable offenses in the 1891 Immigration Act. Courts and immigration officials tried to define the term, but never settled on a firm definition. A wide array of offenses could theoretically be considered crimes of moral turpitude, from passing bad checks to arson to adultery to bigamy to gross indecency and even murder. The arbitrary nature of the term made it problematic for both officials and aliens. In the wake of the Cathcart case, one academic complained that moral turpitude had become "enshrouded by an impenetrable mist."

Given the murky nature of the charge, it is no surprise that Cathcart was unrepentant, despite the very public moral opprobrium cast down upon her. "I have done nothing in my life that I am ashamed of," she told reporters. The affair with the Earl of Craven had quickly gone sour in South Africa. After promising to marry Vera, the earl left her for another woman. Later, he returned to his wife.

By 1926, Vera had managed to get over her failed amorous adventure with one earl and her failed marriage to another. She turned her unfortunate love life into a thinly veiled autobiographical play entitled *Ashes of Love*, and she seemed to be on the rebound. She was now engaged to a commoner, a young playwright named Ralph Neale, who was waiting for her back in England.

Now it looked as if Vera would be seeing her fiancé sooner than expected. She was ordered deported on the same ship on which she arrived. Meanwhile, her friends and the British consulate appealed to Washington, which granted Vera a three-day stay of deportation.

While Vera stewed at Ellis Island, the Earl of Craven was actually in New York, staying with an uncle on Park Avenue. His wife was sick and had come to New York for medical care and the earl was there to be with her. This only added to the soap opera nature of the case. "Why am I to be deported, if the Earl of Craven is to be allowed to remain here," Vera rightly asked. "He has no more right to be in America than I have. If I am guilty, so is he." Officials argued that since the earl had declared himself married, he did not attract the attention of officials. This explanation did little to quell the complaints of a pernicious sexual double standard.

Ellis Island officials were aware that their decision was being scrutinized and sent an inspector up to Park Avenue to interview the Earl of Craven. Meanwhile, Vera spent her time at Ellis Island writing her next play, entitled *Who Shall Judge?*, an autobiographical account of her detention.

Since officials were adamant that Vera not be let into the country, they had no choice but to order an arrest warrant for the Earl of Craven on the same charge. Anticipating the move and no doubt uncomfortable that his affair was once again fodder for the press, the Earl of Craven fled to the Ritz Carlton in Montreal, but he made sure to make his opinion of the affair known before he left town. "Gentlemen, you must be a bunch of Godforsaken idiots," he wrote immigration officials.

In an interview from Ellis Island, Vera said: "I am not a coward and have not run away, like the Earl of Craven. He has proved himself a coward in many ways." (This was a man who had lost a leg in combat as a young officer during World War I.) She had become a victim, not just of a caddish former lover, but also of insensitive government authorities. American women's groups, like Alice Paul's National Woman's Party, called the deportation order against Vera a case of discrimination.

Many in England saw it as another example of provincial puritanism. The *Evening Standard* went so far as to accuse American officials of bad manners. As one of Vera's lawyers said, "Congress did not intend by the enactment of this statute to translate the Department of Labor into a radio of foreign scandal." Congress, he continued, did not mean for immigration authorities to act "as a censor of international sex morals or to send its agents snooping among the divorce records of foreign countries in order that they might obtain evidence which would enable the department to protect our chaste and puritanical Republic."

The 1920s were a time of greater freedoms for women, personified by the fun-loving flapper. These women challenged Victorian-era notions of the proper place of women. Vera Cathcart was just such a modern woman. "I think all persons should be at liberty to do what they choose," she said. Vera symbolized the sexual liberation and right to self-expression of women freed from the conventions of middle-class morality.

Yet traditional morality still held sway with government officials. Despite the national and international uproar, Cathcart remained at Ellis Island, albeit in a private room. In fact, she claimed to be quite comfortable and was surprised by conditions on the island, compared to the horrors she had read about in English newspapers.

Women's groups enlisted the legal help of Arthur Garfield Hays, one year removed from his work with Clarence Darrow on the defense team at the Scopes trial. Hays argued that there was no reason to deport Vera for a crime of moral turpitude since adultery was not a crime in England, South Africa, or the United States. However, Hays was mistaken. Since the late 1800s, a growing number of states had made adultery illegal.

A federal judge issued a writ of habeas corpus in Vera's case, and she was released from Ellis Island after signing a $500 personal bond, which allowed her to remain free for ten days. Then, another federal judge ordered that Vera could stay in the country as long as she liked. Government lawyers, reeling under the embarrassing publicity of the case, did not put up much of a fight.

Vera could now attend to her theatrical career. The notoriety led a producer to offer her $5,000 for her play, as well as a percentage deal on gross receipts and motion picture rights. *Ashes of Love* premiered in London in mid-March, a month after her ordeal began. Her case brought publicity to a previously unknown talent, but it did not prevent negative reviews. One London critic called the play crude: "The dialogue, with few exceptions, is banal and the characters in the piece are wooden and lifeless as dummies."

After London, the play then moved to Washington, D.C., with Vera taking over the lead role. American critics were no kinder. The *Times* called it a "naïve . . . rather childish and undramatic story." Most of the audience seemed attracted only by the curiosity value of Cathcart's story. The play ran for one week.

Angry at the reception her play received, Vera bought it back from the producer. She vowed to finish her play about her detention at Ellis Island. She was careful to remind the public that despite her title and lifestyle, she was not rich. Her stepfather was a wealthy businessman, but he had not given her any money and she was no longer married to the wealthy Earl of Cathcart. She was an independent woman of dependent means, dependent on her marginal literary talent and even more meager acting talent. Perhaps that is why, when Vera Cathcart sailed back to England at the end of March less than two months after her arrival, she told reporters that her treatment at Ellis Island was kind and generous when compared to what she received from critics. Immigration officials she could forgive; theater critics she could not.

Edward Corsi, who ran Ellis Island a few years after the Cathcart incident, admitted that officials probably were too zealous in "catching these wearers of the cloak of royalty in our immigration net. . . . We have used our democracy as a weapon to allow us deliberately to offend them." Corsi may have been correct that democratically minded immigration officials enjoyed the chance to take down minor celebrities and members of high society, yet had Vera Cathcart been a poor peasant girl from Poland, the press would not have taken notice of her case, ambassadors would not have complained to Washington, and women's groups would not have come to her rescue.

Women of all nationalities fell victim to the prying investigations of immigration officials, whether poor Jewish and Italian women or wealthy Englishwomen. There is little evidence to suggest that officials targeted women from eastern and southern Europe for increased scrutiny. In fact, it seemed that the one group most often profiled as potentially immoral was single French women arriving in first- and second-class passage. For Ellis Island officials, policing the border and enforcing the nation's immigration laws often meant enforcing middle-class ideas of sexual morality.

GIULIA DEL FAVERO SAID she would rather jump into New York Harbor than submit to the medical exam. She did agree to have the male doctor examine her breasts, which he thought showed a peculiar appearance that might suggest pregnancy.

Giulia was taken out for special examination during the initial line inspection because an official thought that she looked pregnant. Through a translator, Giulia adamantly denied she was pregnant and declared herself to be a morally pure young woman. The breast exam was one thing, but there was no way that the twenty-three-year-old unmarried seamstress was going to let a male stranger give her a vaginal examination.

Ellis Island commissioner Thomas Fitchie was sticking to his guns. He declared that either Giulia would submit to the exam or she would be deported. But Fitchie ran into strong opposition from his own staff. A female matron named Regina Stucklen complained that forcing such a procedure on young women ran the risk "of examining pure and good moral girls, and thereby, perhaps, injure them morally for the rest of their lives." Even the doctor agreed, telling Fitchie that he thought the young girl was right to refuse the exam, a procedure he believed was "extremely repugnant to a virtuous woman." Such an exam would say nothing about a woman's condition if she were less than three months pregnant. Fitchie backed down and the young girl was allowed to enter the country.

Giulia was not married and if she had been pregnant that would have cast doubt on her moral fitness to enter the country. There were other concerns. Young women were never set free from Ellis Island unless in the custody of a male relative or missionary or immigrant aid official. To do otherwise, officials feared, would risk throwing these women to the proverbial lions, whereby unsavory men might entrap them, steal their innocence, and start them on a life of prostitution.

Sometimes, though, those vultures worked inside the immigration station. Inspector John Lederhilger seemed to take a certain relish in closely questioning single women who passed through Ellis Island. "Did he sleep with you on the boat?" Lederhilger reportedly asked an unmarried German girl arriving in New York with a male companion. "Now tell me how often did he put it in?" If Fitchie and others exhibited genuine interest in protecting single women and upholding traditional morality, Lederhilger seemed more interested in his own sexual titillation.

Immigration officials continued to find themselves enmeshed in the personal lives of immigrants. In 1907, the solicitor of Commerce and Labor ruled that moral turpitude covered issues of private sexuality such as adultery and fornication. Twenty-one-year-old Swede Elin Maria Hjerpe found this out when she arrived at Ellis Island in early 1909. Five months pregnant and single, Elin arrived in the company of her intended husband, a naturalized American citizen and the "author of her condition," as the records state.

Because of her out-of-wedlock pregnancy, the board of special inquiry voted unanimously to exclude her on the grounds of moral turpitude. Yet when the case reached Washington on appeal, Frank Larned, the assistant commissioner-general of immigration, was not convinced. He noted that Elin's only offense was that she had committed fornication, which he believed, when committed in private so as not to "offend the moral sense of the community," was not a crime of moral turpitude. Without excusing premarital sex, Larned believed the circumstances called for leniency. Elin's boyfriend had told officials he wanted to marry Elin as soon as possible. Elin Hjerpe and her boyfriend were married at Ellis Island and she was allowed to enter the country.

A young Serbian woman named Milka Rosceta arrived at Ellis Island a few days later, accompanied by her three-year-old child. Their ultimate destination was Steubenville, Ohio, where Dana Jezdic, the father of the child, resided. Like Elin Hjerpe, Rosceta was detained on the grounds of fornication. An immigrant aid society representative at Ellis Island sent a telegram to Dana about the situation and he responded with an affidavit stating his desire to marry Milka upon her arrival in Stuebenville. He even had a local Serbian Orthodox priest sign an affidavit that he would officiate at the wedding, but this was not good enough for officials.

So Dana took time off from his job at the La Belle Iron Works and traveled by train to New York. Milka and her child were in their sixth day of detention when Dana arrived. There was some discrepancy in their stories. Dana said his girlfriend was only nineteen and they were too young to be married back home; Milka claimed to be twenty-four years old and said the couple could not marry in Europe because Dana had not served in the army. Officials probed Milka's sexual history, asking her: "What other men, if any, have you been intimate with?" She responded that there had been no other men.

The case wound up in Washington, where Frank Larned ruled on it. He noted that this was also a case of fornication, which was not punished under common law unless it was committed "openly and notoriously." While officials had every right to exclude Milka under the moral turpitude clause, Larned again called for moving beyond a literal interpretation of the law—but with a twist.

Larned argued that officials could not hold Milka to the standards of middle-class American morality. "If this appellant had been reared in environments similar to those existing in the United States," he argued, "the commission of her fornication would necessarily impute to her moral turpitude." She was raised in the Balkans under very different standards. "It is extremely doubtful whether she fell from a higher state of character to a lower when she mated with the man to whom she is now destined," he argued. Her conduct, in his eyes, was "unmoral," not immoral. By that same reasoning, Larned argued, officials could not exclude the "wife of a Zulu chieftain from savage Africa" coming to join her mate, even though "they may have mated in no more ceremonial a manner than is observed by the beasts of that country." Larned ordered that Milka and Dana be married at Ellis Island.

The argument contained many of the contradictory feelings embodied in American immigration law, mixing prejudice with leniency. That Larned would use this rationale for the Serbian Milka and not for the Swedish Elin Hjerpe shows how strongly Americans differentiated between northern Europeans and southern and eastern Europeans. Yet even though the reasoning behind the decisions may have differed, both women were allowed to enter the country after their hasty marriages at Ellis Island.

Young women who transgressed the boundaries of middle-class morality could still feel the long arm of the law even after they had been admitted to the country, since immigrants could be deported within three years of their arrival if they were subsequently found to have violated immigration law. Take the case of twenty-year-old Cecilie Kolb, who arrived in May 1910 and went to live with the family of a German baker in the Bronx. Within a year, the baker wrote to Ellis Island complaining that his young charge possessed an "immoral character and I believe it is useless to try to keep her on the right path." Cecilie was having difficulty keeping a job and was living with a fortune-teller in Manhattan.

So in August 1911, Kolb was brought to Ellis Island for a deportation hearing. At first, she admitted to having had illicit relations with two men, but then quickly denied it, saying she only went with them to dance halls and bars. Ellis Island doctors examined her and declared her a virgin. Though Assistant Commissioner Uhl wanted to deport Kolb as likely to become a public charge, Washington officials ordered that the girl be freed because of insufficient evidence.

Immigration officials also showed little compunction about enforcing the moral turpitude clause against men, and wealthy Anglo-Saxons at that. Commerce and Labor Secretary Oscar Straus described the case of an immigrant mill manager from Lawrence, Massachusetts, who was married with children. The man went to Canada on a trip and returned with a woman who was not his wife. Inspectors held him at the Canadian border, where he admitted that he had had "improper relations" with his traveling companion. Officials ordered him excluded on the grounds of moral turpitude.

"I had approved exclusion simply to teach the fellow a lesson in morality," Straus wrote in his diary. When a former governor of Massachusetts lobbied Straus on behalf of the adulterer, Straus relented and ordered him admitted, saying that he did it for the man's family, "not because he deserved it." At the next cabinet meeting, Theodore Roosevelt told Straus he would not have let the man in. Straus figured as much and told the president he had thought of him when deciding the case. "That is a nice affair," the happily married Roosevelt said jokingly. "You think of me when adultery is committed."

In another case, a forty-year-old English businessman named Louis Fairbanks arrived in Boston in December 1908. Although he first claimed to be single, Fairbanks later admitted he had a wife in England who suffered from consumption and bronchitis. He also admitted he had taken up with another woman, with whom he had a child. In contrast to Vera Cathcart's case eighteen years later, immigration officials declared that since ecclesiastical courts in England had declared adultery a crime, they were justified in excluding Fairbanks on the grounds of moral turpitude. Straus agreed and Fairbanks was deported back to England.

Sometimes women could use the moral turpitude clause for their own benefit. Sarah Rosen had married Julius Rosen in Russia in the mid-1880s. A few years later, Julius left for America, and by the late 1890s he was joined by Sarah and their three children, Becky, Mary, and George. According to Sarah, four days after she arrived with her children, Julius deserted the family and left for England. There, Julius married again and had two more children. Julius claimed he was forced into his marriage with Sarah by an uncle, and that the marriage was illegal in Russia because Julius was under the age of eighteen.

More than a decade after he abandoned his family, Julius Rosen returned to the United States. A few months later, Sarah Rosen wrote a plaintive letter to William Williams at Ellis Island regarding her husband. Her children were now fourteen, twenty, and twenty-three. She had managed to raise them by herself and attain some measure of prosperity. The family lived in Brooklyn and ran a stationery store. Sarah appeared to own some real estate, and she believed that Julius's return was motivated by money.

She complained that Julius was making her life miserable and bothering her family. "I am not seeking any revenge," Sarah wrote, "all I desire is to be left alone, to continue to support my little family, and not to be interfered with." She wanted Williams to deport Julius on the grounds that he was a bigamist. "It seems to me that my husband is not a proper person to enjoy the liberties in this country, and I ask that you take steps to force him to return to the place he came from," she asked Williams. She even provided Williams with the addresses that Julius was known to frequent.

A few weeks later, Julius was taken to Ellis Island. He continued to claim that his first marriage was illegal and that he had done nothing wrong by remarrying. Augustus Sherman, acting in place of William Williams, argued that the legality of Julius's marriage to Sarah was a moot point. "If legal, he has committed bigamy; if illegal, he is the father of three illegitimate children," Sherman wrote. Either way, Julius was guilty of a crime of moral turpitude. He was ordered deported.

Officials in Washington upheld the decision to deport Julius Rosen. However, Rosen hired former congressman William Bennet as his lawyer. Bennet took Julius's case all the way to the Supreme Court, which ruled against him. Julius was finally deported in February 1914. While living in Canada, Julius would request permission, through Bennet, to enter the United States many times over the next decade. Even though Julius's second wife had died, the government still considered him a bigamist and he was barred from ever entering the country and bothering Sarah or their children.

These cases show immigration officials struggling with how to enforce the moral turpitude clause. They tried to interpret it in a broad manner while upholding community standards that encouraged marriage, rather than cohabitation, especially when children were involved, and discouraged extramarital or premarital sexual relations.

Oftentimes, the moral turpitude clause covered more than just sexual relations and could sometimes place Ellis Island in the middle of international intrigue. Known as the "Lion of the Andes," Cipriano Castro had ruled Venezuela as a military dictator from 1899 until 1908, during which time he plundered the nation's wealth and executed political enemies. Castro, a cross between Napoléon, Boss Tweed, and P. T. Barnum, with a little bit of Nero thrown in, was worth $5 million, much of it stashed in European banks. Secretary of State Elihu Root referred to him as a "crazy brute." Castro's regime led to the creation of one of the most famous American foreign policy statements: the Roosevelt Corollary to the Monroe Doctrine.

When Castro refused to honor the debts his country owed to European banks, England and Germany erected a naval blockade of Venezuela. Theodore Roosevelt feared this would be a backdoor to allow European colonization in the Western Hemisphere and declared in 1904 that "chronic wrongdoing" on the part of the Latin American nations would lead the United States to intervene in those nations' affairs, so as to prevent the meddling of European powers in its own backyard.

In 1908, Castro left Venezuela for kidney surgery in Germany, leaving the country in the hands of General Juan Vicente Gomez, who wasted little time in declaring himself ruler and expropriating Castro's properties. With that, Castro was a man without a country. To make things worse, the American government was still mad at him and feared he was planning to return to power. French and English authorities made it clear that Castro was not welcome at any of their Caribbean colonies. The U.S. Navy followed Castro's every move and American officials kept him under constant surveillance. He finally ended up in the Canary Islands.

In December 1912, Castro decided to visit the United States, but the State Department ordered William Williams to hold Castro at Ellis Island. Like Vera Cathcart, Castro was only coming for a short visit, not to settle permanently. Having caught wind of the State Department's efforts to bar him, Castro fired a wireless telegram to the *New York Times* complaining about the effort. "That you should insult me simply because I visit you is inconceivable," Castro complained.

He arrived on the last day of 1912 and was taken to one of Ellis Island's hospitals for examination. Doctors could find no medical reason to exclude the former dictator, although Assistant Commissioner Uhl remembered that Castro's body was covered with scars and saber wounds. He described the former dictator as a "blackguard and a cutthroat," but still said he admired the man he described as a "little runt."

At his hearing, Castro told his inquisitors: "At present I have no profession. I am traveling for pleasure." However, because of the inconveniences he was being put through, he decided he wanted to go back to Europe. Then Castro changed his mind and demanded to be admitted to America. While officials in Washington decided his fate, Castro would spend more than a month at Ellis Island, in a detention area reserved for nonsteerage detainees, with a private room, bed, washbasin, and nightstand.

Officials had little with which to hold Castro. He was not sick or diseased, had never been convicted of a felony or other crime, and was not, in the words of a government attorney, "accompanied by a lewd woman." There was one thing that officials hoped they could use to bar him from the country. The Gomez government in Venezuela had implicated Castro in the execution of a rebel general named Paredes.

Castro had a number of hearings while on Ellis Island and proved increasingly uncooperative. When asked about his actions as president and the source of his wealth, Castro refused to respond. When asked about General Paredes, he replied that since he was not in a criminal court, he would refuse to answer. Byron Uhl remembered Castro as "vociferous" and "obstreperous" during the hearings, the most picturesque alien he encountered in his over forty years at Ellis Island. Despite the stress, Castro lived well at Ellis Island. He paid for his own meals and ate voraciously, while dressed in a skullcap of black velvet trimmed with gold, and gold-embossed cloth slippers.

After more than two weeks of detention and hearings, a board of special inquiry denied Castro the right to land. It called him an unreliable witness whose refusal to answer questions, along with his manner and demeanor, constituted an admission to the crime of killing Paredes and therefore a crime of moral turpitude.

William Williams, who had spent hours personally interviewing Castro, was uneasy about the decision. To exclude Castro, there would either have to be a conviction for the crime or an admission of the crime, and officials had neither in this case. Another hearing was held, this time in Castro's room, while he was having breakfast. Castro would have none of it. He threw a fit and locked himself in the bathroom. The board then held its hearing in the adjoining room and again voted to deport Castro.

One month after Castro's arrival, Commerce and Labor Secretary Charles Nagel upheld the decision to deport Castro. Admitting that it was an unusual and difficult case and that Castro would not have been detained had it not been for the request from the State Department, he nevertheless argued that Castro's refusal to submit to the hearings on Ellis Island was cause enough for exclusion. Since entry to the country was a privilege, it was incumbent that aliens submit to a hearing.

Meanwhile, New York Democrats took on Castro's case and provided him with legal help, arguing that the death of Paredes was a political act and therefore did not qualify as grounds for exclusion. With this support, Castro was freed on bail after a month in detention. Two weeks later, a federal judge allowed Castro to remain in the country as long as he wished. The judge ruled that the government needed more proof of his crime than just his lack of cooperation and evasiveness.

In the spring, Castro left for Havana and would later settle on the island of Trinidad, hoping that revolutionaries would prevail against Gomez and return him to power. The revolution never materialized and Castro continued to live in exile.

Castro returned to America in 1916 and the State Department again demanded his exclusion. This time, Byron Uhl noted a different Castro. Unlike the proud and difficult man he had seen three years earlier, Uhl found that Castro's "spirit seemed broken." All hope for returning to power had vanished. Castro, traveling with his wife, now only wanted to land in America temporarily while waiting for a boat that would take him to Puerto Rico. He answered the questions of the board of special inquiry and denied that he had anything to do with the killing of Paredes. The board was still not happy with his answers and ordered his exclusion on the grounds of moral turpitude. His wife was excluded on the grounds that she was likely to become a public charge.

This time, however, officials in Washington sustained Castro's appeal and ordered him released. After spending two days at Ellis Island where they were given a suite of rooms with a private bathroom and complete freedom of access to the entire island, the Castros were released and made their way to Puerto Rico, where the former dictator lived out the rest of his days. He never returned to his native Venezuela and died broke and alone in San Juan from a stomach hemorrhage in 1924. The *Times* remembered him not too fondly as "one of the most remarkable adventurers who ever strutted on the stage of Latin America." The term "moral turpitude," the *Times* editorialized, "fitted him beyond a doubt, for he had never had any principles."

THE NUMBER OF MILITARY dictators attempting to enter the United States was fairly small, but immigrants who violated middle-class sexual mores were more abundant. In 1911, Daniel Keefe, commissioner-general of immigration, argued that adultery was a crime of moral turpitude and therefore an excludable offense. "That offenses that are contrary to chastity and decency, or so far contrary to the moral law, as interpreted by the general moral sense of the community," argued Keefe, "that the offender is no longer generally respected or is deprived of social recognition by good living persons, involve moral turpitude is so well established as to be axiomatic." This was in direct contradiction to the orders of Frank Larned from just two years earlier.

The solicitor of the Department of Commerce and Labor overruled Keefe, going back to the more lenient standard set out by Larned. Officials should not regard "specific instances of sexual immorality as necessarily amounting to crimes of misdemeanor involving moral turpitude," the solicitor concluded, as long as the alien was "clearly not of an essentially immoral character." This hardly cleared up the problem, but it did give officials room to allow immigrants to enter the country despite previous moral lapses.

The case of Marya Kocik, a married Polish woman, showed the difficulty of measuring immoral character. Marya's husband was already living in America and was now able to bring over Marya and their three children. However, Marya was five months pregnant, even though she had not seen her husband in over a year.

After her husband left Poland, she and the children were placed in the home of a friend of Marya's husband. Soon after, Marya began having sexual relations with this man and became pregnant. Now that she was arriving in the United States, Marya's husband freely accepted her and agreed to raise the other man's child as his own. Although this was a clear case of adultery, the department's chief lawyer argued that officials were not bound to exclude Marya, considering it best for the family to reunite with the father.

These debates might seem like the work of excessively prudish male officials, but like the rest of the immigration bureaucracy, the regulation of middle-class sexual morality was one of balancing various interests. Officials often showed leniency to immigrants who had committed adultery or engaged in premarital sex, while still upholding the middle-class sexual norms that held that marriage was the ideal institution within which to deal with human sexuality and raise children.

Officials became concerned where sexual promiscuity verged into prostitution. A twenty-two-year-old Croatian woman named Jelka Presniak, who had recently arrived in the country, was arrested on the grounds that she was a prostitute. She admitted to officials that she had sex with a number of men, but denied ever accepting money. The Labor Department's solicitor ruled that the term "prostitute" could be used for any woman who "for hire or without hire offers her body to indiscriminate intercourse with men." Jelka was ordered deported, but managed to elude authorities. She traveled from upstate New York to Pennsylvania to Ohio, living in different Slavic communities under various aliases, working in restaurants and as a prostitute. She was never found.

Eva Ranc provided officials with a similar dilemma. Like many cases, Ranc's troubles began when Ellis Island officials received an anonymous note in early 1916 warning that a Frenchwoman named Eva Vigneron, traveling under the name of Ranc, was coming to America for an "immoral purpose" with funds provided by a wealthy American businessman named Sig Tynberg.

Ranc arrived in New York Harbor on March 1 and was taken from her second-class cabin on the SS *Rochambeau* for a hearing at Ellis Island. A divorcée and mother of a teenage daughter, the thirty-six-year-old Ranc claimed to be a designer of ladies' dresses in Paris. This was her third trip to the United States.

At her hearing, inspectors asked Ranc where she had lived when she was previously in New York and whether she had received any male visitors then. She swore to officials that she had not, although she did admit that Tynberg had given her money in the past. They wanted to get married, but Ranc claimed that Tynberg's father did not approve of his son marrying a Gentile. Officials asked whether Ranc had a sexual relationship with Tynberg or any other men, which she answered in the negative.

The board then interviewed Tynberg, asking him whether he had had any "immoral connection" with Ranc. "No," he responded, "I have the highest respect for that woman." Tynberg was the owner of an insurance company and president of the North American Fuel Company, with an office in lower Manhattan. He vouched that even though he had occasionally paid Ranc's rent during her past visits to New York, there was "nothing, absolutely, immoral about her." He would marry her if it were not for his eighty-year-old father, an observant Jew, who opposed the idea of intermarriage.

Later in the day, the case against Ranc became clearer. A woman named Myrna Light testified and stated her purpose bluntly: to make sure that Eva Ranc could not enter the country. Light had been engaged to Tynberg for over four years. The anonymous letter warning officials about Eva had come from Myrna. When she asked Tynberg why they could not marry, he told her it was because he was afraid of his mistress and what she might do to Myrna. He told Myrna that Ranc was the stumbling block, the "kind of woman who comes into every bachelor's life and as soon as he got rid of her everything would be settled with us." Myrna claimed that Tynberg was scared of Ranc, telling her once that the "French hooker would tear you to pieces if I married you."

Myrna found out about Ranc's most recent arrival from Tynberg's secretary, who was also romantically interested in her boss. The secretary had wired money to Ranc for her trip to New York and, perhaps out of jealousy, told Myrna of the deal. Sig Tynberg had been stringing Myrna Light along for over four years, and now a scorned Myrna was having her revenge. She had once filed a $25,000 suit against Tynberg for breach of promise, but dropped the suit. Going after Eva Ranc seemed a better strategy.

For two nights, Ranc was held in detention at Ellis Island, while investigators interviewed Tynberg's father, as well as the building superintendents of the two buildings where Ranc had previously stayed in New York. Tynberg's father told the investigator that he did not object to his son marrying a Gentile, only "a colored girl or a girl upon whose reputation there is any stain." The senior Tynberg claimed that a relative went to Paris to seek information on Eva and learned that the woman was "something awful." The supers told investigators they had seen Ranc and Tynberg in bed together and that Tynberg had stayed most nights with Ranc.

With this information, the board ordered Ranc deported. Tynberg again appeared to plead for Ranc. He said he had never slept with Ranc, that he loved her and was going to marry her. Because of the emotional strain, Tynberg made his case in peculiar language. "I think she should be given to me," he pleaded in front of the board, "that woman belongs to me and there is nothing about her I am ashamed of."

Tynberg claimed that his good name and character was known throughout New York's business community. He even brought his friend and business associate, a finance professor at the University of Pennsylvania, to testify on his behalf. That testimony, along with proof of Eva's divorce from her first husband in France and the stated desire of both Ranc and Tynberg to marry, led a majority of the board to overturn its earlier decision and allow Ranc to enter the country. One dissenting member of the board, however, believed there was something fishy about Ranc. Under the rules, the dissenting member could appeal the decision to his superiors.

Ellis Island commissioner Fred Howe looked at the evidence and agreed that while it appeared that Tynberg and Ranc had probably lived together, they had showed genuine love for each other and would get married. Howe complained of the "humiliation—and to my mind unnecessary cruelty" of deporting Ranc. He upheld Ranc's admission, a decision affirmed by his superiors in Washington. Eva Ranc entered the country and married Sig Tynberg shortly after her arrival. Immigration officials had delved deeply into the personal lives of these individuals. Although Ranc was ultimately admitted, the experience was no doubt painfully embarrassing for both Tynberg and her.

At first glance, Ranc's case appears to be a triumph over immigration officials eager to punish a woman for her sexual behavior, but there was more to the story. Two years later, a letter signed only by "A Loyal American" arrived at the State Department in Washington. Apparently, the marriage of Eva Ranc and Sig Tynberg had been short and unhappy, with Eva soon fleeing back to France. This anonymous letter warned that Eva was seeking to return again to America "to make trouble for those she wronged."

Officials at Ellis Island sent an investigator to interview Tynberg, who told his sorry tale. He had truly believed that Eva Ranc was a good woman, but found out shortly after their marriage that she was bringing men back to their apartment and meeting others at the Ritz Carlton. Tynberg also believed Eva had fallen in with a group of blackmailers. After less than four months of marriage, Tynberg served divorce papers on Eva, who left the country before the divorce was finalized. Newly remarried, Tynberg now said that he would do everything in his power to prevent her from returning. There is no evidence that she ever tried to return.

Eva Ranc's case shows just how engaged immigration officials were in the regulation of sexual morality. As Commissioner Keefe noted in 1909, "the purpose of the immigration act is to prevent the introduction into the United States not only of innocent girls who have been seduced into a life of prostitution, but of all girls and women of sexually immoral class." Ranc would have been classified under the category of "sexually immoral," but authorities were also on the lookout for those "innocent girls" forced into prostitution.

There was a term for this: white slavery. Americans believed that unscrupulous men—pimps, "cadets," and "mackerels"—were forcibly trapping thousands of innocent young women into sordid lives of sexual slavery. *The Outlook* warned in 1909 that there was an "extensive traffic in white slaves . . . who are bought, sold, and used as instruments for the gratification of men's lust."

The imagery implied in the term was forceful, as anti-prostitution activists positioned themselves as the new abolitionists. No less a reformer than Jane Addams made the connection between the "social evil" of young women being forced to sell their bodies and the enslavement of blacks. Like the battle against race slavery, Addams thought that the fight against white slavery would "claim its martyrs and its heroes." Reformers were ready to take to the battlefield against this newest injustice. "Few righteous causes have escaped baptism with blood," Addams prophesized.

Reports began to filter into the press about unimaginable horrors inflicted upon innocent women. "There are some things so far removed from the lives of normal, decent people as to be simply unbelievable by them," U.S. Attorney Edwin Sims claimed with melodramatic flair. Americans would come to believe that there was a vast and organized system enslaving young women into sexual service, with immigrants at the center of that system, as both victims and victimizers.

ELLIS ISLAND INSPECTOR MARCUS Braun would never pass up a trip at the government's expense, especially if it allowed him to avoid the mundane duties of work at Ellis Island. So Braun spent five months in 1909 traveling throughout Europe investigating white slavery for the federal government.

In Paris, Braun visited the hangouts of pimps and prostitutes and, in his words, "simply played the part of a traveling tourist who is curious enough to make, once in a while, foolish inquiries and to spend his good money to satisfy his curiousity." He collected the names of suspected European pimps and prostitutes, as well as their mug shots.

After five months, Braun reported back to his superiors that he could find "no such thing as an organized traffic for the shipment of alien women for the purpose of prostitution or any other immoral purpose in existence." Nor did he find "any organized effort of bringing innocent and virtuous women into this country for such purposes of prostitution or other immoral purposes." However, he did find that many European prostitutes were making their way to America, either by themselves or with the help of someone in the business, but Braun believed there was nothing forced about it.

French authorities complained to the American embassy about Braun's investigation. A member of the French ministry told Braun that his country would not assist the United States in its fight against white slavery and prostitution. He said it was outrageous that American immigration laws excluded not only prostitutes, but also those women who were guilty of having committed adultery or premarital sex. To the French, American attitudes towards sex were prudish and provincial.

The Hungarian-born Braun even wanted to expand the categories for exclusion, suggesting that "pederasts and sodomites" be added to the list. He seems to have been traumatized by the thousands of young male prostitutes he saw in Berlin. Not only were these *Puppenjungen*, as they were called, practicing prostitution in the open, but many would blackmail their customers and some got into the business of procuring female prostitutes. It was a menace, Braun warned, that needed to be stopped at the border.

Ellis Island officials had always been worried about forced pros-titution. As early as 1898, Edward McSweeney warned Terence Pow-derly about allegations that some immigrants were selling children into prostitution. Lurking in the coffee houses of the Lower East Side, Mc-Sweeney believed, were nefarious individuals "luring to lives of shame children of innocent years and that their down-fall, once they enter into this course, is incredibly rapid."

McSweeney focused on the case of thirteen-year-old Bertha Hondes, who arrived from Buenos Aires with a woman named Rosa Seinfeld, who claimed to be her aunt. Rosa took Bertha to a brothel in New York where, in McSweeney's words, "the woman had attempted to sell her for immoral purposes." Rosa was not Bertha's aunt, but a prosti-tute, and Bertha's mother was a madam in Buenos Aires. The United Hebrew Charities intervened and took the girl away.

In 1907, Congress banned the "importation into the United States of any alien woman or girl for the purpose of prostitution, or for any other immoral purpose," as well as the pimps and procurers who im-ported these women. The law gave officials more tools with which to clamp down on those who violated middle-class sexual norms.

In 1908, the case of John Bitty made its way to the Supreme Court. He was accused of bringing his British mistress to the United States. The woman was excluded and Bitty was arrested. The woman was not a prostitute, but since her sexual relationship with Bitty was outside the bounds of marriage, the government argued it fell under the "for any other immoral purpose" clause. The Supreme Court agreed, with Justice Harlan concluding that the clause was designed "to include the case of anyone who imported into the United States an alien woman that she might live with him as his concubine."

More serious than the case of Bitty were the procurers who traf-ficked in human flesh and imported women against their will. In the years before World War I, as many as twenty-two white slave narra-tives were published, with titles such as *Fighting for the Protection of*

Our Girls: Truthful and Chaste Account of the Hideous Trade of Buying and Selling Young Girls for Immoral Purposes. These lurid books told of innocent young women lured into degrading lives of prostitution by sinister male pimps.

Former New York police commissioner Theodore Bingham published his own exposé, entitled *The Girl That Disappears: The Real Fact About the White Slave Traffic*, warning that at least two thousand immigrant white slaves came to America each year, "brought in like cattle, used far worse than cattle, and disposed of for money like cattle."

Newspapers and magazines further fanned the flames. S. S. McClure's eponymous magazine had helped give birth to the classic American journalistic tradition of muckraking, publishing Ira Tarbell's exposé of Standard Oil and Lincoln Steffens's attack on corrupt city government. Tarbell and Steffens had left the magazine in 1906, and McClure had to find other writers and crusades.

He found that talent in George Kibbe Turner and that crusade in white slavery. Turner's 1909 article "Daughters of the Poor" explained how Tammany Hall allowed New York to become one of the world's leading centers of the white slave trade. Turner focused on Jewish prostitutes on the Lower East Side and immigrant aid societies such as the New York Independent Benevolent Association and the Max Hochstim Association, which procured women for prostitution rings under the protection of Tammany. Turner thought the political machine was the biggest culprit and showed the evolution of the prostitution trade. "The trade of procuring and selling girls in America—taken from the weak hands of women and placed in control of acute and greedy men—has organized and specialized after its kind exactly as all other business has done," he wrote.

The fight against white slavery was about more than nativism, repressed sexuality, or mass hysteria. It embodied many of the themes of Progressive reform. In the eyes of antivice activists, prostitution and white slavery stood at the intersection of greedy business interests, corrupt political machines, and degraded immigrant masses. Women were exploited by male pimps, selfish businessmen—the owners of bars, cafés, hotels, theaters—who profited from the sex trade, and corrupt ward bosses who skimmed their share of the prostitute's income while providing political and police protection.

Some, like Theodore Bingham, blamed Ellis Island officials for failing to pay adequate attention to the importation of prostitutes. "There seems to be very slight difficulty in getting women in this country," he wrote in his annual report, "and the requirement of the immigration authorities were easily met by various simple subterfuges."

In response, the government did more than just send Marcus Braun to Europe to investigate the sex trade. It stepped up enforcement at Ellis Island, keeping an eye out for prostitutes and pimps entering the country. More importantly, officials actively sought out foreign-born prostitutes operating in New York and beyond. If an immigrant woman was found to have engaged in prostitution within three years of her arrival, she could be deported. Inspectors Anthony Tedesco and Helen Bullis put together a list of over eighty cafés, music halls, and hotels in Manhattan frequented by prostitutes.

Despite the increased vigilance, efforts to bar immigrant prostitutes were often stymied, as in the case of Hermine Crawford. Detained at Ellis Island for prostitution, Crawford became friendly with Roland Colcock, a watchman there. Crawford charmed the humble Colcock, who was in the process of being transferred to the immigration station in El Paso. Crawford was released on bail while the courts decided her habeas corpus petition. While out on bail, Crawford married Colcock, making her ineligible for deportation no matter what the courts or immigration officials decided.

Two months after the wedding, Colcock was at his new job in El Paso and Crawford was soliciting sex on Broadway. She told a policeman she had no interest in moving to Texas with her husband. He did not make enough money for her, and she hoped he would stay in Texas and leave her alone. To make matters worse, Colcock was charged with violating his oath of office for his relationship with Crawford. Acknowledging that his interest in his wife was ill advised, Colcock admitted that he was "impetuous by nature and no one has ever accused me of being of a reasoning disposition. A proposition appeals to me and I enter into it without going into details." A month later, Colcock resigned from the immigration service.

The 1911 Dillingham Commission attempted to determine the extent of immigrant prostitution, as well as assess how well immigration officials detected prostitutes at ports of entry. On one hand, the commission found that many immigrant women were admitted who listed addresses of well-known brothels as their destination or claimed to be heading for known red-light districts in San Francisco or Seattle.

Investigators set out to discover whether the addresses given by a random selection of sixty-five women who had arrived at Ellis Island in January 1908 matched up. Thirty women were found to be living at the same address they listed on their ship's manifest. Not surprisingly, many of the other women could not be found because they had moved or the address provided was incorrect. Of the sixty-five, only three were found to be living under suspicious conditions: two appeared to be prostitutes and the third was married to a man who already had a wife.

On the other hand, the Dillingham Commission found that Ellis Island authorities had improved their enforcement of the law against prostitutes and procurers. Between 1904 and 1908, only 205 prostitutes and 49 procurers were barred at the gate. By 1909, officials had grown more vigilant. They arrested 537 people for prostitution, of whom they deported 273. Much of the work was done after landing, as Inspectors Tedesco and Bullis investigated alien prostitutes working in the city and beyond. Suspected prostitutes from as far away as Utah were brought to Ellis Island for deportation.

Single French women, especially those traveling alone in first- or second-class, were always looked upon with suspicion. Of new immigrant groups, however, Jews were most often linked to prostitution. Even Marcus Braun found that a majority of the procurers he came across in Europe were Jewish. Helen Bullis described the workings of the Independent Benevolent Association, which included "practically all the Jewish disorderly house keepers of prominence in New York." This included the owners of cafés frequented by pimps and prostitutes, clothes dealers who sold their wares in brothels, saloonkeepers, bondsmen, and even doctors who attended to the residents of brothels.

The charge of Jewish involvement in the sex trade could easily descend into anti-Semitism, but the actual numbers show a more complicated picture. In one study in New York, Jewish women made up a little less than half of the 581 prostitution arrests, followed by French, German, and Italian women. In another study, of the ninety-eight women deported for prostitution from Ellis Island in 1907 and 1908, only thirteen were Jewish. Half of these women were French.

The link between prostitution and immigration was a persistent one, even if officials had trouble nailing down exact figures. Marcus Braun estimated that there were 50,000 foreign-born prostitutes and

10,000 foreign-born male pimps in America. He also thought there were around 10,000 immigrant prostitutes in New York, while reformer James Bronson Reynolds argued that the number was three times higher. On the more conservative side, a federal grand jury investigation led by John D. Rockefeller looking into white slavery put the number at only 6,000. The Dillingham Commission had to admit that it was "impossible to secure figures showing the exact extent of the exploitation of women and girls in violation of the immigration act."

Were most prostitutes foreign-born? The Dillingham Commission examined over 2,000 prostitution cases in New York courts between November 1908 and March 1909, and found that only about one-quarter were foreign-born, in a city that was over 40 percent foreign-born. Three other surveys from this time show similar results, showing that an average of around 75 percent of prostitutes were native-born Americans.

Were large numbers of women forced into lives of prostitution as white slaves? Officials could not make up their minds. Commissioner Keefe warned that an "enormous business is constantly being transacted in the importation and distribution of foreign women for purpose of prostitution." One year later, he had changed his mind and now believed that "women and girls are rarely imported into this country for purposes of prostitution."

The Dillingham Commission mixed alarmist rhetoric with data that told a more nuanced tale. "The importation and harboring of alien women and girls for immoral purpose and the practice of prostitution by them," the report began, "is the most pitiful and the most revolting phase of the immigration question." Yet later in that same report, the commission admitted that "the majority of women and girls who are induced to enter this country for immoral purposes have already entered the life at home and come to this country," of their own free will.

William Williams also believed that most prostitutes were not forced into the profession. Even so, he noted that male pimps were increasingly dominating the profession and controlling the earnings of female prostitutes, but he did not think this was white slavery. As he saw it, while there might be some "incidental slavery, particularly at the outset," for the most part women were "usually glad to place themselves under the control of and receive their direction from men."

Williams was probably close to the truth. As one historian has put it, "the vast majority of women who practiced prostitution were not dragged, drugged or clubbed into involuntary servitude." By one estimate, less than 10 percent of American prostitutes were victims of white slavery. At the height of the white slavery scare, slightly more than a thousand individuals were convicted of white slavery.

Many women chose to become prostitutes. Economic necessity and a poor home life were more often greater recruiting tools than physical force and enslavement. Yet it was easier to believe that passive and virtuous women could only become prostitutes at the hands of greedy men. Eva Ranc and Hermine Crawford show that women were often willing participants in the sex trade. They were smart, shrewd, and savvy, often outwitting immigration authorities, the police, and male suitors.

The public may have overreacted to the white slavery scare, but for those women forced into the profession it was a harrowing experience. After her arrest for prostitution, a young Swiss girl named Jeanne Rondez told her story at a deportation hearing at Ellis Island. She had been brought to America at age nineteen to work as a servant. She told inspectors about a few photographs she had made in France, which a friend of hers had given to a man named Lucien Baratte. The photos were likely nudes, and it appears that Baratte was trying to blackmail Jeanne.

While searching for Baratte in New York, Rondez ended up at the home of Mrs. Eloy Miller, who invited Jeanne to dinner. After dinner, the woman refused to allow Jeanne to leave and made her spend the night. Then Baratte entered Jeanne's room and demanded sex. Jeanne refused and was kept in the room for two days before she succumbed to Baratte's advances. She had been a virgin, and the shame of her situation allowed Baratte and Miller to force her into prostitution. For the next six weeks, Jeanne was made to receive men, who paid her $2 for sex. Six weeks after her ordeal began, Jeanne was arrested for prostitution and taken to Ellis Island.

Miller and Baratte were soon arrested, while Jeanne was released from Ellis Island into the care of the Jeanne D'Arc Home. The stress of her ordeal caused Jeanne to fall ill for the next two months, after which time she found a job in the home of Mr. J. Dreyfus in Staten Island. After her release, Inspector Tedesco went to see how Jeanne was progressing. Dreyfus informed him that Jeanne had admitted her past to him and he had no doubt that she was trying to "become a respectable woman." In August 1911, five months after her arrest, the deportation order was canceled. Jeanne Rondez's ordeal was over, but her experience as a white slave no doubt lived with her for the rest of her life.

ON JUNE 9, 1914, twenty-one-year-old Giulietta Lamarca arrived at Ellis Island. Though most of her fellow passengers had embarked at Palermo, Lamarca began her journey from Algiers, an unlikely starting point for most immigrants. Lamarca listed her profession as a domestic and declared she was heading for her intended husband, Marco Giro, in Brooklyn. As a young woman arriving at Ellis Island alone, she was temporarily detained, but eventually discharged when Giro came to escort her off the island.

Lamarca's stay in Brooklyn lasted less than a year. In May 1915, an Italian immigrant named Vincenzo Palumbo was arrested for running a gambling house and brothel at 116 Van Brunt Street, the same house where Giulietta Lamarca resided. As it turned out, Palumbo had brought Lamarca to America to work as a prostitute; his brother had originally recruited Lamarca from Italy to work in a brothel in Algiers. Marco Giro was merely an associate of Palumbo, not Lamarca's fiancé. Palumbo was convicted of procuring prostitutes and sentenced to seven and a half years in an Atlanta jail. Giulietta Lamarca was brought to Ellis Island and detained.

At her hearing, Lamarca began to spin a tale about her life. Despite her Italian ethnicity, she claimed to have been born in Algeria. She maintained that she had a husband in the United States, but that she left him. She vehemently denied she was a prostitute.

Witnesses claimed otherwise. One testified that Lamarca had purchased a watch from him and offered to pay him with sexual favors. The most damning testimony came from Ellis Island inspector Frank Stone, who called Lamarca's case commercialized vice "in its most vicious forms." He claimed that she was infected with syphilis and that she charged men 50 cents for sex. Lamarca's home was along the Brooklyn piers and her clientele was almost exclusively sailors. Stone also hinted at greater evil. Obstetrical instruments were found in the rooms at the brothel and a vaginal speculum was discovered hidden in the springs of a couch. Furthermore, Stone noted, obscene photos "depicting the most revolting sexual and carnal scenes" were found.

With her pimp in jail, the case against Lamarca was clear-cut. Having been in the country for just a year, she came within the statute of limitations for deportation, but much in the world had changed since her arrival. Archduke Franz Ferdinand, the heir apparent to the throne of the Austro-Hungarian Empire, had been assassinated in Sarajevo just two weeks after Giulietta arrived. Officials could no longer deport immigrants back to war-torn Europe, as steamships were now in danger from German U-boats. Lamarca would have to be detained on Ellis Island until further notice.

ALTHOUGH ATTITUDES ABOUT SEXUALITY changed dramatically as the twentieth century progressed, the concept of moral turpitude has remained a viable tool in immigration law.

The Supreme Court failed to clear up the ambiguity of the phrase when it ruled in the 1950s that the moral turpitude clause was not unconstitutionally vague. Between 1908 and 1980, almost 62,000 aliens were deported for moral turpitude, one-quarter for immoral behavior and the rest on criminal charges.

The reach of the moral turpitude clause extends into the twenty-first century. When visitors to the United States fill out an entry form at the border, one of the questions they are asked is: "Have you ever been arrested or convicted for an offense or crime involving moral turpitude?" Vera Cathcart would be amused.

Part IV

DISILLUSION AND RESTRICTION

Chapter 14

War

———————

We must not forget that these men and women who file through the narrow gates at Ellis Island, hopeful, confused, with bundles of misconceptions as heavy as the great sacks upon their backs . . . these simple, rough-handed people are the ancestors of our descendants, the fathers and mothers of our children.
——Walter Weyl, 1914

I have seen so much of the human wreckage of Europe pass through Ellis Island during the past two years.
——Frederic C. Howe, 1916

AT A FEW MINUTES PAST 2 O'CLOCK IN THE EARLY MORNING of July 30, 1916, Peter Raceta, the captain of a barge docked at a Jersey City pier, found himself tossed some twenty feet in the air by the force of a blast he could only describe as something akin to the explosion of a Zeppelin and what others likened to the sound of the firing of a large cannon. Raceta landed in the waters of New York Harbor many yards from his boat, which was now on fire. He was stunned, but unhurt, apart from a severe burn on the back of his head. The other two men on Raceta's small boat were missing.

Just a few miles away, on Central Avenue in Jersey City, the very same explosion threw two-and-a-half-month-old Arthur Tossen from his bed. Unlike Raceta, little Arthur did not survive. He died from shock.

On Manhattan's Lower East Side, Jewish immigrants were jolted from their sleep by the explosion and streamed out of their tenements in panic and fear. Amid the chaos, a young mother named Dveire, who had recently escaped the war in Russia, calmly took her family into the cellar of their East Broadway tenement to ride out the confusion. Accustomed to the noise of battle, Dveire stayed calm, but many of her neighbors did not, as the sounds of shells exploding in New York Harbor made them fear that war had followed them to the New World.

Throughout the New York metropolitan area and extending as far south as Philadelphia, people were awakened by what they thought was an earthquake. Residents of northern Maryland called their local police to complain. But this was no earthquake.

The epicenter of the explosion that had disturbed the sleep of so many people was a place called Black Tom Island. Though once a small island in New York Harbor, Black Tom had since been connected with the mainland of New Jersey by landfill, making it a peninsula that jutted out nearly a mile into the harbor. Piers and warehouses were built along its shoreline and railroad tracks connected them to points west.

A fire had started sometime after midnight on board one of the barges docked at the National Dock and Storage Company's facility at Black Tom. Dozens of these boats were lined up along the piers at Black Tom, while locomotive cars waited at the terminal, their contents to be loaded onto those boats the following Monday. Some were filled with sugar and tobacco, but most were stocked with dynamite, ammunition, shells, and other tools of war headed for Britain, Russia, and France.

Two hours after it began, the fire eventually reached one of the ships filled with munitions, setting off the great explosion that had awakened so many people for miles around. For three hours after the first blast, more explosions followed and the fire spread to other ships. Huge towers of flames lit the early morning sky. Shrapnel dug huge pits in the Statue of Liberty on Bedloe's Island, some two hundred yards from Black Tom.

Thousands of windows in the skyscrapers of lower Manhattan were blown out; the buildings looked as if "they had been targets for scattering handfuls of rocks from some great giant." The Brooklyn Bridge swayed. Smoldering embers continued to explode shells as late as twenty-four hours after the first explosion, causing firemen and others surveying the wreckage to duck for cover. Twelve people in Manhattan were taken to local hospitals to be treated for cuts from shattered glass.

Almost the entire Black Tom facility was reduced to rubble. Warehouses became piles of large splinters, stacked almost a hundred feet in the air. Railroad cars from the Lehigh Valley Railroad Company were now burning hulks. Rails that helped speed goods to the dock were now gnarled and twisted pieces of metal pointing in all directions. Six piers had become smoking ruins, along with thirteen warehouses, eighty-five fully loaded railroad cars, and over one hundred barges.

The explosion was felt on Ellis Island, just a few hundred yards northeast of Black Tom. The *New York Times* described it in the wake of the blast as "a war-swept town." Almost every window on the island was shattered by the concussive effect of the explosions. Shrapnel and other debris were strewn across the island. The terra-cotta ceiling of the main hospital had caved in. The iron-bound door of the main building was jammed inward, as if hit by a direct dynamite blast. An Ellis Island doctor, watching the fire on Black Tom, was thrown fifteen feet against a wall by the power of the blast.

The few barges filled with explosives that did not blow up from the fire had been set loose from their moorings and drifted threateningly toward Ellis Island. Two of them hit the pier there, but softly enough to prevent another explosion. Workers at the island doused the ships with water.

Over three hundred immigrants spending the night at Ellis Island were evacuated to Battery Park, but the mentally ill detainees were kept on the island. They were brought out to the east side of the island, where they were treated to a pyrotechnic extravaganza as rocket shells continually shot over the island like flares, exploding in a large arc of fire. These patients, not aware of what had happened or the danger involved, "clapped their hands and cheered, laughed and cried, thinking it was a show which had been arranged for their particular amusement."

It took Jersey City authorities little more than twenty-four hours to make their first arrest. The city's commissioner of public safety, Frank Hague, ordered the arrests of the head of the National Dock and Storage Company and the local agent for the Lehigh Valley Railroad on charges of manslaughter. Hague was upset that the blast had killed one of his own men, Jersey City patrolman James Dougherty, who died when a warehouse collapsed on top of him while he was investigating the original fire.

Authorities were adamant that there was no evidence that foreign plotters were to blame. Officials from the Lehigh Valley Railroad went so far as to chalk up the fire to spontaneous combustion. Never mind that the destruction of so many military explosives would have cheered the German kaiser. Americans were cozily snug in their cocoon, secure in the thought that the vast Atlantic Ocean would buffer them from Europe's deadly storms. The war in Europe, already two years old, was a distant event for most Americans.

But the nearly $50 million worth of damage caused by the explosion was not a mere accident or spontaneous combustion; rather, it was the deliberate act of human hands. Just before midnight, two German saboteurs, Lothar Witzke and Kurt Jahnke, arrived by rowboat at the lightly guarded Black Tom facility. A third man, Michael Kristoff, joined them by land. The three then lit several small fires and set a number of timed explosives in the boxcars and barges filled with ammunition and shells. Within fifteen minutes, the watchmen at Black Tom began to see fires throughout the complex, which would soon burn out of control. Two hours later, these fires would set off the massive explosions that rocked the New York area.

Witzke, Jahnke, and Kristoff were part of a larger plot by the German government to sabotage the Allied war effort. Though technically neutral, the United States had been aiding its friends in Europe, and now Germany responded by waging a quiet war of sabotage against the United States.

Although the explosion's immediate effect on Ellis Island was measured mostly in broken windows, its long-term effect would be felt with grave consequences for the way that America viewed immigrants. Americans of English stock would be dismayed by the reaction of German- and Irish-Americans who sympathized with Germany against England. Alien immigrants would morph into alien enemies.

The road to the Japanese internment camps of World War II began at Black Tom Island and continued right through Ellis Island. Franklin D. Roosevelt, who was serving as assistant secretary of the navy in 1916, reportedly told an aide after the 1941 attack on Pearl Harbor: "We don't want any more Black Toms."

IT WAS GOOD FRIDAY, April 6, 1917, when Congress declared a state of war with Imperial Germany. After three years of avoiding ethnic squabbles that were ripping apart Europe, and less than a year after the devastation at Black Tom, the United States was officially at war. Two million American soldiers would soon be heading for France, many of them country boys from small-town America who had never ventured far from home.

However, President Woodrow Wilson went beyond simply declaring Germany the enemy of America. More than half of his war proclamation dealt not with affairs in Europe but with a class of individuals he termed "alien enemies."

Any male over the age of fourteen born in Germany, residing in the United States, and not a naturalized U.S. citizen, overnight became an alien enemy, part of a potential fifth column ready to strike America on behalf of the kaiser. Such individuals were banned from possessing any weapons or operating a plane. Alien enemies were barred from living within half a mile of any military base, aircraft station, navy yard, or munitions factory. Such aliens could not "write, print or publish any attack or threat against the Government or Congress of the United States." Above all, no enemy alien could give aid or comfort to Germany, assist its war effort, or disturb the "public peace or safety of the United States." Anyone suspected of violating these orders was subject to summary arrest and confinement. No trial or hearing was necessary.

This action was not unprecedented, but was based on the 1798 Alien Enemies Act, which stated that if the United States was ever at war with a foreign nation, all adult males from that country residing in the United States who had not become naturalized citizens were deemed "alien enemies" and "shall be liable to be apprehended, restrained, secured and removed." The legislation was part of a series of laws known collectively as the Alien and Sedition Acts, pushed through by members of the Federalist Party as the nation prepared for a possible war against France. While the other parts of the Alien and Sedition Acts either expired or were repealed, the Alien Enemies Act still remains law more than two hundred years later.

After Wilson's proclamation, the government wasted little time in exercising its prerogative. America's first wartime action took place not in Europe but on American soil, using agents from the Immigration Service. On the night of Wilson's war proclamation, federal agents began rounding up German alien enemies and taking them to Ellis Island for indefinite detention. Literally overnight, Ellis Island's role changed from an immigrant inspection station to a military detention facility.

As for targets, officials did not have to look far. Less than a mile up the Jersey coast from Black Tom stood Hoboken: the "Mile Square City," the self-proclaimed birthplace of baseball and the hometown of Frank Sinatra, who was just a sixteen-month-old toddler when America declared war on Germany. In 1916, the tiny city had a large German population, thanks in part to the fact that the North German Lloyd and Hamburg-American steamship lines docked at Hoboken.

With war declared against Germany, steamships owned by German companies docked at American piers on April 6, including the *President Lincoln* and *President Grant*, were seized by the federal government. All German nationals working on those ships or on the docks were rounded up and taken to Ellis Island.

These men who made their living bringing immigrants to the United States now found themselves in detention. The ships whose steerage sections once carried immigrants would soon be shuttling American troops to the European front. A year later, German torpedoes would sink the *President Lincoln* off the coast of France.

The German officers and crew members were not prisoners of war and did not receive trials. Deemed alien enemies, they were rounded up and detained using the administrative apparatus of immigration law. The almost 1,500 Germans caused little trouble during their stay on Ellis Island, although they complained that they could not get beer. The men filled their days with calisthenics, games, and reading. The commissioner of Ellis Island found the men "obedient to discipline" and resigned to their situation.

One exception was the case of George Begeman, an officer on the North German Lloyd's steamship *George Washington*. Begeman, along with three other colleagues, was granted a leave to visit a dentist in Hoboken. While the guard was getting a sandwich, Begeman fled the dentist's office. He was last seen in a Hoboken bar downing huge schooners of beer and "calling down the curse of the ghost of Mohammed's black dog on all prohibitionists." All that was left by the time police arrived was a line of empty beer steins on the bar. "Ach Himmel," the bartender told police of Begeman, "he vas a great drinker." Begeman would never be captured.

Another detainee at Ellis Island was thirty-seven-year-old William Hausdorffer, the acting captain of the steamship *Bohemia*. Hausdorffer, his wife, and two small children lived in nearby Bayonne. The Hausdorffer children were born in the United States and therefore citizens, but their parents had not yet become naturalized. Hausdorffer had lived in America since 1906, and his wife since 1899, and the family considered themselves American. Hausdorffer's crew had even derisively nicknamed him "the American" because he sympathized with the United States over the land of his birth. His wife told officials that her husband was even willing to enlist in the U.S. Army. Nevertheless, Hausdorffer was not a naturalized citizen, and his position with a German company was enough to make him an alien enemy.

Not everyone felt the same way as Hausdorffer. William Koerner, who served as a machinist on the *Vaterland*, was also taken to Ellis Island. When questioned as to which country he sympathized with in the war, his answer was Germany. Koerner, like many of the steamship company workers, also served in the German naval reserve. Though not officially in the military, the status of Koerner and his comrades was enough to convince American officials to hold them as alien enemies.

Over 1,500 German detainees would spend some time at Ellis Island. For some, their detention would be short. Albert Meyer, who worked as a cook on the steamship *Vaterland*, had been caught up in the dragnet of April 6. He was detained at Ellis Island for two weeks before he could prove to authorities that he was a citizen of Switzerland and therefore not an alien enemy.

Most were not so lucky. In early June, government officials began transferring the German detainees to an internment camp at Hot Springs, North Carolina. First went 470 officers, the captains, engineers, and chief warrant officers. The rest, 1,100 or so crew members and sailors, would follow their officers to North Carolina later. One could have easily mistaken the place for a summer camp, with tidy cabins in a rustic setting, but it was a militarized facility where detainees were not allowed to leave unless given permission. In total, some 2,300 Germans taken into custody throughout the country would be interned at Hot

Springs during the war. Thirty-six Germans accused of being spies remained at Ellis Island and would later be removed to Fort Oglethorpe in Georgia.

Some wives of the detainees, not considered alien enemies because of their gender, petitioned the government for the freedom of their husbands. William Koerner's wife, Paula, was five months pregnant when her husband was taken to Ellis Island. Not only did she lose her husband, but she also had to give up her job making handbags. To help with their situation, Paula and the other wives of steamship crew members received a monthly stipend from their husbands' employers. Thanks to his wife's pleadings, Koerner received a three-week parole in August 1917 to be with his wife as she gave birth.

Other cases were more tragic. Herman Byersdorff, the chief engineer of the *Kaiser Wilhelm II*, had been caught in the first roundup of Germans on April 6 and taken to Ellis Island. War had already touched the Byersdorff family. His only son had been killed in battle in France while serving in the German army in 1914, driving Herman's wife to a nervous breakdown. She was then brought to America to join her husband, which seemed to calm her nerves.

Byersdorff's detention once again sent his wife down an emotional spiral. A doctor in Hoboken diagnosed her with severe mental depression bordering on melancholia. From his internment camp in North Carolina, Byersdorff asked to be paroled to be with his distraught wife. As the paperwork for his parole made its way through the federal bureaucracy, Mrs. Byersdorff moved to Hot Springs to be closer to her husband.

Finally, in February 1918, the stress and grief proved too much for Mrs. Byersdorff, who committed suicide almost a year after her husband was taken into custody. Herman received a temporary leave to attend her funeral in Brooklyn, but had to return to North Carolina after two weeks. In June 1918, four months after his wife's suicide, Herman Byersdorff's paperwork finally landed on the right desk and he was granted a parole, a small consolation with his wife and only son dead.

It is unclear why Byersdorff's parole should have taken so long. As early as February 1918, William Hausdorffer was paroled. That spring, more paroles followed as the fear of German sabotage subsided. William Koerner left Hot Springs in April 1918.

The militarization of Ellis Island continued after the German detainees were gone. With immigration from Europe slowing to a trickle because of the war, the army took over the island's hospital for wounded troops, while the navy took over the baggage and dormitory building and used them to quarter sailors waiting for their assignments. At times, as many as 2,500 military men were stationed at Ellis Island, most for no longer than two weeks. At the same time, American soldiers wounded at the European front were also sent to recover at Ellis Island's hospital. Young American doughboys who had survived the trenches of the Western Front, often at the cost of an arm or a leg, could be seen wandering the grounds of Ellis Island as part of their convalescence.

The man in charge of Ellis Island during this turbulent period was Frederic C. Howe. He knew little about immigration before assuming the job and later admitted that the topic did not interest him greatly. Unlike Ellis Island's first commissioner, John Weber, whose life was forged in the combat of the Civil War, Howe's formative experience as a young man was graduate school at Johns Hopkins, where he studied under Professor Woodrow Wilson. Although Howe later became a lawyer, his graduate years instilled in him an idealistic temperament and a restless intellectual curiosity. His job prior to coming to Ellis Island in 1914 was head of the People's Institute, a debating society for liberal intellectuals in New York.

Whereas William Williams was comfortable, if not smug, with his position in society and his relationship with his ancestors and background, Howe spent most of his life, in his words, "unlearning" the values of his childhood. Raised in a comfortable middle-class, church-going, Republican family in western Pennsylvania, Howe worked to rid himself of the lessons and values of his small-town childhood as he moved up in the world and became engaged in politics.

Howe was a Progressive, a man driven to public service to reform a society reeling from the effects of industrialism, mass immigration, and urbanization. Both Williams and Howe possessed a moralism that stoked the engines of reform. Both men saw the world divided between good and bad. For Williams, the good consisted of people of his class and background, the descendants of the Puritan forefathers. The bad were the undesirable new immigrants whose presence brought crime, disease, and political machines and threatened the Republic that Williams's ancestors had built.

Howe's heroes were those liberals who also had unlearned the values of their youth and committed themselves to changing the world. His villains were selfish and narrow-minded people who pursued economic self-interest at the expense of the public interest. Unlike William Williams, whose progressivism was based on ideas of efficiency, Howe was a humanist who defined his type of reform as "sentimentality, or the dreaming of dreams." No one would have ever accused William Williams of being a dreamer.

Howe sought to humanize Ellis Island, a not-too-subtle dig at his predecessor, and saw his new job as "an opportunity to ameliorate the lot of several thousand human beings." He sought to spruce up the Great Hall, mixing in some Americanization with beautification. Potted plants were placed throughout the grand, yet sterile hall. Photos of American presidents and paintings of important events from American history hung from walls and large American flags from the balcony. Howe also placed suggestion boxes around the station where immigrants, visitors, or employees could voice their complaints.

"I was struck by the dreadful idleness of these poor people," Howe said of the detainees. "Some three hundred of them were detained here, compelled to sit hour after hour on hard benches in a bare room." Instead, Howe ordered that benches be brought out of storage and placed on the lawn outside so that immigrants previously cooped up in indoor cells could now enjoy the outdoors. A playground was created for detained children, with an adult supervisor in charge of ball games and jump rope. Sewing materials, periodicals, and toys were now available. English classes were offered, as was schooling for children. One day, an Italian group brought over Enrico Caruso to entertain the detainees for a Sunday afternoon concert.

Despite their differences, Howe and Williams agreed on one thing. Both sought to end discrimination between steerage passengers and those traveling in first- and second-class cabins. The former were always sent to Ellis Island for inspection, while the latter were inspected aboard ship and were only in rare circumstances ordered to Ellis Island.

"Aliens traveling in the cabin are no more exempt from the immigration laws (which apply to *all* aliens) than they are from the customs laws," Williams wrote. "Some of the most objectionable of the prohibited classes are likely to have means sufficient to enable them to buy a first-class ticket." Criminals, pimps, and prostitutes were sometimes found in first-class cabins, and steamship officials sometimes listed aliens as citizens, which meant, to the cost-conscious Williams, that the government coffers were being deprived of its $4 immigrant head tax.

In January 1912, ninety-two first- and second-class passengers on the *Carmania* were having a pleasant dinner when immigration officials boarded their ship. They were ordered to stop eating, form a line, and answer questions. The inspection lasted forty-five minutes and netted six people who were sent to Ellis Island for further hearings, including four suspected prostitutes and one notorious embezzler.

News of this inspection provoked outrage. A letter signed "One of the Upper Class, Newport, Rhode Island," complained to the editor of a New York newspaper. "We of the better class consider the action of the immigration authorities a gratuitous insult," the indignant writer protested. "There is nothing to my mind that strikes a more violent blow at our 'position' and 'caste' than . . . the intimation that 'first-class' passengers are not one whit better in the social scale than those horrid people who cross the Atlantic in the nauseating and ill-smelling steerage."

Fred Howe would go a step further and ask for permission to send all second-class passengers through Ellis Island, along with steerage passengers. Steamship companies complained and forced a public hearing on the matter. "There has always been maintained in this country that distinction between the cabin and the steerage," said a representative of the steamship companies. "Most of the people who travel second-cabin are most self-respecting people."

Steamship passengers paid a premium for that distinction. The average cost in 1915 for a first-class passage was between $85 and $120; for second-class passage, $50 to $65; and for steerage, $35 to $46. "A man in the first-cabin might consider it almost a joke to be, as he would express it, put with immigrants," said the man from the steamship company. "A person in the second cabin would regard it as a very serious protest in his own mind." The fear that such a measure would cut into the profits of steamship companies, as well as ingrained class prejudice against steerage passengers, meant that the reforms of Williams and Howe went nowhere, and nearly all first- and second-class passengers would continue to bypass Ellis Island.

By this time, however, there were more pressing matters. Most of Howe's reforms came during a unique period in Ellis Island's history.

For most of his five-year tenure, war raged in Europe. During 1914, 878,000 immigrants came through Ellis Island; the following year the war had brought that number down to 178,000. During Howe's entire administration, only about half a million immigrants passed through Ellis Island.

While the pressure to inspect large numbers of immigrants had subsided, war created other problems. Those denied entry and ordered deported could not be sent back because of the war. Many of Howe's reforms were meant to ease conditions on Ellis Island for these men and women stranded because of the violence and destruction at home, yet blocked from legally entering the United States.

Whereas a strict segregation of sexes had been the rule at Ellis Island, Howe allowed men and women to mingle throughout the day on the grounds and in the common detention hall—with matrons keeping an eye out for any illicit activity.

With Ellis Island overflowing with detainees, Howe took a more liberal approach to enforcing the law. He released on bond a number of immigrants designated as feebleminded. A special report by the New York State Department of Labor condemned the move from both an economic and a eugenics standpoint. "The precipitation of feeble-minded females or [sic] marriageable age without restrain into the community, is to be condemned in the strongest possible terms," the report complained.

With the controversy over detainees swirling around the island, Howe also tried to reform the operations of Ellis Island. He believed that the money exchange, the railroads, and the food concession all exploited immigrants. When the contract for the food concession expired, he pushed for the federal government to take over the responsibility of feeding immigrants. Deeply suspicious of private enterprise, Howe had previously championed the public ownership of railroads and utility companies. Here was a chance, on a much smaller scale, to push his ideas about the inherent justice and efficiency of public ownership of business. Private business, argued Howe, should not make money off immigrants on government property.

Like many reformers, Howe had a tin ear for politics. He was apparently unaware that New York congressman William S. Bennet was the lawyer for Hudgins & Dumas, the food concessionaire. Bennet was no reactionary; he was the lone dissenting voice on the Dillingham Commission to oppose the literacy test and was an opponent of immigration restriction. However, Howe had touched a nerve—or Bennet's pocketbook—and the congressman used an amendment to a House bill to block Howe's plan.

At Bennet's urging, Congress began an investigation of Howe in the summer of 1916. It focused heavily on the issue of the sexual morality of female immigrant detainees. Bennet charged gross immorality under Howe's watch, calling him "a half-baked radical" who supported free love.

Even more pointed was Bennet's charge that Howe was lenient toward immigrant women of questionable morality. The white slavery hysteria meant that more suspected prostitutes were taken to Ellis Island and stranded because of the war. Bennet complained that prostitutes were allowed to mingle during the day with other detainees and that detained Chinese sailors were gambling and cavorting with detained prostitutes.

One case that aroused congressional interest was that of a suspected prostitute named Ella Lebewitz. Officials accused her of having sex with a nineteen-year-old Brazilian male, also in detention. Lebewitz denied the charge and argued she would be foolish to ruin her chances of being allowed into the country. A Labor Department official called her "absolutely incorrigible," "subnormal or abnormal," and "positively a degenerate." Though Howe claimed he had been aware of Lebewetz, he doubted any sexual impropriety.

At the hearing, Texas congressman James Slayden asked Howe: "What percentage of the people who are detained at Ellis Island are downright immoral people?" Howe responded that the figure was around 20 to 50, out of the 400 to 600 detained at any one time.

Not only was Howe determined to ease the pain of detention; he was also willing to reconsider deportation orders. He sent a team of female social workers to investigate some of the cases. Howe believed that "the great majority of women were casual offenders who would not have been arrested under ordinary circumstances. In many instances their misfortunes were the result of ignorance, almost always of poverty." In his autobiography, he mentions the case of an immigrant named Sarah, who lived in St. Louis and whose drunken husband abandoned her and her infant. In despair, Sarah sold herself to a man on the street, was arrested, and sent to Ellis Island to be deported.

Alice Gouree was not a prostitute, but she still encountered problems at Ellis Island. Having lived in New York since 1906, the thirty-one-year-old Frenchwoman returned to New York from France the same day that Congress declared war on Germany. Gouree was preceded upon her arrival by an anonymous letter to Ellis Island warning officials that she had had an affair with a married man.

Thanks to the letter, Gouree was detained at Ellis Island and questioned about her sexual past. She admitted to having had sexual relations with the man and that he had paid for her apartment, although she claimed not to know he was married. She also admitted to an affair with another man years earlier who had also paid her rent, as well as a third relationship with another married man. After her hearing, the board ordered Alice excluded as an immoral woman. With deportations suspended until the end of the war, Howe was not interested in keeping Alice detained indefinitely and advised that she be admitted. His superiors in Washington ruled against Howe, calling Gouree a "self-confessed courtesan with very warped ideas of moral uprightness," and ordered her detained until she could be deported. However, the number two person in the Labor Department, Louis Post, agreed with Howe and ordered Alice paroled to her sister.

But Gouree was not free from the long arm of immigration officials. Investigators monitored her situation, reporting that she had worked as a maid in a hotel after her release from Ellis Island, but that she was fired after a few months for improper behavior with a married waiter named Muhlenberg, who had left his wife and three children to live with Gouree. Seven months later, Gouree was back at Ellis Island to answer for her sexual promiscuity. She admitted to the affair with Muhlenberg, but said she believed he was going to leave his wife to marry her.

In tune with the anti-German hysteria sweeping the country and faced with deportation back to France because of the affair, Gouree told officials that she broke up with Muhlenberg not because he was married and wouldn't leave his wife, but because he was German. She also informed officials that she believed that Muhlenberg, a German citizen, had not properly registered with the government as an enemy alien. The newly patriotic Gouree begged officials to allow her to stay, admitting her mistake and saying that she had found another job as a maid for a Park Avenue matron.

When it looked as if patriotic, anti-German appeals were not going to win her the right to stay in America, Gouree lashed out. "I have done nothing wrong and was brought back to Ellis Island for no good reason," she said. "Why should I be kept here?" One can only imagine Gouree's humiliation at having to discuss her sex life in front of male authorities. It was all too much. Although she was released on bond again in February 1918, she told officials that when the war was over, she would return to France at her own expense. "This sort of treatment will make a bad woman out of any good woman," Gouree wrote. In 1919, when officials sought Alice's deportation, they were informed that she had already kept her promise and left America.

Then there was Giulietta Lamarca. Sent to Ellis Island for prostitution in the summer of 1915, Lamarca remained there for months, unable to be deported back to her native Italy. Her case was one of those that attracted Howe's attention. "This woman has conducted herself with propriety," Ellis Island matrons informed Howe. "She has kept away from the men. She has a son in Italy and she wants to make a little money in order to bring him over here." Howe believed that an abusive husband had forced Giulietta into prostitution and decided to give her a chance.

"I have, I admit, thought of the poor, ignorant, immoral women detained at the Island as human beings entitled to every help to a fair start in the world," Howe wrote in response to his critics. Working with charitable groups, he sought to find homes that would help rehabilitate these women. Giulietta was released on bond to work as a servant in the home of an Ellis Island doctor who lived in New Jersey.

Giulietta seemed to be a good worker. A year after she left Ellis Island, she was working for another government official living in New Jersey, a man named S. L. Norton. Lamarca only worked for Norton for four days before leaving. Inspector Frank Stone was sent to look into Norton's complaints against his former employee.

Norton was angry that Lamarca had left his employ early. Giulietta claimed she was hired to be a cook for Norton, but instead had to clean up after Norton's wife, who was suffering from an ailment that forced her to wear a diaper. Norton took after Lamarca with a vengeance. He told Stone that Giulietta had had indecent contact with his two dogs.

His proof: when the dogs left Giulietta's company, they were panting and excited, which to Norton showed "that she had committed some crime against nature with them."

Norton also complained that the former prostitute was corrupting the morals of the decent young women of Cranford, New Jersey. Because of her past, Lamarca's relationship with men was open to investigation. Stone found that although Giulietta had had some conversations with an Italian chauffeur and an Italian garbageman, she had "conducted herself properly while in Cranford." He concluded that Norton's charges were "inspired by malice and vindictiveness" and anger at Howe's policy of releasing prostitutes from detention at Ellis Island.

Ellis Island officials allowed Giulietta to remain free. She continued to live and work in New Jersey. Howe argued that he had found that out of the hundreds of women paroled, "not more than a dozen" had reverted to their former lives of prostitution. Howe possessed a positive view of human nature, that men and women were victims of their environment and that rehabilitation was an exercise in humanity, not futility.

With cases like Giulietta's seemingly to have turned out so well, Congressman Bennet's hearings went nowhere. The specific complaints were dropped, the former prostitutes were out on parole, the food concession remained in private hands, and Howe remained in office. At least one of Bennet's criticisms, though, was on the mark.

Bennet charged that Howe spent less than half of the working week at Ellis Island, making him "the most absentee commissioner" in the station's history. As if to prove the point, Howe could not be reached for comment on Bennet's charge because he was vacationing for a week in Nantucket.

Howe described his daily schedule for the congressional committee. He would arrive at Ellis Island sometime between 8:30 and 10:00 A.M., depending on which ferry he caught. Usually, he got to his office around 9:30 A.M. His days on Ellis Island would end around 4:15 P.M., but he admitted that "many days I leave before that when I clean up all the work and there is nothing more to do."

Howe's inattention was due less to laziness than to overextension. Howe still spent a great deal of time dabbling in personal intellectual and political pursuits, few of which directly related to immigration.

He was more likely to make news for his views on unemployment, the nationalization of railroads, or public ownership of utilities than on immigration policy. Most of his letters to Woodrow Wilson dealt with recommendations on everything from who should serve on the new Federal Trade Commission to what kind of peace Wilson should seek when the war in Europe ended.

Howe spoke out about the conflict in Europe, giving a speech in lower Manhattan in 1915 in which he warned against rushing into war, since he believed that "wars are made by classes and privileged interests." This was a far cry from what his boss, President Wilson, was saying.

Even the *Times*, a defender of Howe against attacks from Bennet, called Howe "a glib spokesman of glittering and ignorant theories, a thinker of vealy thoughts, an individual whose public utterances are often of the half-baked kind." It encouraged Howe to continue his humanitarian work at Ellis Island, but "stay off the lecture platform."

There was a deeper issue at work. Not only did Howe know little about immigration, but he was also growing increasingly disillusioned with government. Whereas William Williams wielded the powers of his office comfortably—perhaps too comfortably—Howe seemed uneasy with his role at Ellis Island. More a thinker than a doer, he had difficulty administering the station and admitted that his superiors in Washington often ignored his suggestions and left many of his letters unanswered.

Howe had also grown disillusioned with government workers, finding them nothing more than petty clerks. "The government was their government," he wrote. The great success of the Progressive Era was the creation of the administrative state that would regulate private business in the public interest. In theory, civil service reform helped staff that bureaucracy with professionals instead of hack politicians. Yet Howe found that this bureaucracy "moved largely by fear, hating initiative," caring only about "its petty unimaginative salary-hunting instincts." He felt that his position at Ellis Island was not just irrelevant, but unnecessary. Howe had no desire to preside over what the *Times* called the "petty Czarship" of Ellis Island commissioner and saw little need to weed out the desirable from the undesirable.

Howe's career indicated a steady change in American liberalism, an evolution from the progressivism of earlier years to a more modern form of liberalism. The Great War only brought more disillusionment with the state, as liberals increasingly emphasized individual rights and humanitarianism.

Even Howe's choice of a home made a statement. When they arrived in New York in 1910, Howe and his wife, Marie Jenney Howe, chose to live in Greenwich Village. There the couple mixed with a growing band of bohemians and political radicals. Marie became active in the Women's Suffrage Party and helped found the Heterodoxy Club, a debating society for women that served as an incubator for early feminism. Despite his continual "unlearning" of the conservative values of his childhood, Fred Howe could never fully come to grips with his wife's feminism, which put a strain on their marriage. When Marie read her husband's autobiography, she reportedly asked him, her voice dripping with sarcasm: "Why, Fred, were you never married?"

The 1910s were an exciting time in Greenwich Village. One man who helped give the area its bohemian feel was a hunchbacked dwarf named Randolph Bourne, who walked the streets dressed in a black cape. Bourne, who had suffered from spinal tuberculosis as a child and whose difficult delivery as an infant left his face misshapen and disfigured, became a prominent voice among liberal intellectuals.

Much of the immigration debate had been fought over the idea of the melting pot, a phrase made popular by Israel Zangwill in his 1908 play of the same name. Whether immigrants could be absorbed into American society was the question that divided Prescott Hall from Oscar Straus, a dividing line that the politically agile Theodore Roosevelt danced along for his entire political career.

Randolph Bourne believed that the melting pot had failed, but he turned the idea around on the Prescott Halls of America. No group, Bourne argued, clung more tenaciously to the virtues of the old country than Anglo-Saxons who worshipped everything British and whose allegiance to England was getting the United States dangerously close to participating in the faraway war in Europe.

Bourne also complained that assimilation was a one-way street, accomplished only on the terms set by Anglo-Saxon Americans. Bourne feared that assimilation would take the distinctiveness of the nation's ethnic communities and wash them "into a tasteless, colorless fluid of uniformity." Assimilation, Bourne argued, was bad for immigrants and turned them into people "without a spiritual country, cultural outlaws, without taste."

For Bourne, America's strength was that it was a "world-federation in miniature." He was the prophet of multiculturalism, a hunchbacked John the Baptist laying out arguments that would not gain currency for more than sixty years. Perhaps only a misfit like Bourne could have foreseen this trend in American society. In 1916, these ideas found few adherents beyond the streets of Greenwich Village. Bourne's colleagues at *The New Republic* argued that if America continued to be fractured ethnically, "we cannot expect to attain the homogeneity of feeling and action essential to our position of power with international rights and obligations." The editors of this newest liberal magazine argued in favor of stricter regulation of immigration to end the "wholesale transplantation upon our soil of alien communities."

The war raised questions about ethnic loyalty. Were German-Americans going to support the kaiser? Were Irish-Americans so hateful toward the British that they would side with Germany? Such concerns led to a new enemy on American soil, one so tiny and seemingly insignificant, yet fraught with peril for the entire nation. It was not a person or an organization, but a lowly punctuation mark, a short horizontal line used to connect two words: the hyphen. Irish-Americans, German-Americans, Polish-Americans. "There is no room in this country for hyphenated Americanism," Theodore Roosevelt warned in 1915.

War in Europe and fears of ethnic disloyalty at home recharged the case for immigration restriction. Since the 1890s, the literacy test had been the gold standard for restrictionists. Congress again took up the cause in the waning months of 1916. Both chambers overwhelmingly passed the bill. A week after Wilson gave a speech to Congress calling for "peace without victory" in the war in Europe and only days before Germany resumed its submarine warfare in the Atlantic, the president vetoed the literacy test. It was Wilson's second veto of the bill since becoming president and the fourth veto of a literacy bill since the 1890s.

The president had few strong feelings on the issue, but had promised ethnic groups during his 1912 campaign that he would veto any literacy test to make amends for his earlier anti-immigrant writings. Wilson called the test a radical departure from traditional policy. Unlike other justifications for the exclusion of immigrants, Wilson argued, the literacy test was not a "test of character, of quality or of personal fitness," but instead penalized those who lacked opportunity in Europe. Wilson's arguments were moot. Within days, both the Senate and the House had easily overridden Wilson's veto. The literacy test was finally law.

The new law would require all immigrants over the age of sixteen to be able to read a short text in their native language. In a nod to America's traditional role as an asylum for refugees, those fleeing religious persecution were exempt from the literacy test. To give a sense of how far restrictionist sentiment had evolved, the 1917 Immigration Act contained twenty-six different exclusionary categories for aliens. In contrast, the 1891 law contained only seven.

The literacy test consisted of about forty words from the Bible in the immigrant's native language. The decision to use the Bible had little to do with evangelizing and more to do with the fact that the Bible was the most translated book in the world. Assistant Commissioner Uhl said that the biblical verses were a "non-controversial matter in every case and are practically all from the Old Testament."

Instead of rejoicing at victory, Prescott Hall believed that the work of restriction had just begun, hinting that a literacy test would have little effect on immigration. Since, between 1908 and 1917, some 1.6 million illiterate immigrants had entered the country, many had assumed that the new law would bar a large number of aliens. Yet in its first five years, a mere 6,533 people were barred by the literacy test. After a quarter-century of political agitation, this was at best a tepid victory for restrictionists.

The enactment of the literacy test coincided with America's entry into the Great War, when hostility toward immigrants was channeled toward German-Americans. War propaganda painted the murderous and rapacious Hun as a virulent enemy. Anti-German hysteria spread across the continent as schools stopped teaching German, and German-language newspapers folded. Anything remotely German was suspect: Americans went so far as to rename sauerkraut as "liberty cabbage."

The war greatly strengthened the hand of prewar restrictionists. Charles Warren served as assistant attorney general during the war. He had been a founding member of the Immigration Restriction League and though not as prolific a pamphleteer as some of his colleagues, he perhaps had a greater influence in the long term.

At the Justice Department, Warren began work to resuscitate the

Alien Enemies Act of 1798 to give the government greater control over German alien enemies. Warren was also the architect of the Espionage Act, which passed Congress in 1917 and was designed to go after domestic opponents of the war effort. While Prescott Hall could merely fulminate against the inferiority of the new immigrants, Warren quietly changed the law affecting thousands of people.

The targeting of Germans was also tied to German-owned steamship companies, which had been responsible for much of the immigrant traffic to America in the past quarter century. Otto Wolpert, superintendent of the Hamburg-American docks in Hoboken, and Paul Koenig, the chief detective for Hamburg-American, were accused of assisting German saboteurs and bomb makers. Though only a small percentage of steamship employees were involved in Germany's covert war effort, it was enough of a link to reinforce negative views of steamship companies and tie wartime sabotage directly to immigration.

Then there was the case of the increasingly hapless Marcus Braun, the former head of the Hungarian Republican Club in New York and sometime friend of Theodore Roosevelt. Braun had pushed his way into a patronage job at Ellis Island in 1903, which led him to his native Hungary to investigate the causes of immigration. After leaving the immigration service, he started his own newspaper, *Fair Play*.

Braun's career took a strange turn during the war. In 1915, he was discovered carrying documents from the Austrian consul general in New York to the Ministry of Foreign Affairs in Vienna. Though not illegal, Braun's activities reinforced notions that foreign-born Americans still held loyalties to their mother countries and were willing to assist them in wartime.

It was later revealed that Count Johann von Bernstorff, Germany's ambassador to Washington, had secretly purchased Braun's newspaper. It became apparent that Braun had been a shill for the German government since 1915. Von Bernstorff's activities went beyond buying up American newspapers and extended to overseeing the whole operation of German propaganda and sabotage—including the Black Tom explosion.

Braun somehow escaped punishment, but found his reputation and career in ruins. His name came up a number of times during 1918 congressional hearings looking into the relationship between German-American brewers and German propaganda during the war. The man who had dined with President Roosevelt, inspected immigrants at Ellis Island, and investigated white slavery in Europe had been publicly disgraced. After the war, he moved to Vienna, bought another small newspaper, and passed away unnoticed in 1921.

SITTING IN HIS OFFICE after Armistice Day in 1918, Fred Howe must have thought that the worst of his troubles had passed. The war was now over. Detained immigrants could be released from their Ellis Island imprisonment. But the end of the Great War would not bring peace either to Ellis Island or America.

Chapter 15

Revolution

———•———

The worst dump I have ever stayed in.
  —Emma Goldman, referring to Ellis Island, 1919

ON THE MORNING OF FEBRUARY 6, 1919, SOME 65,000 workers in the city of Seattle began a general strike that would shut down the city for the next five days. Mayor Ole Hanson feared that his city was in the grip of a political and social revolution. Tensions ran high, but revolution never came. The strike ended five days later, after federal troops arrived to restore order.

Even before the strike began, government officials had their eyes on the immigrant radicals of Washington State. On the day that the strike began, some forty-seven suspected radical aliens from Seattle, Spokane, and Portland found themselves on a train headed for Ellis Island instead of manning the barricades. Most were Wobblies, members of the radical Industrial Workers of the World (IWW), but a few belonged to the Union of Russian Workers. Newspapers eagerly dubbed the train the Red Special.

As the train approached Montana, some one thousand Wobblies were waiting for it in Butte in hopes of freeing their comrades, but the Red Special bypassed the city by way of Helena and avoided any problems. In Chicago, the train picked up seven more suspected radicals headed for deportation. The *Times* called the group a "motley company of I.W.W. troublemakers, bearded labor fanatics, and red flag supporters."

The train arrived at Hoboken, New Jersey, and its fifty-four passengers were hustled onto a waiting barge for Ellis Island, where a melee erupted after an argument between a guard and one of the prisoners. This was one exception to a fairly peaceful trip, although the radicals did heap abuse and insults upon their guards throughout. As the guard in charge of the Red Special explained to his boss in Washington, it "went against my grain, as well as every guard aboard the train, to handle them without force, as they were very insulting at times." One guard said the detainees needed gags, not handcuffs. "This is a musical gang," he told a reporter. "They sing foreign songs for hours. Some of 'em wake up in the night to do it."

When the Red Special radicals arrived at Ellis Island, Fred Howe was not there to greet them. He had been away since December accompanying President Woodrow Wilson at the Paris Peace Conference. In his absence, Byron Uhl was acting commissioner, faithfully carrying out deportation orders from Washington. The fifty-four suspected radicals were held incommunicado. Neither their relatives nor their lawyers could see them. The headline from the socialist paper *New York Call* read: "Mystery Thick Around Exiles in Ellis Island: Keepers of New Bastille Terribly Fussy About Even Relatives Seeing Inmates." Officials soon relented and allowed lawyers to review the cases.

Attorneys Caroline Lowe and Charles Recht led the fight to free the detainees. However, they were unfamiliar with immigration law. "A sovereign state has the right to deport every alien, under any laws or rules it pleases," an astounded Recht later remembered in his autobiography. "An alien deportee cannot invoke the Bill of Rights or the Constitution, for these do not apply to him."

In contrast to the depiction in newspapers, Lowe saw her clients as admirable citizens of high character, "clean cut, upright, intelligent, educated." All were literate and could speak English. Americans had been so worried about the pernicious effect of illiterate immigrants that it had enacted a literacy test, yet these radicals would have had no problem passing such a test. The stereotype of the anarchist and radical was usually the Jewish Socialist or the Italian with a bomb, but most of the immigrants on the Red Special were English or Scandinavian.

The detainees were a random lot of IWW organizers, political radicals, and eccentrics. Among them was thirty-four-year-old E. E. Mc-

Donald, who had been born in Denmark and had come to the United States when he was eight. A local newspaper called the picturesque McDonald the poet laureate of the Ellis Island detainees. He even composed a poem there called "Song of the Alien Deportees."

> In the shadow of the statue
> That Bartholdi's hand designed
> We are waiting for the mandate
> That will make us leave behind
> All the friends and kin and loved ones
> We have here on this fair shore
> We are waiting to be exiled
> From this land forevermore.

McDonald and the other passengers of the Red Special were still at Ellis Island when Fred Howe returned from Europe. When Howe complained about the status of the detainees, his superiors told him to mind his own business and follow orders. Howe was also dismayed that his colleagues at Ellis Island were "happy in the punishing power which all jailers enjoy, and resented any interference on behalf of its victims."

Howe was swimming against the tide when it came to the country's attitude toward radicals. Congress had added anarchists to its list of excluded groups back in 1903, and the 1917 Immigration Act expanded the definition of excluded or deportable immigrants to include not just anarchists, but also "persons who believe in or advocate the overthrow by force or violence of the Government of the United States." The following year, Congress gave officials more latitude to define alien radicals. Undesirable immigrants were now defined as those "opposed to all organized government," who advocated or taught "the unlawful destruction of property," and who belonged to an organization that advocated any of the above measures.

This expansion of the law allowed Anthony Caminetti, commissioner-general of immigration, to launch a personal crusade against foreign-born, nonnaturalized radicals living in the United States. One of his first targets in 1918 was the Home Colony, a radical commune on the west side of Puget Sound some forty miles from Seattle. An investigation by government officials showed that the Home Colony was a kind of utopia-turned-sour whose middle-aged members seemed more interested in free love than revolution. It was small potatoes when compared to the IWW.

Though he was out of step with American anti-radical laws, Fred Howe did have one trump card at his disposal. He simply postponed all deportations, allowing the IWW lawyers to present their case to Washington. With additional time to hear the cases, the acting secretary of labor, John Abercrombie, overruled Caminetti and issued a memorandum to all immigration officials stating that the department had never declared the IWW to be an anarchistic organization and therefore its members could not be deported. In all future cases, he declared, a Wobbly's actions, and not simply his membership, would be the basis for deportation.

Using this new standard, the department took up the case of James Lund, an immigrant from Sweden and a member of the Seattle IWW. The Labor Department found, contrary to earlier findings, that there was little evidence that he advocated the overthrow of the U.S. government. Therefore, it ordered Lund released on his own recognizance, or in effect paroled. The cases of eleven others were deemed to be similar to Lund's and they too were paroled on March 17. In the next six weeks, eleven more alien radicals were set free. One suspected radical escaped from custody, while four others were discharged outright and one was found to be an American citizen.

For those Red Special radicals still in custody at Ellis Island, attorneys Lowe and Recht pursued a round of habeas corpus writs to free the detainees. Judge Augustus Hand ruled in the case of Sam Nelson that he could only find that the detainee believed in an "irreconcilable conflict between employer and employee." This was not enough, in the eyes of Hand, to justify Nelson's detention or deportation. Using Nelson's precedent, more Red Special detainees were paroled. Later in June, Judge Hand ruled on seven more cases, allowing the deportation of six men while freeing one: Ellis Island's poet laureate, E. E. McDonald.

Martin de Wal was one of the unlucky ones whom Judge Hand ruled against. After three months of the tedium of detention, de Wal sent a letter to the editor of The Survey asking readers to send books and other reading material to them. By June, de Wal again wrote to thank readers for the apparently large number of books and pamphlets they had sent, although de Wal noted that they could not tell how many books had actually been sent, since officials at Ellis Island withheld material so as "not to spoil our morals further by allowing us radical or truthful books." Hopefully, de Wal had enough reading material, for he was to remain at Ellis Island until the end of September.

As de Wal and his colleagues whiled away their time in detention, more than thirty bombs were being mailed to prominent Americans like J. P. Morgan and John D. Rockefeller. Postal officials intercepted most of them, but one package they missed arrived at the home of former Georgia senator Thomas Hardwick where it exploded and blew off the hands of Hardwick's maid. In June, another bomb exploded in front of Attorney General A. Mitchell Palmer's Washington home, damaging the house and killing the man who planted the bomb. To Americans, these seemed like dangerous times.

Meanwhile, despite the hoopla surrounding the big roundup of radicals from the Red Special, only nine of the detainees had actually been deported. Most of those taken from Seattle to Ellis Island were eventually released on parole. The big Red roundup had actually been a bust.

In the middle of this was Fred Howe, a public servant with impeccably bad timing. Not only were suspected radicals being released from his custody, but just a few days before the bomb exploded in front of Palmer's home, Howe had presided over a Justice to Russia rally at New York's Madison Square Garden. His presence attracted the attention of Senator William King of Utah, who demanded that Howe be fired. "I don't think a man who has sanctioned Bolshevism, as he did by presiding at that meeting, is fit to remain in office," King said. "If there is any hint of Bolshevism at Ellis Island, through which the immigrants of the world pour into the United States, it must be wiped out." Howe was unapologetic and denied that the meeting was "pro-Bolshevist" or "pro-Soviet." Yet a *Times* article claimed that participants cheered for Trotsky and Lenin, while booing the mention of Woodrow Wilson's name.

Howe was not a Communist, but a Labor Department report showed that he had been solicitous of the comfort of the Red Special detainees. When the radicals complained that they had to get up at six thirty in the morning, but could not eat breakfast until eight thirty, Howe ordered that their mandatory wake-up be postponed closer to breakfast. Howe also allowed detainees to receive such IWW periodicals as *The Rebel Worker* and *The Red Dawn*.

The attacks on Howe also came from an unexpected source: Fiorello La Guardia. The former Ellis Island translator, who had recently returned to his House seat after serving in the army during the Great War, lashed out at Howe on the floor of Congress. La Guardia was a liberal and sympathetic to the plight of immigrants at Ellis Island, but was disgusted enough to condemn Howe as a radical and complain that he had allowed anarchist literature to be available to detainees. While Senator King proposed impeaching Howe, La Guardia merely wanted to cut his pay by 50 percent.

In the midst of the criticism, Howe resigned in September 1919. He was mostly guilty of a political tin ear, a victim of political naïveté and poor judgment. The irony is that despite his sympathy for the radicals, one of the intercepted explosive packages sent by anarchists in the spring of 1919 was addressed to Howe, perhaps because he was technically in charge of the detention of IWW radicals.

This did not prevent Congress from initiating three days of hearings in November 1919 at Ellis Island to look into charges that Howe's administration was lax, especially regarding suspected radical detainees. Howe had already weathered one congressional hearing dealing with his alleged lenient treatment of alien prostitutes.

During this second hearing, Congressman John Box of Texas called the inspection of immigrants at Ellis Island a farce, a characterization that Howe's deputy, Byron Uhl, did not dispute, saying that it had become "largely a matter of checking names." The committee chairman, Albert Johnson, asked Uhl whether it was Howe's desire to turn Ellis Island into "a place of individual government, letting everyone do as he pleased." Uhl had been fairly taciturn in his responses, but answered that that had been his impression. In addition, he admitted that nearly all employees at Ellis Island were of the opinion that Howe's policies were "utterly improper." Uhl admitted that under Howe each detainee at Ellis Island could just about do as he or she pleased.

The committee also released at letter from anarchist Emma Goldman to Howe in 1915, addressed to "My Dear Fred." Critics argued that the letter implied a friendship between the two, yet another piece of evidence that Howe was soft on radicals.

Howe was present during the hearings at Ellis Island, but was not on the witness list. At a number of points during proceedings, he tried to answer charges but was silenced by the chairman. Later, Howe made his case to the press outside the hearing, explaining that he had never released anyone from Ellis Island without the explicit order from the Labor Department. In a literal sense, what Howe said was true. The decision to parole or release detained radicals was made by his superiors, but it was Howe's intercession that stalled the proceedings and allowed the radicals a second chance to make their case to Washington.

Back inside the hearing, the congressmen seemed particularly bothered that not only were the Red Special detainees, as well as others held at Ellis Island, released on their own recognizance, but the government had no idea where they were. "Whereabouts now unknown," was the phrase that attached itself to name after name of suspected radicals. In the course of the hearings, it came out that 697 warrants of arrest had been issued for the deportation of suspected radicals between February 1917 and November 1919. Of that number, only 60 had actually been deported.

The press had a field day with the revelations. The most colorful, if overwrought, description came from the *Cleveland News*, which described Ellis Island as a "government institution turned into a Socialist hall, a spouting ground for Red revolutionists . . . a place of deceit and sham to which foreign mischief-makers are sent temporarily to make the public think the Government is courageously deporting them." The *New York World* complained that Ellis Island was in danger of becoming a "perpetual joke," where a workforce of guards consisting of "one-legged, one-armed or decrepit old men" was in danger of losing control to anarchists.

The case of the Red Special detainees was a false start in the government's battle against suspected alien radicals. The next round of arrests and deportations, which were already underway during the Howe hearings, would be much different.

The next series of roundups had their genesis on the desk of A. Mitchell Palmer, the attorney general. But there was a problem: the power of deportation lay not with the Department of Justice but with the Department of Labor. William B. Wilson, who headed the newly created department, reminded Palmer of this fact in a letter, temporarily derailing Palmer's crusade. But Wilson had become increasingly disengaged from his job and was in no position for bureaucratic infighting. His wife had recently suffered a stroke, so he took an extended leave from his job to care for her. Adding to his burdens,

Wilson himself fell sick and was rarely seen in his office throughout most of 1919.

With Secretary Wilson turning over effective control of his department to subordinates, Palmer saw an opportunity. Commissioner-General Caminetti had already shown that he was committed to the idea of rounding up alien radicals. In Secretary Wilson's absence, Caminetti made an end run around his superiors and worked directly with Palmer and the Justice Department. His liaison was the twenty-four-year-old head of the General Intelligence Division, J. Edgar Hoover, whom a congressman referred to as a "slender bundle of high-charged electric wire." A direct phone line to Hoover's Washington office would be installed at Ellis Island.

If the earlier roundups of suspected radicals consisted of mostly obscure figures from the West Coast, the main targets of the fall 1919 campaign were the country's most notorious radicals: Emma Goldman and her former lover Alexander Berkman. Goldman was a notorious nonconformist known for her fiery rhetoric and anarchist beliefs. The U.S. attorney Francis G. Caffey referred to her as a "continual disturber of the peace." Berkman's claim to fame was his attempted murder of industrialist Henry Clay Frick, for which he served fourteen years in jail.

Both Berkman and Goldman had been born in Russia. Goldman had arrived at Castle Garden in 1886. Berkman was not a citizen, but Goldman claimed citizenship via a brief 1887 marriage to Jacob Kershner, a naturalized Russian immigrant. Goldman's citizenship should have left her immune from deportation, but that was not to be.

Beginning in 1907, immigration officials began to monitor Goldman closely. Commerce and Labor secretary Oscar Straus began going after anarchists, in part to compensate for the criticism he was receiving from restrictionists. "There is no doubt about it that Emma Goldman, who is a woman of the French Revolution type, is dangerous by reason of her incendiary ability," Straus wrote in his diary. Secret Service agents monitored Goldman's public speeches.

For two years, Straus vacillated on the Goldman case. At one point, he ordered Robert Watchorn to take her into custody at Ellis Island for an administrative hearing. Yet that never happened. Straus claimed that Goldman's speeches were "very skillfully worded so as not to be actionable." He argued that although she was an anarchist, arresting her would only add to her prestige among radicals.

As Straus continued to debate action against Goldman, a federal judge revoked her ex-husband's citizenship as fraudulent. It was a peculiar move. By 1909, Kershner was dead and it was not readily apparent why the government thought it necessary to pull the citizenship from a corpse. The move was not really about Kershner, who had been little more than a poor factory worker. The real target was Emma Goldman. In revoking Kershner's citizenship, the government also revoked Goldman's. By this dubious legal move, Goldman was now subject to deportation under the immigration law.

This did not put Goldman in immediate jeopardy, although she was more than capable of getting into trouble on her own. Before World War I, she was arrested for lecturing on birth control. Her real problems began after the United States entered the war, as officials continued to monitor her speeches for criticisms of the war effort. In 1917, Goldman and Berkman were arrested under the Espionage Act for speaking out against the draft. They were sentenced to two years in prison.

As Julius Goldman was about to find out, the mere attendance at an Emma Goldman speech could place one in legal jeopardy. No relation to Emma, Julius was a nineteen-year-old deli clerk on Manhattan's Lower East Side. He had been in the country since 1913. One night after seeing a movie, he walked down East Broadway for dinner when he saw a large crowd at Forward Hall, the headquarters of the city's Yiddish-language, socialist paper. Emma Goldman and Alexander Berkman were speaking. After their speech, all men in the audience were stopped by police and asked to show their draft registration cards. Having been caught up in the crowd, Julius Goldman was interrogated by police. Was he an anarchist, one policeman asked? More questions followed: "Do you believe in the overthrow of law and government by force? Do you believe in organized government? Do you believe in free love?" Because he had admitted to being an anarchist and since he was a nonnaturalized immigrant, Julius was sent to Ellis Island.

Officials quickly realized that Julius was hardly a bomb thrower. His lawyer argued that Julius's appearance "does not stamp him as one who has been given over to too much study." He had simply wandered into the meeting and mistakenly said he was an anarchist out of fear. Caminetti called Julius a "somewhat unsophisticated lad" with no knowledge of anarchism and he was released on bond. Julius had come to find that even a random association with Emma Goldman could be dangerous to one's liberty. A short time later, with government officials and policemen monitoring their every utterance, Emma Goldman and Alexander Berkman were arrested and convicted of obstructing the draft by speaking out against the war. The two were each sentenced to two years in jail.

When released from jail in September 1919, Goldman, stripped of her citizenship for a decade, knew that deportation was a possibility. She was ordered to appear at Ellis Island for a hearing on October 27 to answer charges that she was actively advocating anarchy and the violent overthrow of the government. At the hearing, Goldman asserted her citizenship, going so far as to state that her name was Emma Goldman Kershner. She submitted a long statement for the record, denouncing the "star chamber hearing," and then proceeded to refuse to answer most of the questions officials put to her. To question after question, Goldman responded: "I refuse to answer." A subsequent hearing in November produced much the same result, and officials recommended deportation.

Goldman and Berkman were asked to arrive at Ellis Island on December 5 to await their imminent deportation to Russia. There they joined eighty-eight other suspected radical aliens. For more than two weeks, Goldman and Berkman would remain in detention, to be joined by more radicals rounded up by the government. After thirty-three years here, Ellis Island was to be Emma Goldman's last home in America.

Detained at Ellis Island alongside Goldman and Berkman was Joseph Poluleck, who had already been there for almost a month. While Goldman was famous or infamous, Poluleck was an anonymous figure. A packer at the American-European Distributing Company on the Lower East Side, he had arrived in America from Russia six years earlier. He was arrested in early November while attending math classes at the People's House night school run by the Union of Russian Workers, one of the radical organizations targeted by government officials.

At his hearing, Poluleck adamantly denied being an anarchist and claimed to like the United States and support the country. "There is not a word of truth in the charges," he told immigration officials, "I am not an anarchist and I am not affiliated with any organization of that kind." He had only been taking classes at the People's House since

September and the only organization he belonged to was the Methodist Episcopal Church.

The case against Poluleck was weak. Even Byron Uhl admitted there was no evidence to substantiate the main charges against him. The government's case rested on the fact that each student at the school received a book from the Union of Russian Workers, which implied membership in the organization. Though Labor Department officials had declared earlier that mere membership in a radical organization was not grounds for deportation, by late 1919, the Justice Department reversed the policy and Poluleck was ordered deported.

Meanwhile, Goldman called the conditions at Ellis Island "frightful" and argued that little had changed in the treatment of immigrants since she had arrived at Castle Garden more than thirty years earlier. While in detention, Goldman suffered an attack of neuralgia, a painful condition affecting her jaw and teeth. The Ellis Island doctor could not help with the pain and, as she later put it, for "forty-eight hours, my teeth became a federal issue." Eventually, officials allowed her to visit a dentist in New York, accompanied by a male guard and female matron. Goldman called her ailment "very timely," since the visit allowed her friends a chance to visit her. At Ellis Island, detainees were allowed only occasional visits conducted behind screens and with the oversight of guards.

Apart from the minidrama with Goldman's dental pain, there was little else for detainees at Ellis Island to do but wait for their day of deportation, which was kept secret from them. To pass the time, Goldman did something she was especially good at. She wrote. Most of her efforts were directed toward a pamphlet she was writing with Berkman entitled "Deportation: Its Meaning and Menace," further subtitled, accurately but melodramatically as the "Last Message to the People of America by Alexander Berkman and Emma Goldman."

Afraid that officials would confiscate their material, they wrote in their cells at night while their roommates kept watch for guards. On their morning walks, the two would discuss the material and trade suggestions for the next night's writing. The pamphlet included an introduction by fellow radical and political cartoonist Robert Minor, who called the impending deportation, the first effort of "the War Millionaires to crush the soul of America and insure the safety of the dollars they have looted over the graves of Europe." A mixture of melodrama, grandiosity, and conspiratorial history pervaded the pamphlet. Goldman and Berkman saw their tribulations as nothing less than another form of czarism. "Now reaction is in full swing," they wrote. "Liberty is dead, and white terror on top dominates the country. Free speech is a thing of the past."

While Goldman was angry at her detention, she was especially saddened to find out that Assistant Secretary of Labor Louis Post had signed her order of deportation. The seventy-year-old Post had been a noted liberal journalist and possessed none of the traditional starchy appearance of most public men of the time. With his thick, unkempt head of hair, bushy, gray Van Dyke beard, and thin wire-rimmed glasses, at a quick glance Post looked a little like an American-born Trotsky. More philosopher than bureaucrat, he called himself a rational spiritualist and had been an early supporter of Henry George's single-tax theory, a plan popular with utopian thinkers disheartened by the vast accumulations of wealth in the industrial age.

Years earlier, Post had come to Goldman's defense when she was accused of involvement in President McKinley's assassination. Not only did he defend her in the pages of his magazine, but Goldman had also once been a guest in his house.

In his waning professional years, Post went to work in the Wilson administration. Like Howe, he was uncomfortable having to enforce laws that went against his beliefs. Coming to the Labor Department in 1914, Post had hoped to work on issues dealing with the condition of workers, but instead found that some 70 percent of the department's appropriations and more than 80 percent of its staff went toward enforcing the immigration laws. One of the nation's few advocates of an open-door policy for immigrants, Post had little interest in this work, which put him in a depressive mood for the rest of his tenure. "I found myself moving about in a cloud of gloom from the beginning to the end of my service in the Department of Labor," Post later wrote.

Post complained about the administrative nature of immigration law. While serving as assistant secretary, he published an article arguing that the exclusion or deportation of aliens "should not be determined finally by administrative decision." It was unusual for a serving political appointee to write in an academic journal criticizing policies he was bound to uphold, but Post had few good options.

Still in office in late 1919 and taking on more responsibility with the continuing absence of his boss, Labor Secretary Wilson, Post was faced with the cases of the radical detainees. The decision to deport Goldman rested in his hands. He spent a great deal of time contemplating her case and came to the conclusion that the only issue that mattered under the law was whether or not Goldman was an anarchist, not whether she had ever participated in revolutionary or violent actions. On that question, Post could only answer yes; the result had to be deportation. Post signed the order.

Post found that he had to enforce the law even if it clashed with his own beliefs. To do otherwise would be a violation of his oath of office and, as he wrote, "essentially repugnant to the developing democratic principles of our Republic." Such thinking did not impress Emma Goldman, who thought Post had another option open to him: resignation. Since he chose to remain in office and carry out the deportations, Goldman "felt that Post had covered himself in ignominy."

Post, however, could do something for Goldman. The deportation called for her to be brought back to Russia, where the civil war was raging. To send Goldman back to areas controlled by White Russians would have been a death sentence, so Post ordered her deported to Soviet-controlled Russia.

In the early morning hours of December 21, Emma Goldman was in her cell, which she shared with two other female detainees. She was doing what she had been doing for most of her detention: writing. At the sound of guards approaching their cell, Goldman hid her notes under her pillow and pretended to be asleep. The guards were there for another reason. The hour of deportation—that inevitable, yet carefully guarded secret—had finally arrived.

Collecting their things, the three women were marched into the Great Hall, where they joined 246 men, including Alexander Berkman, shivering in the cold. In a short time, the group would march single-file through the main building and outside to a waiting ferry that would take them on the first leg of their journey. Walking through the bitter air of an early December morning with snow covering the ground, the band of ragged, sleepy, and dispirited radicals made their way to the ferry under the watchful eyes of armed soldiers and a group of federal officials, including J. Edgar Hoover and Congressman Albert Johnson, chairman of the House Committee on Immigration and Naturalization. "Scores of cruel eyes staring us in the face," was Goldman's recollection of the event. As Goldman was boarding the ferry, someone yelled sarcastically: "Merry Christmas, Emma," to which the anarchist thumbed her nose.

Colorado congressman William Vaile was also on hand. He described the deportees as having "rather stupid faces" and being "degraded and brutalized men." Vaile believed that the deportations were perfectly justified. "Deportation is merely the act of ridding ourselves of foreigners who are not eligible for residence here under our laws," he wrote. Though the government could not expel citizens for holding anarchist views, he believed that "a nation has the right to refuse its privileges and protection to any class of aliens whom it may consider undesirable residents."

Vaile shared his cigarettes with a few of the deportees as they waited to board the ferry, but stopped after listening to their conversations, filled with a "bitter sneer." Disgusted with these radicals, Vaile was overwhelmed by feelings of loathing and decided that "the rest of my tobacco should go to Americans."

From the Ellis Island pier, the 249 deportees were first taken to Fort Wadsworth in Staten Island. Goldman and the other two female deportees were segregated from the men during the two-hour ferry ride. As the ferry passed the Statue of Liberty, it crossed paths with another ferry crowded with incoming immigrants headed for Ellis Island, who let out a cheer upon seeing the other boat, not realizing the destination of its passengers. Goldman, with her typewriter case beside her and holding a few sprigs of holly, engaged Hoover in conversation. America's time was coming to an end, she told him matter-of-factly. Just as a new day was dawning in Russia, Goldman believed, so too would revolution come to the United States.

It must have been an odd sight, with the middle-aged anarchist and the young federal agent engaged in political conversation. The thin veneer of civility between Goldman and the authorities was a sign of the anarchist's defeat. Goldman was still bitter at Hoover for not informing her lawyer about the deportation, and she let the young government official know it. "Haven't I given you a square deal, Miss Goldman?" a defensive Hoover responded. "Oh, I suppose you've given me as square a deal as you could," she replied. She could not refuse one final dig at her adversary: "We shouldn't expect from any person something beyond his capacity."

Upon arrival at Fort Wadsworth, the passengers were transferred to the *Buford*, a thirty-year-old army transport ship that had been in use during the Spanish-American War. Only 51 of the *Buford*'s passengers were deemed anarchists, including Berkman and Goldman. Some 184 of the deportees were members of the Federation of the Union of Russian Workers, a group designated as advocating the overthrow of the U.S. government. This included Joseph Poluleck, the Methodist whose major offense was that he took math classes at the wrong place. Finally, 9 of the passengers were excluded as likely to become a public charge, while 5 others had violated other parts of the immigration law.

The press was quick to give the *Buford* a new name, one that would stick throughout history: the Soviet Ark. The *Pittsburgh Post* called Goldman and the other passengers "the unholiest cargo that ever left our shores." Because of the supposedly dangerous nature of the *Buford*'s human cargo, the army provided a contingent of sixty-four soldiers and officers to provide protection and prevent a mutiny, joined by nine officials from the Immigration Service.

Goldman and the others elicited little sympathy from Americans. Contrary to what Goldman and Berkman wrote, their deportation did not signify the beginnings of czarism or the end of freedom in America. Rather it was one of the many big and small events that, when taken as a whole, helped break apart the national consensus on immigration and herald a new era when Ellis Island—and the immigrants who once streamed through its doors—were less relevant to America.

"One could not imagine a more quiet movement of so many people," Commissioner-General Caminetti reported the next day.

FROM THE BLACK TOM explosion to the deportation of Emma Goldman, Ellis Island found itself witness to the traumas of the Great War and its aftermath. The war was now over, but the debate over the power of exclusion, detention, and deportation remained.

A few years before Goldman was expelled, Justice Oliver Wendell Holmes succinctly summarized the government's view on deportation.

It is not a punishment, Holmes wrote, but instead "simply a refusal by the government to harbor persons whom it does not want."

The sailing of the Soviet Ark, which forever banished the country's number one anarchist, emboldened the Justice Department to make further arrests. While the *Buford* was still on the high seas, hundreds more suspected alien radicals were rounded up as part of the Palmer Raids and brought to Ellis Island for deportation, many of whom belonged to the Communist Party. At the Labor Department, Louis Post tried to rein in the Justice Department's excesses. With Secretary Wilson still ill, much of the burden fell on Post's shoulders. He did not save Emma Goldman, but now, at the end of his career and with little to lose, Post ordered the release of over two thousand suspected radicals across the country, although he did uphold the deportations of a few hundred individuals.

Post made enemies with his actions, not the least of whom was J. Edgar Hoover. The young Justice Department official had dug up an affidavit that Post had signed in 1904 in support of anarchist John Turner. In Hoover's files was a poem entitled "The Bully Bolshevik," which was "disrespectfully dedicated to 'Comrade' Louie Post." It is not clear whether Hoover wrote the ditty, but it certainly summed up his views:

> The 'Reds' at Ellis Island
> Are happy as can be
> For Comrade Post at Washington
> Is setting them all free.

The anger toward Post extended to Congress. Six months earlier it had been Fred Howe who was being grilled for his sympathy toward radicals. Now it was Post's turn. In May 1920, the House Rules Committee began impeachment hearings against him. By then, the Red Scare had petered out almost as quickly as it had begun. When Palmer's dire warnings of a May Day revolution failed to come true, the public lost interest in the crusade. Congress quietly dropped its proceedings against Post.

At the height of the Red Scare, between November 1919 and May 1920, warrants were issued for 6,350 aliens suspected of radical activity, leading to around 3,000 arrests. Of that number, only 762 were ordered deported and only 271 were actually deported, including the

249 who left on the *Buford*. In the year after May 1920, an additional 510 alien radicals were deported.

The roundup and deportation of alien radicals were merely a continuation of longstanding immigration policy. For years, immigrants safely landed in the United States were at risk of deportation if they were subsequently found to qualify under one of the categories of exclusion. Between 1910 and 1918, almost twenty-five thousand immigrants already residing in the United States found themselves rounded up by authorities and deported back to their homelands for various reasons. After World War I, the government focused its attention more closely on radical aliens, but the mechanism it used was largely the same as had been used to deport immigrants before the war.

While the deportation process that characterized the Red Scare had long been part of the immigration law and would be used for decades more to come, the emotions that fueled this particular spasm of antiradical sentiment quickly died out. In hindsight, this period was a disjointed blip, a hiccup of tension and conflict. To the American mind of 1919 and 1920, however, the world seemed ablaze with danger.

A global flu outbreak had erupted before the armistice and continued into 1919. The worldwide death toll has been estimated at anywhere from 20 million to as high as 100 million. Many in the United States referred to it as the Spanish flu, reinforcing the alien nature of the disease and the danger of foreign entanglements. Some one-quarter of all Americans came down with the flu, and 675,000 died in less than one year, including Randolph Bourne, who passed away in December 1918. To many Americans, war and pestilence seemed their grim reward for becoming a world power.

During 1919, Americans were on edge. Some 4 million workers across the country went out on nearly 2,600 strikes. Steelworkers, miners, even Boston policemen walked out on their jobs during that tumultuous year. The American Communist Party was formed that year. And it was not just the United States that was in turmoil: following the lead of the Russian Bolsheviks, Communist uprisings occurred in Bavaria and Hungary.

The Great War turned the world upside down and dashed the optimism of a generation. Modern civilizations tore each other up on the battlefield as new technologies like airplanes, machine guns, and poison gas made the traditional destruction of war that much worse.

The number of military dead was staggering: around 2 million Germans and Russians each, and around 1 million English, Austrians, and French each, not to mention the wounded, maimed, or shell-shocked. In a little over one year of war, America lost more than 115,000 men, with more than 200,000 wounded.

When the war ended, people on both sides of the Atlantic began to ask why and received few answers. The victorious Allies carved up the map and took their war booty, while Woodrow Wilson's romantic vision of a League of Nations that would end war forever would have to function without the participation of the United States, when the Senate failed to ratify the Treaty of Versailles. When Americans asked what the war had been for, some answered that it had been fought only to fatten the pocketbooks of big business.

The scars of war remained on the American psyche and disabused many of their positive feelings for government. For liberals, the disillusionment was even more pronounced. They were the ones who had built up the federal government, who hoped to use it to counteract the power of corporations and provide protections for workers and consumers. The government, run by educated, middle-class professionals, was supposed to rescue America from an orgy of commercialism and ignorance, but instead it bumbled into a bloody European war for no apparent reason, stirred up ethnic hatred at home, and used its new police powers to quash dissent.

No one felt this disillusionment more than Fred Howe. "I hated the new state that had arisen, hated its brutalities, its ignorance, its unpatriotic patriotism, that made profit from our sacrifices and used its power to suppress criticism of its acts," he wrote in his autobiography. The man who once argued that government should take control of public utilities now changed his tune. "I became distrustful of the state," he complained, "And I think I lost interest in it, just as did thousands of other persons . . . who were turned from love into fear of the state and all that it signified."

To Howe, the brutality of the state was on display at Ellis Island. A few weeks after the *Buford* left New York Harbor, Howe penned a scathing critique of U.S. immigration laws, the same laws he had been sworn to carry out for five years. The article's title said it all: "Lynch Law and the Immigrant Alien." He condemned deportations as cruel and criticized the secret hearings held at Ellis Island to determine the fate of immigrants. He painted a dark picture of European immigrants living in a "state of panic" and "perpetual fear." He ominously pronounced: "We have made Americanization impossible." Of course, in retrospect Howe was wrong. The policies at Ellis Island and the Red Scare had few long-term effects on the attitudes of immigrants to their adopted country, but they certainly scarred Fred Howe.

His disillusionment can also be seen in his shift away from the idea of government control over business and utilities toward the idea of a cooperative "producers' state," where workers participated in the management and ownership of business. After leaving Ellis Island, Howe tried to put this idea into practice as the executive director of the Conference on Democratic Railroad Control.

When Howe left in September 1919, he was at the depth of despair. He had been condemned on the floor of the House of Representatives. He had survived one congressional investigation, and another one loomed. He despised his superiors and lost faith in his fellow citizens. He had begun work at Ellis Island hoping "to make it a playhouse for immigrants." When he left, he found it a prison for aliens deemed unworthy by the government, but it had also, as Wendell Phillips once said about slavery, "made a slave of the master no less than the slave."

Before leaving Ellis Island for the last time in the fall of 1919, Howe gathered up all of the personal papers that he had been saving to use for a book on his experiences there. Instead of taking them with him, he sent for a porter and the two men carried the materials to the island's engine room where they threw the papers into the flames.

Chapter 16

Quotas

———•—•———

AT EXACTLY MIDNIGHT ON JULY 1, 1923, THE STEAMSHIP
President Wilson rushed across an imaginary line that spanned the Narrows of New York Harbor. Thirty seconds later, the *Washington* crossed
that same line, which stretched from Fort Hamilton on the Brooklyn
side to Fort Wadsworth on the Staten Island side. Within six minutes,
a total of ten steamships had sailed past the line. One more ship would
slip across a few hours later.

Immigration officials stationed at the two forts duly noted the times
the ships crossed this line. When the mad midnight dash was over,
eleven ships had arrived at Ellis Island, containing over eleven thousand passengers seeking entry to the United States. By morning, immigration officials were busy processing the new arrivals.

To anyone awake at that midnight hour, the throng of massive
transatlantic steamers jockeying for position in the middle of the night

in New York Harbor must have been a sight to behold. Why were these ships waiting in the harbor for the tolling of the midnight hour? Why did immigration officials patrol an imaginary line along the Narrows in the middle of the night? And why did these ships race across that imaginary line and have their times recorded as if it were an Olympic track meet?

The exact time a steamship crossed that invisible line held the potential to change the lives of thousands of immigrants aboard those vessels and spoke to the dramatic turn in American immigration laws since the end of World War I. The postwar disillusionment meant that the old way of dealing with the regulation and processing of immigrants—sorting the desirable from the undesirable—was over.

Restrictionists had long thought the process at Ellis Island was too lax, while immigration defenders thought it too strict. Yet the little island kept the concerns of both groups in balance, allowing a generally free immigration while barring those few deemed undesirable. War disrupted that balance, and both sides lost faith that government could weed out undesirables while treating its guests with a modicum of respect. Summing up the nation's disillusionment, the *Saturday Evening Post* complained in 1921 that "the Department of Labor knows no more about immigration than it knows about the habits of the viviparous blenny or the gambling systems in use at Monte Carlo."

Though the hysteria of the Red Scare had subsided, economic concerns deepened. The United States had entered a severe postwar recession. With some 2 million Americans out of work—many of them returning soldiers—the prospect of a postwar revival of European immigration was troubling. While four years of war had drastically reduced the number of immigrants, more than 430,000 people arrived between July 1919 and June 1920, and almost double that number would arrive in the following twelve months.

Americans feared that was just the tip of the iceberg. When they looked to Europe, they saw a continent teeming with people living amid the rubble and destruction of war. To those poor souls, America looked more and more attractive. Anthony Caminetti investigated conditions in Europe in late 1920 and reported back that some 25 million Europeans were ready to emigrate. Steamship officials told immigration authorities that some 15 million Europeans were "vociferously demanding immediate passage." Lothrop Stoddard, author of *The*

Rising Tide of Color Against White World-Supremacy, feared as many as 20 million.

"The influx of aliens will be limited only by the capacity of the steamships," a *New York Times* editorial warned of this potential deluge of war-displaced Europeans. "Our equipment for handling the alien flood, meanwhile, has pitiably broken down. . . . Ellis Island is a chaos."

This was all too much for Albert Johnson, the chairman of the House Committee on Immigration and Naturalization. Johnson returned from another visit to Ellis Island in November 1920 and announced that what he found there was so bad that he was sure "the country does not realize the menace of immigration." He promised that on the first day of the new session of Congress he would offer a bill to restrict immigration.

That is exactly what he did. At first, Johnson pushed for a two-year suspension of immigration, but his colleagues could only be convinced to support a one-year moratorium. Had the legislation passed, it would have marked the first time in American history that the gates of the nation were closed completely. Suspending immigration was a tactic that not even Henry Cabot Lodge or Prescott Hall had ever suggested in their darkest, most pessimistic moods.

The plan went nowhere. In the Senate, William Dillingham, former chairman of the U.S. Immigration Commission, had other ideas. He resurrected a plan that emanated from his 1911 report: institute a quota on new immigrants of 5 percent of the number of foreign-born for each nationality in the United States as counted by the 1910 Census. The plan would also impose a limit of six hundred thousand immigrants per year, well above the wartime figures but half the number that had arrived in the boom years of 1905–1907 and 1913–1914. The House dropped its immigration moratorium plan and signed on to the Senate's efforts, although Johnson and his allies managed to shrink the quota down to 3 percent and lower the overall ceiling.

The bill came to the desk of Woodrow Wilson for signature in his final days in office in 1921. His body withered by a stroke and his soul embittered by the failure of the Senate to accept his beloved League of Nations, Wilson did not act on the bill, thereby effecting a pocket veto. No public reason was given.

Congressman Johnson was not finished. A new president, more sympathetic to immigration restriction, was about to enter the White House. Less than two months after Wilson's pocket veto, President Warren Harding signed a nearly identical bill. More surprising than the drastic change in policy was its relatively uncontroversial nature. The bill passed the Senate with only one negative vote, and it passed in the House with only thirty-three nays. Ethnic groups opposed the measure, but their arguments found little traction in those unsettled postwar years.

As Congress moved rapidly toward restriction in the spring of 1921, Prescott Hall lay ill in his bed in Brookline, Massachusetts. He had devoted the previous twenty-eight years of his life to the ideal of an Anglo-Saxon nation. The sickly Hall used the one weapon at his disposal—his pen—to rail against undesirable immigrants from southern and eastern Europe and in favor of the literacy test. The Immigration Restriction League actually had its own version of immigration quotas introduced into Congress in 1918. The organization admitted that its goal was to "discriminate in favor of immigrants from Northern and Western Europe, thus securing for this country aliens of kindred and homogeneous racial stocks." That bill went nowhere.

Hall lived long enough to see Congress pass the new quota law, then passed away that May at the age of fifty-two. Joseph Lee, the Boston reformer and IRL member, eulogized Hall in the *Boston Herald*. "Mr. Hall's work was unknown, unpaid, unrecognized," Lee wrote, noting that without Hall, "the gates would have still been unguarded."

The new law setting quotas by nationality went into effect at the end of June 1921 and limited immigration to a total of 355,000 quota immigrants per year. (Immigrant children and wives of American citizens, naturalized or native-born, could enter outside of the quotas.) The bill was passed as a one-year measure, but Congress would reauthorize the legislation for 1922 and 1923 as well.

The quotas severely restricted immigration from eastern and southern Europe; only 43 percent of immigrant slots were allotted to those regions. On a country-by-country basis, the effect of the quotas was even more startling. Although 296,414 Italians came to America in 1914, the last year in the prewar immigration boom, under the new quotas only 40,294 would be allowed to enter. In addition, no more than 20 percent of a nation's yearly quota could be filled in any given month. That meant that the yearly quota for most nations would be filled in the first five months of the fiscal year.

If one of those ships on the night of June 30, 1923, had passed the imaginary line before midnight, it would have been marked as having entered in June 1923, the final month of the fiscal year, and all of its passengers would have been counted toward that year's quota, which by then had most certainly been filled. Such a miscalculation, even by one minute, would mean that most of those immigrants would be barred from entry and sent back to Europe. The steamship race across the Narrows would be repeated at midnight on the first of the month for the next few months.

What had caused this drastic change in immigration policy? America's unhappy experience in World War I helped turn the nation inward and soured its citizens. By 1920, Europe meant destruction, disease, and pointless ethnic conflict, and Americans sought once again to use the Atlantic Ocean as a barrier to the wretched influence of decayed Europe.

The link between immigration and radicalism further poisoned American attitudes—formerly ambivalent, yet relatively open—toward immigration. The fear of alien radicals caused many in the business community, usually in the forefront of the pro-immigration lobby, to acquiesce to the new restrictive legislation.

A major backbone of pro-immigrant sentiment had been the German-American community, which never fully recovered from the suspicions brought on by the Great War. In 1910, there had been 634 German-language newspapers in the country; by 1920, that number was down to 276.

The National German-American Alliance, one of the largest German-American organizations in the country, had been a staunch supporter of immigration and opponent of restriction. The organization—and especially two of its leaders, Henry Weismann and Alphonse Koelble—had been a fierce critic of William Williams. The Great War destroyed the NGAA. By 1916, Weismann and Koelble were charged with trying to set up an office in Washington to lobby on behalf of the German government. By 1918, Congress voted to revoke the charter of the NGAA. The cumulative effect was that the strongest, loudest, and most fearless pro-immigration voice in the country was now eager to prove its "100 percent Americanism" and would never fully regain that voice.

The growing popularity of eugenics also contributed to the success of the quotas. After the war, Prescott Hall called immigration restriction "a species of segregation on a large scale, by which inferior stocks can be prevented from diluting and supplanting good stocks." A number of eugenicists linked their work to immigration restriction. Congressman Johnson, a leading proponent of quotas, was deeply influenced by eugenics. Harry Laughlin, director of the Eugenics Record Office, served as a researcher for the House Committee on Immigration. However, as Stephen Jay Gould has noted, "Restriction was in the air, and would have occurred without scientific backing."

Madison Grant's *The Passing of the Great Race*, a paean to Nordic supremacy, was originally published in 1916 and received little notice. The early 1920s, however, provided a more welcoming environment for his views. Grant noted how the Great War seemed to shift public attitudes toward immigrants, since "Americans were forced to the realization that their country, instead of being a homogeneous whole, was a jumbled-up mass of undigested racial material." He also worried that immigration was affecting the national stature of Americans—literally. He complained that the Army had lowered its height requirement to allow the conscription of soldiers from "newly arrived races of small stature."

The fact that many immigrants and their children fought in the U.S. military was surely a positive sign of assimilation. For Grant, assimilation was a false god. This was one of the few areas where he agreed with proto-multiculturalists like Randolph Bourne. Grant mocked the famous war propaganda poster with Miss Liberty paying homage to an honor roll of names from Du Bois to Smith to Levy to Chriczanevicz. "Americans All!" shouted the poster, an idea that Grant found difficult swallow.

"These immigrants adopt the language of the native American; they wear his clothes; they steal his name; and they are beginning to take his women, but they seldom understand his ideals," Grant bemoaned. The problem was not the lack of assimilation, but rather that the melting pot was being "allowed to boil without control." He painted a bleak future where assimilation would "produce many amazing racial hybrids and some ethnic horrors that will be beyond the powers of future anthropologists to unravel." The question for Grant was: Was it too late?

Such views were not just isolated to cranky Manhattan snobs. The *Saturday Evening Post*, the nation's most widely read weekly magazine and best known for its Norman Rockwell covers that embodied Middle American values, became one of the leading voices of restriction. A 1921 article warned middle-class Americans that "immigration must be stopped. This is a matter of life and death for America."

America's postwar attitude toward immigrants had a substantial effect on the fortunes of immigrants such as the accused prostitute Giulietta Lamarca. The war had meant a reprieve from being returned to Europe, but since their deportation orders had never been rescinded, peacetime meant they were again vulnerable. The turmoil of war and the Red Scare briefly pushed these deportation cases into a bureaucratic black hole, but by 1921, as the American mood toward immigration grew darker, the government once again turned its attention to immigrants like Giulietta.

In the summer of 1921, officials reopened her case. Byron Uhl, the assistant commissioner of Ellis Island, noted that Giulietta had been living in open adultery in New Jersey for a few years, despite having a husband and child in Italy. This, coupled with the outstanding deportation order for prostitution from 1915, was enough to warrant another stay for Lamarca at Ellis Island.

This time, her boyfriend, Dana E. Robinson, the son of the Ellis Island doctor to whom Frederic Howe had paroled Giulietta in 1916, wrote officials to plead for mercy. He was very much in love with Giulietta (whom he referred to by her Americanized name, Juliette) and wanted to marry her. There had been no further charges against her in the last five years and Robinson found it "hard indeed to believe that the old charges are true as she has been under the careful and kind attention of my mother for the past three years." Despite her documented past and abandoned family in Italy, Robinson stated that his beloved Juliette was "as good a girl morally as any" and promised that their mutual love would keep them morally pure.

On the word of Robinson and his mother, Paula, Giulietta was once again released. It appeared that Howe had been correct that she could turn around her life and there appears no evidence that Giulietta had fallen back into a life of prostitution. But the happy ending that Frederic Howe, Dana Robinson, and many others had hoped for never materialized. Within three months, Paula Robinson wrote to the Labor Department. "I have to confess," she wrote in anguish, "that when I asked for clemency in the case of Juliette Lamarca I made the gravest mistake of my life."

It is hard to tell what went wrong in those few months, but something clearly did. According to Paula Robinson, Juliette threatened that neither the government nor Paula would "have anything further to say about what she does and that if the Government does anything to her, she will show them what she can do." Juliette vowed that if she were turned over to immigration officials, she would take her story to the newspapers and ruin the Robinson family by publicizing the fact that Dana was going to marry a former prostitute. She also threatened to have the Black Hand kill both mother and son if they turned her over to immigration authorities.

Did Juliette Lamarca finally have enough of the harassment of immigration officials and the threat of deportation that lingered over her head for five years? Was she merely exerting her independence from a meddling future mother-in-law? Or was she a scheming conniver who had latched onto a prosperous American fiancé and, once married, was going to kick her mother-in-law out of her house, as Paula Robinson feared? Juliette was clearly not a naïve woman, having seen the world from the brothels of Algiers and the Brooklyn docks. Perhaps her intentions were less than admirable, or perhaps she had just snapped under the pressure of such prolonged and intrusive scrutiny.

What we do know is that less than two weeks after receiving Mrs. Robinson's letter, immigration officials rescinded Juliette's stay of deportation. Four days later, she was taken to Ellis Island for the third time in five years, and on December 3, 1921, she was deported back to Italy.

HENRY H. CURRAN, THE new commissioner of Ellis Island, took office on July 1, 1923, the morning after the mad dash of steamers at midnight. A feisty and irreverent New York politician who had spent his adult life working in politics as an outnumbered Republican in a Democratic city, Curran had run for mayor in 1921, losing to his Democratic opponent by a margin of more than two to one. No wonder the reserved Calvin Coolidge found Curran "a little peppery."

His new job at Ellis Island seemed only slightly less quixotic than his mayoral campaign. When first approached for the job, Curran responded: "My God, but . . . that stuff is all over." He was correct that the best days of Ellis Island were behind it, but after witnessing the mad rush of steamships, Curran knew things were not entirely done.

Curran referred to Ellis Island as a "red-hot stove," something with which his predecessors would have agreed. The facilities, operations, and morale at Ellis Island were at their lowest since the days of the Mc-Sweeney-Powderly feud two decades earlier. Part of the problem rested with the weak administrative talents of Fred Howe, but the larger problem had to do with the wartime use of Ellis Island. Detaining German sailors and IWW radicals and housing wounded doughboys had taxed the island's infrastructure. With immigration at a near standstill, the workforce at Ellis Island was severely reduced in a cost-saving measure. Even after the war, the government showed little desire to spend more money on its operations.

"It was a poor place to be detained," Curran thought to himself when he began work. The waters surrounding the island were thick with sewage. Rats and mice made the buildings their home, and bedbugs nested in the sleeping quarters of the detainees. Curran's greatest reform was convincing Congress to appropriate money to replace the wire bunks, stacked three high with a stretch of canvas serving as a mattress, with real beds for the detainees.

There was little that Curran could do to silence the never-ending criticism of Ellis Island. In 1921, *The Outlook* magazine had called Ellis Island "one of the most efficient factories in the world for the production of hatred of America and American institutions." Another magazine warned that the "hatred that Ellis Island breeds is spreading like a plague to increase the discontent which menaces our institutions and the Government itself." Such criticism had been a constant since the facility opened, but by the early 1920s, the cries of one ethnic group in particular had reached a crescendo. While many ethnic and religious groups complained about poor treatment or exclusionary policies, British citizens had another grievance entirely.

Complaints by the British were not new. Back in 1903, a Protestant missionary working at Ellis Island told an investigative commission that the English had a reputation as proverbial "grumblers," although the missionary noted that most of the complaints centered on British detainees being forced to sleep with blankets that had been used by non-British foreigners. One of Ellis Island's most famous grumblers was the Reverend Sydney Herbert Bass, whose brief 1911 detainment made headlines.

Even Fred Howe noted that the British gave him the most trouble during the war years. When detained, an Englishman would rush to the telephone to complain to the British Embassy. When deported, "he sizzled in his wrath over the indignities he was subjected to." English citizens were indignant at being forced to endure inspection by immigration authorities. "All Englishmen seemed to assume that they had a right to go anywhere they liked," Howe remembered with some exasperation, "and that any interference with this right was an affront to the whole British Empire."

The British seemed especially perturbed by being forced to interact with other, seemingly inferior, immigrants. British subjects held at Ellis Island considered other immigrants to be foreigners and refused to sleep in the same room as them. Britain's undersecretary of state for foreign affairs, Roland McNeill, complained that the facilities at Ellis Island were basically for people "of a low standard of conduct" and a hardship for those of "any refinement, especially women."

A female British journalist named Ishbel Ross traveled through Ellis Island to report on conditions for the *New York Tribune*. She seemed quite animated by the prospect of mixing with the "steerage hordes," those poor immigrants who not only lacked the proper social graces, but who had also gone without a bath for a long time. "It must unquestionably shock immigrants of any degree of refinement to come into intimate and enforced contact with the strange assortment of humanity that seethes into the country through the gates of Ellis Island," Ross noted.

There had been a long litany of complaints by British subjects at their treatment at Ellis Island, but now the issue reached the British Parliament. Speakers there likened Ellis Island to "the Black Hole of Calcutta." As the *Literary Digest* put it: "Ellis Island a Red Rag to John Bull."

The British continued to argue that they were entitled to special privileges, including the right not to be mixed with uncouth and less cultured immigrants from southern and eastern Europe. But despite the ideas of Nordic and Anglo-Saxon superiority that floated through the air, most American officials had little compunction about subjecting the British to the immigration laws. To the Americans, most of the British aliens coming through Ellis Island were just that: aliens.

What the British wanted was to be segregated from others at Ellis Island. There was already some segregation by class at Ellis Island. While all detainees ate at common tables in the dining hall, sleeping accommodations were structured like steamships. First-class and second-class passengers, noted Ishbel Ross, possessed smaller rooms with fewer people; first-class passengers were even allowed to sleep in individual beds. Both classes received mattresses instead of canvas, with clean sheets and pillows with pillowcases. Detainees who arrived in steerage received more spartan accommodations.

Yet this was not enough for the British. In late 1922, the British ambassador, A. C. Geddes, made a tour of Ellis Island and reported his findings to Parliament. Contrary to some of the criticisms of his fellow Englishmen, Geddes's report was moderate in tone and sympathetic to the plight of immigration officials. Like many British critics, Geddes blamed other immigrants for much of the problem. "Many of the immigrants are innocent of the most rudimentary understanding of the meaning of the word 'clean,'" he reported. "If they were all accustomed to the same standards of personal cleanliness and consideration for their fellows, Ellis Island would know few real difficulties." This "pungent odor of unwashed humanity" mixed with more general odors to give Ellis Island a "flat, stale smell" that lingered with Geddes for thirty-six hours after he left.

"I should prefer imprisonment in Sing Sing to incarceration on Ellis Island awaiting deportation," wrote Geddes, clearly affected by what he had seen. He provided a list of suggested improvements, including fresh paint, better ventilation, and a thorough cleaning of the facility. Geddes thought Ellis Island was too small to handle large numbers of aliens. Rather than just build a new and larger facility, Geddes suggested a number of separate and smaller inspection stations for different classes of aliens.

It soon became clear just what kind of segregation Geddes had in mind. "After considering the matter with some care," Geddes concluded, "I have come to think that it might be feasible to divide the stream into its Jewish and non-Jewish parts." The report complained about Ellis Island doctors examining immigrants for veneral diseases.

"I saw one nice, clean-looking Irish boy examined immediately after a very unpleasant-looking individual who, I understood, came from some Eastern European district," Geddes reported. "The doctor's rubber gloves were with hardly a second's interval in contact with his private parts after having been soiled, in the surgical sense at least, by contact with those of the unpleasant-looking individual."

Curran dismissed the report and nothing came of its recommendations. When he arrived at Ellis Island, Curran was sympathetic to immigrants and proved willing to bend the rules on occasion. When a Hungarian girl was ordered deported because the quota had already been met, Curran noticed that she was carrying a violin and asked her to play. When she was done, Curran declared her an artist, a category that was exempt under the quotas, and she was allowed to enter.

Curran admitted that restricting immigration was the last thing on his mind when he took office, but he was soon arguing that America would be better off with fewer immigrants, or none at all—at least for a time. "Take again the intelligence, honesty and cleanliness of the average immigrant of today," Curran warned. "Those who have served at Ellis Island for thirty years and more will tell you that he is below his predecessor of a generation ago—far below, by all three counts." That would have been news to Americans in the 1890s who claimed that the immigration of *that* era was significantly inferior to what had arrived thirty years earlier.

Though this made Curran sound like William Williams, Curran's heart was not in the job of restricting immigrants. When he received another job offer, he dropped his position at Ellis Island "like a hot cake." "I have never seen such concentrated human sorrow and suffering as I saw at Ellis Island," Curran later wrote. "Three years were enough."

Congress had already reauthorized the 3 percent quota twice, but in 1924 it was ready for even stricter measures. Eventually, Congress agreed to a new quota of 2 percent of each foreign-born nationality based on the 1890 Census, with a ceiling for quota immigrants around 287,000. The rationale for using the 1890 Census instead of the 1910 Census was clear. There were far fewer Italians, Greeks, Poles, Jews, and Slavs in the country then. In fact, the new quotas meant that almost 85 percent of the quota allotments would go to northern Europeans. The Italian quota went from roughly 40,000 a year to 3,845; the Russian

342 / AMERICAN PASSAGE

quota from about 34,000 to just 2,248 and the Greek quota went from just over 3,000 to a negligible 100.

There were even more changes. Beginning in 1925, the inspection of immigrants moved from American ports to American consulates abroad. People who wanted to come to the United States sought permission at the nearest American consulate, whose officers were tasked with inspecting the individual and making sure he or she would make a desirable immigrant. Upon successful inspection and the payment of a fee, consular officials would grant the individual a visa.

It was now the responsibility of American consulate officials to make sure potential immigrants met the monthly quota, which was now reduced to 10 percent per month of the yearly quota. This eliminated the mad midnight dash of steamships across the Narrows.

The shifting of inspection to American consulates abroad was a measure sought for many years by Americans on both sides of the immigration debate. Senator William Chandler argued as far back as 1891 that consular inspections, far from the prying eyes of the press and immigrant-aid societies, would be stricter and conducted without the intervention of friends, relatives, and politicians seeking the immigrant's entry.

Fiorello La Guardia was also a proponent. Before his stint at Ellis Island he had served as a consular official in the port city of Fiume, where he conducted his own inspection of potential immigrants. Granting immigrants official permission to land *before* their transatlantic journey meant the end, with rare exceptions, of the heart-wrenching scenes of exclusion and deportation at Ellis Island and other ports. Immigrants who had sold all their property in order to come to America now possessed a visa that practically guaranteed their entry into the country. Of course, with the new stricter quotas, far fewer immigrants would actually experience that luxury and peace of mind.

Though La Guardia may have thought the new overseas inspection process was an improvement, he was no fan of the new quotas. The former Ellis Island interpreter was now representing a Manhattan district in the U.S. House of Representatives. With little actual power in Congress, La Guardia took on the role of gadfly, denouncing restrictive legislation and defending the contributions of immigrants. A child of immigrants, he condemned the quotas as being in the "spirit of the Ku Klux Klan."

These new quotas covered immigrants from Europe, Africa, Australia, and New Zealand. But nearly as many immigrants arrived from the Western Hemisphere, which was exempt from the quota system. Throughout the 1920s, 60 percent came from Canada and 30 percent came from Mexico.

Since the 1890s, more than 70 percent of immigrants entered through the Port of New York; throughout the 1920s, that number was about 50 percent. While the twenty-seven acres of Ellis Island served as the legal border for most immigrants, the new gate of entry became the nearly two-thousand-mile border with Mexico and the even longer border to the north with Canada. The future of American immigration, little grasped at the time, would not be with Europeans, but with those coming from south of the border.

Stricter quotas led to greater efforts to evade the new law. Illegal immigration began to attract the attention of the nation's leaders. In 1923, Labor Secretary James J. Davis warned President Harding that as many as one hundred thousand immigrants were crossing into the United States surreptitiously. Other reports, no doubt exaggerated, put the figure at a thousand a day. After taking office later that same year, President Calvin Coolidge warned the nation's governors of this "seepage over the borders," which he called a "considerable menace" to the success of the new immigration legislation.

Deportations also increased during the 1920s. From 1910 to 1918, an average of 2,750 immigrants were deported each year. By 1921, over 4,500 immigrants were deported annually, and by 1930, that figure had skyrocketed to 16,631, as the nation's mood increasingly soured toward immigrants. As more people were being stopped at the front door by quotas, still more were being kicked out the back door with stepped-up enforcement of the law.

By far the most important change brought by the new law would not go into effect for a few more years. Not happy with the near-complete exclusion of most southern and eastern Europeans, restrictionists saw a gross disparity in these quotas: they were based upon the *foreign-born* population. If the goal was to maintain America as an Anglo-Saxon nation, why not figure the quotas on the ethnic background of the entire population, both native- and foreign-born. In fact, the 2 percent quota based on the 1890 Census had actually reduced the quota on immigrants from the United Kingdom by more than half. The big winners of the 1924 quota law were midnineteenth-century immigrant groups such as the Irish and Germans.

To rectify the situation, Congress authorized a study to determine the precise ethnic makeup of all American citizens living in the country in 1920. The result was a so-called national origins plan. In keeping with the rigid racial boundaries of the era, the study included only white Americans and omitted blacks, Asians, and American Indians.

The commission calculated that by 1920, the United States was no longer a majority Anglo-Saxon nation, as more than 56 percent of the population was descended from non-British ancestors. An optimist like Henry Curran could defend the national origins plan for ensuring "that all future immigration will consist of the same racial proportions as are found in the stock of the hundred millions of us already here." For Madison Grant, however, the future was bleak: Americans of colonial descent were soon to "become as extinct as the Athenian of the age of Pericles and the Viking of the days of Rollo."

The new national origins plan lowered the overall immigration ceiling to 150,000 per year and granted immigrants from the United Kingdom almost half of the yearly quota. The big losers were the Germans, Irish, and Scandinavians, who saw their previous quotas cut by more than half. Ironically, although quotas were originally designed to bar southern and eastern Europeans, quotas for Italians, Greeks, and Russians all went up from the previous ones based on the 1890 Census, but their numbers were still pitifully low. Now, only 307 Greeks and 5,802 Italians would be allowed in each year.

On the surface, the quotas possessed a scientific precision that lent the endeavor the air of authenticity. Unlike the 1921 or 1924 quotas, the national origins plan would not be instituted without a fight. German-Americans, ten years removed from the harrowing effects of the war, began to speak up, as did Irish-Americans. One of those voices was a familiar one.

Edward F. McSweeney had resurrected his professional career and reputation after the imbroglio that led to his departure from Ellis Island in 1902 and the criminal charges for attempting to steal government documents. Now a respected citizen of Massachusetts, McSweeney served as chairman of the Knights of Columbus Historical Commission. The former union man and government official used his new post to call for a home-grown national history of the United States, untainted by what he felt was a creeping British bias in some histories. Anglo-Americans, argued McSweeney, were the real hyphenated Americans who overemphasized the contributions of the English to the exclusion of other groups. "What America needs most," he argued, "is the Americanization of most self-appointed Americanizers."

More substantively, McSweeney's group commissioned a number of books to counter overly pro-British histories, creating something called the Racial Contribution Series, whose monographs detailed the contributions of various racial, ethnic, and religious communities. One product of the series was W. E. B. Du Bois's *The Gift of Black Folk*, for which McSweeney wrote the introduction.

Foreshadowing a trend that would blossom decades later, ethnic groups were beginning to lay claim to their own Americanness, evolving into staunch patriots and defenders of a distinctly American history as they became more assimilated, while more established ethnic groups would often succumb to more critical attitudes toward American history and nationalism. McSweeney's work with the Knights of Columbus was a way to fight Anglo-Saxonism and immigration restrictionists with patriotic fervor.

To a pro-immigration Anglophobe like McSweeney, the whole national origins plan smelled fishy. In his mind, it was an un-American fraud perpetrated by Anglo-Americans. The data on national origins, in McSweeney's words, were an "impudent imposition . . . fabricated for a sinister purpose and are in truth discrimination."

Despite McSweeney's efforts, the National Origins Act went into effect in 1929. However, McSweeney never lived to see the implementation of a plan he believed violated America's traditional attitude of judging immigrants as individuals, not by their ethnic, religious, or national background.

In the late afternoon of November 16, 1928, McSweeney was driving home in Framingham, Massachusetts, when his car stalled at a railroad crossing in the face of an oncoming train. McSweeney's car was demolished by the train, which dragged it some sixty feet. Suffering serious head trauma and many broken bones, McSweeney was rushed to a hospital where he lingered for two days before succumbing to his injuries. He was sixty-three years old.

McSweeney had managed to outlive his former nemesis, Terence Powderly, by four years. He had rebuilt his life to such a degree that senators, congressmen, judges, and other dignitaries turned out for his funeral. In contrast, Powderly died in relative obscurity. He had gone from being the most famous labor leader of the late nineteenth century to an obscure government bureaucrat, a low-level functionary working within the Immigration Service that he once ran.

Powderly had once been a staunch restrictionist who opposed immigrant contract laborers and warned that immigrants posed a menace to the nation's health. By 1920, Powderly changed his tune. In his new position, he was concerned that government was neglecting the needs of immigrants. "We have admitted them as we have received baled hay, bars of pig iron and casks of olive oil," he wrote to his boss, "not a single throb of human sympathy has been extended to them and not a thing has been done to assure them of a welcome." By the time of his change of heart, it was too late. Powderly was little more than a powerless bureaucrat who needed his job for the paycheck that staved off poverty in his old age.

Freed from the burdens of petty political and labor squabbles, Powderly lived out his final years with little of the mental and emotional stress that plagued him in the past. He remarried, wrote his autobiography, and continued his work as an amateur photographer. He died in 1924 at the age of seventy-five.

The immigration work of McSweeney and Powderly belonged to another era. They both passed away during a time when the nation's immigration laws changed dramatically and Ellis Island, the site of their bitter feud a quarter century earlier, had gradually begun to fade in importance.

Powderly was not the only person to have second thoughts. Psychologist Henry Goddard, who coined the term "moron," had done much to buttress beliefs in the mental inferiority of immigrants. By the late 1920s, he had changed course and now believed that most individuals scoring below the mental age of twelve were not morons. Despite his lifetime of work on the subject, Goddard wrote in 1928 that psychologists were "still limited to a definition of feeble-mindedness that is unscientific and unsatisfactory."

Taking issue with supporters of eugenics, Goddard came to believe that feeblemindedness was curable and that environment played just as strong a role in intelligence as genes. In the late 1920s, he even concluded that there was not much evidence to show that feebleminded parents begat feebleminded children. Goddard had never personally been drawn to the racism that infected others associated with eugenics, but by the 1920s he would go so far as to write that the "distribution of intelligence in the different races is probably the same." By this time, immigration quotas were solidly in effect and Goddard's national influence had waned.

Unlike Goddard, University of Wisconsin sociologist Edward A. Ross had been much more heavily invested in the genetic inferiority of immigrants. He had earlier coined the term "race suicide" and complained that many new immigrants resembled prehistoric creatures and were "the descendants of those who always stayed behind." A proud Anglo-Saxon and defender of Nordic superiority, Ross was also a progressive who believed immigrants from southern and eastern Europe retarded the advancement of American civilization by bringing illiteracy, vice, and political corruption.

By the time he wrote his autobiography, Ross had moderated his views. He still professed a belief in eugenics and birth control and was proud that his writings had helped build support for the quota laws of the 1920s. Yet something happened to the man who had once penned articles such as "The Causes of Race Superiority" and "The Value Rank of the American People." Since then, Ross had traveled the world and softened his views toward non-Nordic cultures. A chastened Ross now declared: "Far behind me in a ditch lies the Nordic Myth. . . . Difference of race means far less to me now than it once did." He regretted that it took him more than two-thirds of his life to come to realize the "fallacy of rating peoples according to the grade of their culture."

In 1904, he had referred to eastern Europeans as "beaten members of beaten breeds." More than thirty years later, he recanted. "I rue this sneer," Ross admitted. The change of heart did nothing to change U.S. immigration quotas, but the newfound attitudes of Powderly, Goddard, and Ross foreshadowed the slow and steady abandonment of racialist thinking that would develop in the twentieth century.

NINE-YEAR-OLD EDOARDO CORSI AND his brother Giuseppe Garibaldi Corsi stood on the deck of the steamship *Florida* as it sailed into New York Harbor in November 1906. They were two of the over 1 million immigrants who would pass through Ellis Island in that record year.

Amid the excitement of the end of their journey, they thought they spied mountains rising out of the haze in the distance and wondered why their peaks were not topped by snow. Their stepfather corrected them. Those were not mountains, but the highest buildings in the world, he said pointing to the Manhattan skyline.

The Corsi family—two young sons, two sisters, mother, and stepfather—had arrived from the Abruzzi region of southern Italy. Adding to the sense of confusion brought on by those mysterious urban mountains, the Corsis felt an apprehension about what lay ahead of them at Ellis Island. Their acceptance into America was not assured, although Edoardo's stepfather had spent the rest of the family's money to buy his wife a second-class cabin ticket to ease her entry. "I felt a resentment toward this Ellis Island ahead of us," Corsi later reminisced.

The child who thought the Manhattan skyline was a mountain range would make his adult life within those urban mountains. Edward Corsi became active in the settlement house movement in New York and a progressive Republican in the mold of his congressman, Fiorello La Guardia. His political connections eventually led him to be named commissioner of Ellis Island in 1931 by Herbert Hoover.

Corsi was not the first foreign-born commissioner, but he was the first to have entered through Ellis Island. He presided over a much-diminished station. It had once attracted the attention and ire of many Americans. Presidents had visited the island for an up-close view of its operations. Restrictionists thought the system was too lenient; immigration defenders thought it too strict.

Those days were over. As America became mired in the Great Depression, Ellis Island slipped into the far recesses of the collective American mind. "Only occasionally now does this most famous of national gateways appear in the news," the *Literary Digest* noted in 1934. When Ellis Island was mentioned, it was often in highly negative tones. A 1934 report commissioned by Labor Secretary Frances Perkins began its findings by noting the popular myth that Ellis Island had been a place of misery, "a dungeon from which the immigrant is lucky to escape."

The 1930s would represent a low point in U.S. immigration history. The island's welcoming role continued to shrink, while its more punitive side increased. "An important consequence of restriction has been to make Ellis Island as much an emigrant as an immigrant station," one newspaper noted. "One may even say that its major activities now are concerned with deportation since of course to slam the front door is to challenge entrance through the back."

The combination of restrictive quotas and economic distress meant that by 1932, three times as many people left the United States as came to it. In the following year, only 23,068 individuals made the decision to come to immigrate, the smallest number since 1831. Ellis Island had given up its decades-long role as a "proper sieve" to inspect immigrants. By the 1930s, Corsi noted with more than a touch of sadness, "deportation was the big business at Ellis Island."

With fewer immigrants to process and no longer the nation's primary gate for inspection, Ellis Island increasingly reverted to a role that it had played sporadically in its history: a prison for unwanted aliens. Much would change in the coming years. World War II and the Cold War would highlight the dangers that existed in the world. As Americans concerned themselves with fighting those threats abroad, they also began looking to threats on the home front. The nation's immigration laws became increasingly entangled with national security concerns. Once again, Ellis Island would find itself at the center of controversy.

Chapter 17

Prison

———•◦•———

*I would never go back to Ellis Island. I spent too much time
facing the back of the Statue of Liberty. I always felt that even
though she had welcomed immigrants promising the American
dream, she turned her back on us just because of our ancestry.*
    —Eberhard Fuhr, German enemy alien detainee

*Government counsel ingeniously argued that Ellis Island is
his "refuge" whence he [Mezei] is free to take leave in any
direction except west. That might mean freedom, if only he
were an amphibian.*
    —Justice Robert Jackson, *Shaughnessy v. Mezei*, 1953

"HERZLICH WILLKOMMEN! HEIL." THOSE WORDS ON A
large poster greeted visitors to Room 206 at Ellis Island in 1942. This
was the headquarters of a small clique of pro-Nazi German nationals
who had been detained by the U.S. government as enemy aliens. Even
before the United States entered the war, the administration of Franklin D. Roosevelt was drawing up lists of suspicious aliens to be arrested
and detained if and when the country joined the war effort against the
Axis powers. J. Edgar Hoover's Federal Bureau of Investigation spent
a great deal of time between 1939 and 1941 collecting information on
noncitizens living in the United States who were suspected of sympa-
thizing with Nazi Germany or Fascist Italy. In October 1941, the attorney general warned officials at Ellis Island to prepare for an avalanche
of wartime detainees.

Hoover had run into bureaucratic difficulties during the Red Scare because the power to detain and deport aliens resided in the Labor Department. Now he would have no such problem. The Immigration Service had been moved to the Justice Department in 1940. Immigration was now officially a law enforcement issue.

On December 8, 1941, as the nation was reeling from the previous day's surprise attack on Pearl Harbor, Major Lemuel Schofield, head of the Immigration and Naturalization Service (INS), wrote to Hoover with a list of individuals "considered for custodial detention" because of their views about Germany and Italy. This information gathering had begun before either of these countries had actually been declared enemies of the United States.

More disturbing still, Schofield's list included "American citizens sympathetic to Germany" and "American citizens sympathetic to Italy." In all, over four thousand individuals were under consideration for detention.

Shortly after Pearl Harbor, Roosevelt issued three presidential proclamations declaring nonnaturalized Japanese, Germans, and Italians living in the United States to be enemy aliens. The proclamation against Japanese civilians was issued on December 7; the other two were issued on December 8, 1941, three days before the United States was technically at war with Germany and Italy.

The government wasted little time in rounding up alleged enemy aliens. On December 8, the attorney general ordered Hoover to immediately arrest "alien enemies who are natives, citizens, denizens or subjects of Germany." They were to be arrested and delivered to the INS for detention. Hoover's FBI moved at lightning speed. On December 9, 1941, working off the lists it had been compiling for the past two years, FBI agents arrested and detained 497 Germans, 83 Italians, and 1,912 Japanese enemy aliens. The following day saw more than 2,200 additional arrests. Some of these individuals would be quickly released, but a month later the government was holding nearly 2,700 enemy aliens in facilities across the country.

Some of the internees had belonged to organizations like the German-American Bund. Others made comments, whether to neighbors or in letters to the editor, opposing America's entry into the war. Informants would report to the FBI if they noticed a picture of Hitler in the home of German-Americans or if they overheard comments favorable to the Nazis or opposing the Allies.

This internment of enemy aliens was distinct from the relocation and internment of Japanese and Japanese-Americans on the West Coast, which began in February 1942. Under FDR's Executive Order 9066, certain zones in the United States could be designated as military areas, off limits to any or all unauthorized personnel. Later that spring, military officials ordered everyone of Japanese ancestry who resided on the West Coast moved to camps in the nation's interior. This was accomplished by a new agency called the War Relocation Authority. Unlike the military relocation and internment of Japanese-Americans, enemy aliens were rounded up under the auspices of the INS.

A large number of enemy aliens were initially detained at Ellis Island. Four days after Pearl Harbor, 413 German enemy aliens found themselves in detention at Ellis Island. "For the time being," the *New York Times* wrote of Ellis Island's new role, "New York has a concentration camp of its own."

The Office of Strategic Services (OSS), the nation's newly formed wartime intelligence agency, took an interest in the detainees at Ellis Island. In the summer of 1942, it placed an undercover agent there for three weeks. When the unnamed agent filed his report, he told his superiors of a large chink in America's security. "Ellis Island is undoubtedly a major information spot for the Axis, both for getting it and sending it," the agent wrote. "There is every reason to suppose that they regard Ellis Island as an important transmission center."

The OSS report described a tightly organized and disciplined "Nazi clique" among some detainees at Ellis Island. Their informal headquarters was Room 206. They sang the "Horst Wessel Lied" and other Nazi songs and plastered their rooms with drawings and articles mocking the American war effort. "They act as though it were inevitable that Germany win this war," the report noted. The Nazi sympathizers who congregated around Room 206 "can carry on effective propaganda and intimidate the weak."

Were these few hundred Germans, Italians, and Japanese held at Ellis Island in the summer of 1942 a major threat to the American war effort? The OSS agent certainly thought so, believing that it "would be strange, indeed, if such well-organized and fanatical Hitlerites only carry on harmless activities. The chances for conspiracy are practically limitless." He argued that German detainees kept watch on the shipping activity on the docks of New Jersey and reported this information back to Germany. Yet even the OSS agent had to admit that this was largely speculation and that in his three weeks among the detainees, he had found "no actual instance of this happening."

By the fall of 1942, FBI director J. Edgar Hoover was hearing gossip about this OSS report and demanded that an underling get a copy immediately. What angered Hoover was not the far-fetched claims that Nazis were operating an intelligence gathering operation for the Third Reich from Ellis Island. What really concerned him was that the report criticized, in Hoover's words, "the incompetent and venal custodial practices at Ellis Island." He wanted all such talk of lax security immediately "scotched."

Hoover was right. The OSS report was absolutely blistering in its depiction of the guards. "The system of supervision and control is inadequate to cope with experienced conspirators," the report concluded. The guards were "unpolitical and unobservant." Most were only interested in their weekly paychecks, sports, food, and drink. "Race prejudice, especially anti-Semitism among the guards, is conspicuous," the report noted.

The report painted many of the guards as easily corruptible by the petty payoffs and gifts of the detainees. Some of the guards could be found cavorting with detainees, sharing cigars and drinks. Much of the blame for the corruption of officials was placed at the feet of one detainee: William Gerald Bishop. For the remainder of the war, no detainee would give the government more headaches than Bishop.

One Justice Department official called Bishop "one of the most unreliable individuals with whom I ever came into contact," while another called him one of Ellis Island's "worst sources of mischief-making and corruption of employees." Bishop was accused of encouraging guards to violate rules, leading to the dismissal of a number of them. He constantly bullied uncooperating guards and officials by threatening them with his "political influence." At various times, he incited a hunger strike among the detainees, stole food from the dining hall, and was accused of abusing and cursing Jewish guards. It was reported that Bishop had three white poison tablets hidden in a pencil that he said were meant for Jewish guards. "If I can't make them leave the Island one way, I will make them leave another," Bishop is reported to have told a fellow detainee.

Not only did Bishop enjoy many privileges on the island, but he also spent a great deal of time in Manhattan on leave. A friendly eye doctor would require Bishop to make weekly appointments for exams. Guards would accompany him to the doctor's office, but were easily paid off in food, drinks, and cigars, and would allow Bishop to visit friends and do as he pleased until it was time to return to the island.

Although Bishop was taken to Ellis Island on February 27, 1942, his problems had actually begun back in January 1940, when J. Edgar Hoover held a press conference to announce that the FBI had arrested seventeen members of an organization known as the Christian Front for plotting to bomb various buildings in New York. Hoover claimed that the plotters had hoped these bombings would eventually lead to the overthrow of the U.S. government. "Plots were discussed for the wholesale sabotage and blowing up of all these institutions so that a dictatorship could be set up here, similar to the Hitler dictatorship in Germany," Hoover dramatically claimed. The alleged plotters were going to start their revolution with eighteen cans of explosives, twelve Springfield rifles, and assorted other guns and ammunition. One of their leaders was William Gerald Bishop.

During the spring 1940 trial of the Christian Front plotters, all of his fellow codefendants turned against Bishop, portraying him as a hothead who wanted to commit violence against the government, a man whose rhetoric was so extreme some of them believed he had to be a government informant. It was Bishop who admitted stealing many of the weapons and ammunition from a National Guard armory. In keeping with later government reports, the trial also showed Bishop to be suffering delusions of grandeur. He asserted that prominent politicians, such as Senator Arthur Vandenberg of Michigan, were among his supporters. He also claimed to have fought in the 1930s with Spanish rebels in North Africa, where he served as secretary to General Francisco Franco.

In June, the jury came back with its verdict. In a slap to the government, it acquitted nine of the men, while the cases of five others, including Bishop, ended in a hung jury. (Two men found their cases dropped before coming to trial and one committed suicide.) Shortly thereafter, the government quietly dropped its case against the five remaining defendants.

However, Bishop's troubles had just begun. During the trial, his citizenship had become a subject of debate. At various points, he referred to his birthplace as Salem, Massachusetts; California; Switzerland; and

Vienna, Austria. At trial, he finally admitted that he was born abroad and had entered the country in 1926 as an illegal stowaway, leaving him vulnerable to the much looser rules of immigration law. Immediately after Bishop's legal case ended in a hung jury, officials issued a warrant for his deportation. Because of the war in Europe, the government suspended the order and Bishop remained free.

By February 1942, Bishop faced another threat. He was now considered an enemy alien, since authorities declared his place of birth as Austria, though this was unusual since Austrian citizens were generally not considered enemy aliens. He was now brought to Ellis Island with hundreds of other accused enemy aliens.

Though Bishop was vocal about his support for Nazi Germany, the OSS report was careful to note that many of those imprisoned on the island were not Nazis and several were "on the verge of a nervous breakdown only because of this intolerable Nazi atmosphere." These unfortunate individuals had been caught up in a bureaucratic dragnet based on false accusations.

One of them was the forty-nine-year-old Italian opera singer Ezio Pinza. The leading basso at the Metropolitan Opera, Pinza was arrested at his home in suburban New York in March 1942 as an enemy alien. The news of his arrest made the front page of the *New York Times*. Pinza would spend nearly three months in detention at Ellis Island and feared that his career was over.

The FBI had talked to a number of informants willing to peddle salacious stories about Pinza, including a fellow opera singer who resented him and former girlfriends jealous that he had recently married another woman. The case against Pinza rested on a number of allegations: he had owned a ring with a Nazi swastika on it; he had a boat from which he broadcast secret radio messages to Europe; he was friends with Mussolini and was even nicknamed after the dictator; he sent out coded messages during his performances at the Metropolitan Opera; he had organized a collection of gold and silver at a benefit for the Italian government in 1935. Only the last charge had any merit. Pinza, along with other Italians working at the Met, contributed to a benefit for Italy, but less out of sympathy for fascism than for patriotic support for their homeland. The benefit occurred after Italy's invasion of Ethiopia, which Pinza, like most Italians, supported at the time.

Thanks to a good lawyer and the dogged persistence of his wife,

Pinza was able to prove his innocence. He was even able to enlist the aid of New York City mayor Fiorello La Guardia, whose dentist was Pinza's father-in-law. He was eventually released on parole from Ellis Island in June and had to report weekly to his local doctor, who acted as his sponsor. On Columbus Day 1942, a few months after Pinza's release, the Roosevelt administration lifted the enemy alien designation from Italians living in the United States, but it was not until 1944 that Pinza received his unconditional release.

Ironically, three years after his release, Pinza was invited to sing "The Star-Spangled Banner" at a welcome-home celebration for General George Patton at the Los Angeles Coliseum. In 1950, Pinza won a Tony Award for his role in the Broadway musical *South Pacific*. Yet he never completely got over the heartbreak of his wartime detention. His widow, Doris, charged that his imprisonment worsened his heart condition and helped speed his early death in 1957 at the age of sixty-four.

Ezio Pinza's story was just one of thousands. By September 1942, some 6,800 aliens of German, Japanese, and Italian ancestry had been arrested by the Justice Department. Of those, half were quickly released or paroled, like Pinza. The other half remained in detention, including the Neupert family. In the summer of 1942, Emma Neupert was taken to Ellis Island as an enemy alien. Her husband, George, was a naturalized U.S. citizen and her nine-year-old daughter, Rose Marie, was a U.S. citizen by birth. By December, Rose Marie was taken to Ellis Island to be with her mother. A few months later George had his citizenship revoked, and he too found himself in detention with his wife and child.

Young Rose Marie remembers that the internees spent most of their days in the Great Hall of the main building, which by 1942 had become "dingy, dirty, and grey with age." To make matters worse, "every time anything would be moved, the roaches would scurry about." The food was "almost inedible." At night, Rose Marie shared a small dormitory room crowded with eight women and two children. During the daytime, the female detainees crocheted, knitted, and sewed to pass the time.

Most of the detainees would be transferred from Ellis Island to other internment camps throughout the country. Many would be joined by their spouses and children—some of whom were American citizens like Rose Marie—who voluntarily agreed to be detained with their family. The Neuperts were sent to a camp in Crystal City, Texas. Other Ellis Island detainees, including William Gerald Bishop, were taken to Fort Lincoln in Bismarck, North Dakota.

The end of the war in the summer of 1945 should have meant the release of the remaining enemy aliens. However, that was not to be. In July 1945, President Harry S. Truman issued Presidential Proclamation 2655, which ordered that all enemy aliens presently detained and found "to be dangerous to the public peace and safety of the United States" be deported. Most of these so-called enemy aliens challenged their deportation orders.

By March 1946, Ellis Island was again filled with enemy aliens, as the government closed down other internment camps around the country and shipped the remaining detainees back to New York, where they waited for the resolution of their cases. Officials from the Red Cross and the State Department inspected the facilities at Ellis Island and found them wanting. The old inspection station was inadequate to hold so many people for such an indefinite period. Morale among detainees was low, their futures uncertain, and a growing number were in need of psychiatric help.

One of those not holding up well under the strain was Helene Hackenberg. She had arrived in the United States from Germany in 1926 and married a fellow immigrant named Rudolf in 1937. Both were accused of belonging to pro-Nazi organizations. Rudolf was arrested in January 1943 and Helene in November of that year, and both taken to Ellis Island. From there they were sent to the Crystal City camp and then transferred back to Ellis Island in early 1946. They would remain there for two and a half years while they fought their deportation orders. Although that time was punctuated by a number of paroles to arrange personal affairs, as well shopping trips for female detainees to Fifth Avenue stores, Helene's depression deepened and she began talking of suicide.

Hundreds of these enemy aliens remained at Ellis Island while they petitioned the courts to cancel their deportation back to war-ravaged Germany. As months went by, their cases lingered in the courts. At the beginning of 1947, almost a year and a half after the cessation of all military conflict, over three hundred still remained at Ellis Island, including William Gerald Bishop, who had been transferred back to New York from North Dakota after the war. For some of them, repatriation would have meant living in Soviet-occupied Germany where, one detained couple feared, they would "be placed in a concentration camp where we will be held indefinitely."

The Fuhr family arrived at Ellis Island from Crystal City in 1947. Carl and Anna Fuhr had come to America from Germany in the 1920s, bringing their sons, Julius and Eberhard. They settled in Cincinnati, Ohio, where the family added a third son, Gerhard. The Fuhrs never became U.S. citizens. Like a number of nonnaturalized Germans, they came to the attention of the FBI in 1940, when informants, many of them anonymous, accused Carl of being a member of the German-American Bund and the Friends of New Germany, of being a strong critic of the United States and supporter of Hitler, and, of saying that his oldest son would return to Germany to "fight for Hitler."

The Fuhrs remained free until the summer of 1942, but more reports had filtered into the FBI by that time. Carl and Anna were arrested in August 1942 and sent to an internment camp in Texas along with their youngest son, the American-born Gerhard. Julius and Eberhard joined the family at the camp in March 1943.

While in custody, the family continued to make statements that reinforced the government's decision to hold them. Julius and Eberhard told authorities they would refuse to serve in the U.S. military. The senior Fuhr, according to officials, possessed "the mind of a man who continues to believe in the Nazis," while Julius was found to be "completely Nazi." In keeping with Truman's postwar orders on German detainees, the family was ordered repatriated back to Germany in the spring of 1946.

However, the family had changed its mind about America and decided to fight the deportation order. They argued that their recalcitrance while in custody was due in part to their anger at the internment. They slowly came to discover that they were more American than German and wanted to remain in the country. The uncertainty of returning to war-devastated Germany no doubt also played into the family's desire to remain in the United States.

By 1947, the family was transferred to Ellis Island to await deportation. Eberhard Fuhr remembers the facility as "cramped, dirty and stultifying." Despite the poor conditions, the Fuhrs made a favorable impression on authorities. "A definite reformation has taken place," according to one report. By now, it was more than two years after the end of hostilities in Europe and more than two hundred individuals, including the Fuhrs, were still being held in custody.

These men and women found a champion in the form of Senator William Langer of North Dakota, who convinced Justice Department officials to form a committee to hear the cases of those still stuck in political and legal limbo at Ellis Island. Throughout the summer of 1947, Langer made several trips to Ellis Island with the committee and held hearings for every single German detainee.

Langer introduced a bill in Congress to cancel the deportation orders of 207 German detainees, including Rudolf and Helene Hackenberg, George Neupert, and the Fuhr family. The bill stalled in Congress, but at the end of the summer of 1947 the Fuhrs managed to secure their release from Ellis Island and headed back to Cincinnati to rebuild their lives. They were the exception. Despite Langer's efforts, by the fall of 1947, some two hundred German enemy alien detainees were still stuck at Ellis Island.

One of those not on Langer's list was William Gerald Bishop. In fact, Langer had already introduced a separate bill in April 1947 calling for the cancellation of Bishop's deportation. Not only did Langer believe that Bishop had been deprived of his rights during five years of detention, but he argued that sending enemy aliens like Bishop "to Communist controlled territory would subject them to the purge, enslavement or liquidation, which according to reports being received daily from Europe affect all persons disliked by Communists." Langer's efforts failed and Bishop was finally deported back to Austria in October 1947.

As for the remaining German detainees at Ellis Island, in June 1948 the Supreme Court rejected their petitions for release from custody. Defense attorneys had argued that Truman's proclamation was invalid since the United States was no longer at war with Germany. The Court's majority was not interested in that issue, but instead decided the case on much narrower grounds, concluding that the habeas corpus petitions were invalid since they were filed in Washington, while the detainees were held in New York.

At the end of June 1948, three years after the end of the war, 182 Germans were still held at Ellis Island, including 9 "voluntary detainees," American citizens who had joined family members in detention. One couple, Marie and Eugen Zimmerman had actually conceived a child, George, while in detention at Ellis Island.

In the following weeks, government officials would work to settle the cases of these unfortunate individuals. On July 8, fifty-seven detainees lost their fight and were sent back to Germany. A few, including Helene and Rudolf Hackenberg, avoided repatriation by voluntarily leaving for a new life in Argentina. (They would finally receive visas to reenter the United States in 1960.) Most of the remaining detainees were released or paroled from Ellis Island and allowed to restart their lives in America, including the Zimmerman family and George Neupert, who was now able to rejoin his wife and daughter. By August 1948, the government had disposed of all the cases of detained German enemy aliens at Ellis Island with the exception of Frederick Bauer, a former U.S. Army sergeant arrested in late 1945 and charged with being a German spy.

Although exact numbers vary, the FBI arrested over thirty thousand German, Japanese, and Italian enemy aliens during the war. Roughly one-third were interned in government camps for some period of time, including a few thousand German and Japanese nationals deported from Latin America to the United States for detention.

By 1948, German enemy aliens had become an anachronism. The enemies of the previous war—Germans—were evolving into new allies, while the allies of the last war—Communists—had become the new enemies. Those last few German detainees at Ellis Island in June 1948 found themselves sharing quarters with men like Gerhard Eisler, Irving Potash, and John Williamson, Communists who were detained and ordered deported for their politics. The Cold War had begun, but the intersection of national security and immigration would continue to run through Ellis Island.

WITH AMERICA ONCE AGAIN at war in the fall of 1950, this time on the peninsula of Korea, Congress passed the Internal Security Act. Spearheaded by Senator Pat McCarran of Nevada, the law would force Communists and other subversives in the United States to register with the federal government.

The bill also granted government greater powers to exclude aliens from the United States. Going beyond already existing laws banning anarchists and Communists, the new law would bar all those who not only advocated totalitarianism but were affiliated with any organization that advocated any form of totalitarianism.

President Truman came out strongly against the bill. On September 22, he gave a lengthy explanation of his reasons for vetoing it. He said there was no need for changes regarding the admission of aliens since the present law was already strong enough to keep out suspected subversives and Communists. He also warned that the bill would require the government to bar foreigners from "friendly, non-Communist countries" such as Spain. Refusing to heed Truman's warnings, both the House and Senate overrode his veto by overwhelming majorities.

Embarrassed at having his veto soundly overridden, Truman decided to get even with his congressional opponents. In a fit of pique, the president declared that if Congress wanted such a law, his administration would strictly enforce it. Attorney General J. Howard McGrath ordered that the Internal Security Act be applied not just to members of the Nazi, Communist, or Fascist Parties, but to anyone who had ever been forced to join such organizations, "regardless of whether or not he may now be harmless, anti-totalitarian, pro-American, or the circumstances under which he was a member." Five years after the end of the war, Germans, Austrians, Italians, and other Europeans who may have been forced to join Nazi or fascist organizations were now barred from entry.

Ellis Island once again found itself in the firing line. Twenty-year-old Viennese pianist Friedrich Gulda, who would later become a renowned avant-garde musician, was one of the first detained under the new law because he had belonged to the Nazi Youth as a preteen during the war. He was in New York for his Carnegie Hall debut.

Gulda arrived at Idlewild Airport in Queens shortly before midnight on October 6. After being detained at the airport and questioned, he was taken to Ellis Island in the early morning hours. It was unclear whether Gulda would ever make it to Carnegie Hall. At Ellis Island, he practiced on an old piano until Steinway & Sons received permission to send a concert grand piano to the island. After three days in detention, Gulda was released and was able to perform at his concert. He was given until the end of the month to stay in the country, but left shortly after his concert.

The Metropolitan Opera was concerned that eight of its singers for the fall season would be barred from the country. One of them was Fedora Barbieri, a twenty-five-year-old Italian mezzo-soprano heading for her debut in Verdi's Don Carlo. She was briefly held at Ellis Island because as a young girl she had attended Fascist schools in Italy. Of course every Italian schoolchild in the 1930s and early 1940s had gone to Fascist-controlled schools. Victor de Sabata, the conductor at Milan's La Scala, was also temporarily detained at Ellis Island. Even the great conductor Arturo Toscanini was questioned, though he escaped detention and was allowed to land.

The law also affected average Americans recently married to Europeans. For seven months, Arthur Sweberg, an American serviceman living in New York, had to live apart from his new bride, a German national who had been a member of the Nazi Youth as a child. Josephine Mazzeo, of Evanston, Illinois, married an Italian national in October 1949. Because of the new law, her Italian husband could not enter the country because he had belonged to a Fascist youth organization during the war.

George Voskovec watched this whole scene unfold before him. The forty-five-year-old Czech playwright and actor had been held at Ellis Island since May 1950, before the passage of the new law. Voskovec had lived in the United States from the late 1930s until the end of the war and was married to an American. He had been a vocal anti-Nazi and worked for the Office of War Information during the war. Upon returning to America in May 1950 to apply for citizenship, he was detained at Ellis Island. Authorities were concerned that Voskovec had been allowed to leave Communist Czechoslovakia legally, setting off alarm bells as to his political sympathies. Now he was joined on Ellis Island by hundreds of other suspected subversives.

As Truman predicted, the strict interpretation of the Internal Security Act made the law look foolish, but it was the price he was willing to pay to embarrass Congress into at least tightening the law. And it worked. By late March 1951, Congress amended the Internal Security Act to exempt those who may have been members of a totalitarian organization, but who were under the age of sixteen at the time, were "involuntary members" of the organization, or had joined the group "for purposes of obtaining employment, food rations, or other essentials of living."

George Voskovec's detention at Ellis Island would end shortly after Congress revised the Internal Security Act. After ten months and seventeen days at Ellis Island, he was a free man. Only one witness had come forward to accuse him of being a Communist. On the other side, a number of prominent Czechs and Americans vouched for his character, including the playwright Thornton Wilder.

Upon his release, Voskovec noted that none of the inmates at Ellis Island had been mistreated. However, that did not ease the frustration at his imprisonment. Speaking of his situation, he told a reporter that a detainee "isn't told the particulars of his offense, his accusers are nameless, and the weeks and months pass, as if human beings were no more to be considered than ciphers in a manila folder." Even more bluntly, Voskovec said of Ellis Island: "I want to go on record that it's a disgusting place—a prison."

Voskovec would later dramatize his imprisonment at Ellis Island in a made-for-television play. *I Was Accused* aired in November 1955, the same year he gained his U.S. citizenship. Voskovec's career would later take him to Hollywood, where he made a living as a character actor in movies and television, most famously starring as one of the sweaty and stressed-out jurors in the classic film *12 Angry Men*.

When Friedrich Gulda was taken to Ellis Island in October 1950, he found nearly two hundred people held there, including George Voskovec and a European war bride. Most likely Gulda was referring to Ellen Knauff, already on her second stay at Ellis Island. Her first detention began when she arrived in New York in August 1948.

She was born Ellen Raphael in 1915 in Germany. In the 1930s, she moved to Prague and married a Czech man named Boxhorn. Being Jewish, Ellen escaped from Prague—and the marriage—after the Nazi invasion and made her way to England, avoiding the fate that befell much of her family in Nazi concentration camps. During the war, she worked as a Red Cross nurse and then served in the Royal Air Force. After the war, she made her way back to Germany, where she landed a job as a civilian with the U.S. military government, first working for the Civil Censorship Division and then as a secretary in the Signal Corps.

In February 1948, Ellen married Kurt Knauff, a naturalized U.S. citizen and an honorably discharged army veteran working as a civilian for the military occupation. After the war, the U.S. government passed the War Brides Act, which allowed U.S. servicemen to bring back foreign-born brides without regard to either the strict mental and physical requirements required of immigrants or the national origins quotas.

When Ellen arrived in New York in August 1948, however, she was not greeted with any celebrations. Instead, she was ordered detained at Ellis Island. No explanation was given. When a government official told her, "I am sending you to a place where they will look after you,"

Ellen broke down in tears. She had lost family members in the Holocaust and the detention order caused her to fear that she too was heading to some kind of concentration camp. Ellen was not allowed a hearing, nor was she informed of the charges against her. Therefore, she was stuck at Ellis Island with no apparent way to prove her innocence and gain entrance to America.

The government's case against Ellen was as follows: When she was employed by the army's Civil Censorship Division in Germany, she furnished Czech agents with secret information, including copies of telephone conversations that her department was monitoring. She was also accused of warning the chief of the Czech Liaison Section in Frankfurt against using telephones since they were being tapped by the Americans. She was also alleged to have described to Czech agents the type of decoding machines used by American intelligence.

All of this took place before the 1948 Communist coup in Czechoslovakia, so the charges were not that Ellen was a Communist, although later witnesses would testify that they saw Ellen enter Communist Party headquarters in Frankfurt. The major source of the charges against Ellen was an unnamed "former highly placed Czech official" who had defected and was now assisting the American military. Two other Czechs also provided testimony against Knauff.

Though still flush with wartime victory, Americans were growing increasingly insecure and vulnerable regarding national security threats at home. Ellen Knauff was ordered detained just nine days after Alger Hiss appeared before the House Un-American Activities Committee (HUAC) to deny falsely that he had been a Communist spy.

It would be more than two more years before Ellen would hear these details, since they were kept classified to protect confidential intelligence sources. She would remain in detention at Ellis Island for the next nine months while her lawyers submitted a habeas corpus petition. In her letters to her husband, who was still working in Germany, Ellen told of her "bitter disappointment in the Ellis Island version of American freedom." She called it "a concentration camp with steam heat and running water" and said the food there was only "fit for pigs—if you were not particular about what your pigs ate." For all of her anger, Ellen had nothing but good things to say about the men and women who worked at Ellis Island.

Eventually, her case reached the Supreme Court. While the Court was deciding the case, it allowed Ellen to be released on bond. The Court reached a decision in January 1950. By a vote of 4 to 3, it rejected Knauff's request. Two justices did not take part in the case, including newly appointed Justice Tom Clark, who had been attorney general and was technically the authority who had detained Knauff in 1948.

The Court relied on the plenary power doctrine that had long given the executive branch tremendous latitude in its treatment of aliens. In familiar language, the Court reiterated that "an alien who seeks admission to this country may not do so under any claim of right. Admission of aliens to the United States is a privilege granted by the sovereign United States Government."

The War Brides Act of 1945 may have superseded some aspects of immigrant law, but it did not override national security concerns. Though the plenary power doctrine was well-trod legal ground, the Court also outlined the little-known history of recent presidential proclamations and regulations that led to Knauff's exclusion. The paper trail began with FDR's May 1941 declaration of an "unlimited national emergency" in dealing with the threat posed by the European war, even if America still technically remained on the sidelines.

Congress then allowed the president to impose additional restrictions upon those entering the country during times of national emergency. This was followed by Presidential Proclamation 2533 in November 1941, which ordered that no alien should be allowed to enter the country if his presence was "prejudicial to the interest of the United States." This was followed by a Justice Department regulation that allowed the attorney general to deny a hearing to an excluded alien if the evidence was confidential.

Justice Robert Jackson, who had served as chief prosecutor during the Nuremberg Trials, delivered the dissenting opinion. He called Ellen's exclusion without a hearing "abrupt and brutal." The Court, Jackson wrote, basically told Kurt Knauff, an American citizen and army veteran, that "he cannot bring his wife to the United States, but he will not be told why. He must abandon his bride to live in his own country or forsake his country to live with his bride."

While much of the majority decision followed precedent on immigration law, it showed how much Roosevelt's administration had expanded executive power, both before and during wartime. No one, though, seemed to comment on the oddity that the authority for the denial of a hearing to Ellen Knauff and other aliens was based on the unlimited national emergency declared by FDR in 1941. Was the government implying that this emergency was still in effect nine years later during peacetime?

Having lost in the Supreme Court, Ellen Knauff now headed back to Ellis Island to await deportation. Twice in the spring of 1950, she had her deportation stayed by the courts. The second time, Justice Jackson stayed her deportation just twenty minutes before Knauff's flight back to Germany was set to take off from Idlewild Airport. Yet the reprieve did not mean freedom. Knauff was returned to detention at Ellis Island.

In the meantime, her case had aroused such interest among the public that Knauff was not out of options. A young woman in her thirties, Knauff made a convincing and sympathetic victim. After all, Ellen was the war bride of an American GI, a woman who had lost family members in the Holocaust, and someone who worked for the U.S. military as a civilian in occupied Germany. Newspapers like the *St. Louis Post-Dispatch* took up her case.

Congress also took notice. Her cause was taken up by Senator Langer, who had previously fought for the rights of German enemy aliens, and Congressman Francis Walter, an anti-Communist Democrat who would later chair the House Un-American Activities Committee. Both introduced bills in Congress to free Knauff.

In the spring of 1950, Ellen was invited to Washington to testify before a congressional committee investigating her case. "That whole day was a true American fairy tale," Ellen later wrote. "A prisoner on Ellis Island woke up one fine morning to be flown to Washington, D.C., to be heard before a congressional subcommittee in an effort to make truth prevail." The Justice Department was invited to present its evidence against Knauff, but refused to cooperate on the grounds that it would compromise confidential sources. Despite her treatment as the guest of Congress, by day's end Ellen would find herself on a plane, headed back to confinement at Ellis Island. Following the hearings, the House unanimously passed a bill to allow Knauff to remain in the country, although a similar bill stalled in the Senate.

The press attention in Harry Truman's home-state newspaper helped bring Ellen Knauff's case to the president's attention. In mid-June 1950, Edward Harris of the *St. Louis Post-Dispatch* argued to

Truman that Ellen was the only war bride ever to receive such treatment, and that she was entitled to a hearing at least, "in view of her own war record and her husband's valorous combat service." Harris correctly noted that it was within the power of the president or the attorney general to change the regulations so that every alien was entitled to receive a hearing except in time of "actual warfare." Within days, Truman personally asked his aide, Steve Spingarn, to look into Ellen's case "and see if anything can be done to straighten it out."

The Justice Department stalled in releasing the Knauff file, but finally relented. In September 1950, Spingarn detailed his findings to the deputy attorney general. He was deeply unimpressed by the case against Knauff. "It seems to me most meager despite the seriousness of the allegations," he wrote. "Indeed it all boils down to a few paragraphs in an Army Intelligence report repeated several times in Immigration Service and FBI reports." He recommended that the Justice Department grant Knauff a hearing in camera, which would allow her to answer the charges but would preserve the confidentiality of the intelligence sources.

However, the Justice Department ignored Spingarn's suggestion and Truman, preoccupied with the war in Korea, seemed disinclined to interfere. The Justice Department also ignored congressional calls for her release. Instead, it pushed ahead with her deportation. Edward Shaughnessy, the district director of immigration in New York, summed up the rationale for debarring Knauff. The attorney general, he said, "decided she is not the type of person wanted in this country and that is all there is to it. It was felt prejudicial to the best interests of the country to let her stay here." Understandably, Ellen did not want to return to postwar Germany. Instead, she told her lawyer: "I am ready to stay on Ellis Island till doomsday."

In January 1951, almost a year into Ellen's second stint at Ellis Island, Kurt Knauff received a leave from his job in Germany, where he worked as an assistant chief of supply at an American army base, and arrived in New York. The Justice Department granted Ellen temporary parole into her husband's custody and she was free again, but her troubles were far from over.

In March 1951, more than two and a half years since she was first detained at Ellis Island, Ellen Knauff received her first hearing in front of immigration officials. The government was under no obligation to hold such a hearing, but the publicity forced the issue. Ellen finally was able to see the evidence against her. She vigorously denied all of the charges, stating that she never gave away secrets and went to the Czech mission only to get an extension on her passport.

It took the board members only an hour to deliberate the case. Ellen Knauff was to be barred from entry into the United States because her presence was deemed "prejudicial to the national security." Ellen's parole was revoked and she was once again sent to Ellis Island to await the results of her appeal to Washington. If it failed, she would again face deportation.

By the end of August 1951, a board of immigration appeals made its ruling. By a two-to-one vote, the board overturned the decision to exclude Ellen Knauff and recommended that she be allowed to enter the country. "There is no charge that Mrs. Knauff is or has been a Communist," the majority concluded. "There is not the faintest thread of traditional party line thinking or Marxist philosophy apparent in her background." In fact, they determined that Knauff's politics were conservative. She was a supporter of Churchill and an opponent of England's Socialist Party and believed that Soviet Russia was as evil as Nazi Germany. The dissenting member of the board continued to argue that the testimony against Knauff was sufficient to exclude her from the country.

Now Ellen Knauff's fate rested in the hands of Attorney General McGrath. He was well acquainted with the case and had previously showed no inclination to admit Knauff. But in November 1951, McGrath ordered that Ellen Knauff be admitted to the United States. It is unclear why he changed his mind.

McGrath released his decision at 6:00 P.M. on November 2. Fifteen minutes later, the phone at Ellis Island rang with the good news. Ellen quickly gathered her belongings in time to make the 7:30 ferry to Manhattan. The media was waiting for her at the Manhattan pier, snapping photos of a jubilant and beaming Knauff standing on the ferry. First, she wanted to call her husband with the news. Then, she told reporters, "I want to have a lobster dinner." Kurt and Ellen had spent more time apart than they had together and now had to decide whether they would make their home in New York or if Ellen would join Kurt while he remained in Germany working for the military.

In total, Ellen Knauff spent nearly twenty-seven months imprisoned on Ellis Island while fighting for her right to become an American. During that time, she penned a book about her case, which was published a few months after her release. Although it contained no new information, it helped to solidify the public impression that she had been a victim of a security-obsessed nation.

However, there were serious charges against Knauff. Though she vigorously denied the accusations and no further evidence was produced to corroborate her accusers, it is still unclear why three Czech refugees would deliberately lie about her. Ellen theorized that the refugees testified against her in an effort to receive U.S. citizenship. She also believed that rumors of her alleged espionage were being spread in Germany by one of her husband's old flames, who in a fit of jealousy reportedly said that she would do her best to ruin Ellen's arrival in the United States.

Though she ultimately won her battle against the U.S. government, the victory came at the cost of her marriage, which did not survive the 1950s. With Ellen detained at Ellis Island and Kurt working in Germany, the first three and a half years of their marriage could hardly be termed a honeymoon. After her divorce, Ellen remarried. With her new husband, William Hartley, she cowrote a number of children's books. Ellen Raphael Boxhorn Knauff Hartley lived a quiet life in America until she passed away in Florida in 1980.

Ellen Knauff's plight had gained nationwide publicity. But her case also brought attention to the fact that individuals could be detained and deported without benefit of an official hearing and without any knowledge of the evidence against them.

The widespread sympathy that Ellen Knauff's case elicited did not mean any slackening of the nation's anti-Communism. Ellis Island would continue to serve as a detention center for suspected Communists and other political radicals. One of them was a middle-aged Trinidadian writer named Cyril Lionel Robert James. He was arrested and taken to Ellis Island in June 1952 for his political affiliations and because the government alleged that he had entered the country illegally in the 1930s. Immigration authorities had spent a number of years trying to sort out his immigration status and his political proclivities. Now he was at Ellis Island awaiting deportation.

Following in the footsteps of Emma Goldman and Ellen Knauff, who both used their detention at Ellis Island to write about their plights, C. L. R. James also devoted his time as a prisoner to writing. His unlikely topic was Herman Melville's *Moby-Dick*. James's experience at Ellis Island profoundly influenced his reading of Melville's classic. He would sit at his desk and write, sometimes for twelve hours a day, all the while suffering from painful ulcers made worse by the stress of his confinement. Within a few weeks of his detention, James was living on milk, boiled egg yolks, soft bread, and butter. He was then taken to the U.S. Marine Hospital on Staten Island (in a cost-saving measure, the hospitals on Ellis Island had recently been closed), where he would recuperate under twenty-four-hour guard.

In the final chapter of his Melville book, James wrote what he called "A Natural but Necessary Conclusion." It was in part the story of his detention at Ellis Island, but more importantly it was James's attempt to convince the government that he was not a dangerous subversive and should be allowed to remain in the country. James was not in fact a Communist, but a Trotskyite, and a harsh critic of Stalin and the Soviet Union. "I denounced Russia as the greatest example of barbarism that history has ever known," James wrote. When he arrived at Ellis Island he was placed in a room with five Communists. Because of his past criticisms of Stalin and the Soviet Union, James feared for his life among these men, "conscious of their murderous past, not only against declared and life-long enemies, but against one another."

The U.S. government was not interested in parsing the internecine battles among Marxists, sorting out Trotskyites from Stalinists. As far as it was concerned, James was a Marxist critic of capitalism and author of books such as *World Revolution 1917–1936: The Rise and Fall of the Communist International* and *A History of Negro Revolt*. There was enough revolution there to expel him from 1950s America.

As much as he played up his anti-Soviet and anti-Stalinist views in the hopes of being allowed to stay in the country, James pulled no punches when it came to government officials. "Hence on Ellis Island, in particular, the arbitrariness, the capriciousness, the brutality and savagery where they think they can get away with it," James wrote, "the complete absence of any principle except to achieve a particular aim by the most convenient means to hand." For his guards, however, he had nothing but kind words. "They were a body of men in a difficult spot," James wrote, "yet they remained, not as individuals but as a body of men, not only human but humane." Although the government continued to refer to individuals like him as detainees, James thought it "a mockery for me to assist them in still more deceiving the American people." He and the others at Ellis Island were nothing less than prisoners.

James was freed on bail in October 1952 after four months in detention. His Melville book, *Mariners, Renegades, and Castaways: The Story of Herman Melville and the World We Live In*, was released the following year. Despite the fierce anti-Communism that only a Trotskyite could muster, James was eventually deported to England in 1953. There, he made a living as a writer on cricket. He also traveled back and forth to his native Trinidad, where he became involved in local politics. James eventually returned to the United States for extended visits in the 1970s, when Ellis Island was a dim memory and the Cold War a growing embarrassment for Vietnam-fatigued Americans.

C. L. R. James died in relative obscurity in 1989. Posthumously, James's reputation would grow as one of the leading black social critics of the twentieth century. *Mariners, Renegades, and Castaways* would be republished after his death and garner attention in academia and beyond. Ellis Island inspired millions of true-life sagas of joy and heartbreak among the many who passed through there. Few could imagine that it also inspired a major work of literary criticism.

At least C. L. R. James had a home country to which he could be deported. The same could not be said for fifty-two-year-old cabinetmaker Ignatz Mezei. Arriving in February 1950, after a visit to Europe, Mezei was detained at Ellis Island and refused readmittance to the country he had called his home for over twenty-five years. Like Ellen Knauff, Mezei was also refused a hearing because the charges against him were based on confidential information.

Mezei was not a random immigrant to America. He had made his home in Buffalo for a quarter century before returning to Europe in 1948 to visit his dying mother in Romania. However, he ended up detained in Hungary and never managed to make it to see his mother. While in Hungary, his common-law wife, Julia Horvath, arrived from America and the two of them officially married. They then returned to the United States in 1950. While Horvath was allowed to return to Buffalo, Mezei was detained at Ellis Island and ordered excluded. He was denied a hearing and not allowed to see the specific charges against him. The basic accusation was that he had been a member of a Communist-affiliated group while residing in America.

Mezei was ordered deported, but to where? As a court would later declare, "there is a certain vagueness about [Mezei's] history." He had arrived in the United States illegally in 1923, having gone overboard in New York Harbor from a ship on which he served as a seaman. He was born in 1897 on a ship off the Straits of Gibraltar, but raised in Hungary and Romania. In his twenty-five years in the United States, Mezei had never become a naturalized citizen. All of this left his actual citizenship uncertain.

This was a dilemma for U.S. officials deciding where to send Mezei. When the government deported him back to France, that nation turned him away. The same thing happened when Mezei was sent to England. The State Department then asked the Hungarian government to take him, but it refused. Mezei wrote to twelve Latin American countries asking for entry, but not one would accept him. Ignatz Mezei was stuck at Ellis Island, a man without a country.

The next step was for Mezei to file a habeas corpus petition. Eventually, his case reached the Supreme Court. While the judicial process unfolded, Mezei was released on a bond in May 1952, after nearly two years imprisoned at Ellis Island. He returned to Buffalo and tried to earn a living as a cabinetmaker while the courts untangled his case.

In March 1953, the Court came to a decision. In a 5-4 ruling that relied heavily on *Knauff*, it declared that the exclusion without a hearing and subsequent detention of Ignatz Mezei at Ellis Island was constitutional. The Court agreed with the Justice Department that Mezei was not actually imprisoned at Ellis Island, since he was free to leave at any time to any country that would accept him. "In short, respondent sat on Ellis Island because this country shut him out and others were unwilling to take him in," wrote Justice Tom Clark.

The Court again reiterated the plenary power doctrine that recognized that "the power to expel or exclude aliens" was "a fundamental sovereign attribute exercised by the Government's political departments largely immune from judicial control." Even though Mezei had previously lived in the United States and was currently on American soil, the Court recognized the legal fiction that Mezei had not formally and legally "entered" the United States and was therefore not eligible for constitutional protections such as due process. "Neither respondent's harborage on Ellis Island nor his prior residence here transforms this into something other than an exclusion proceeding," Clark wrote.

In his dissent, Justice Hugo Black complained that Mezei was being excluded at the "unreviewable discretion of the Attorney General," noting that such powers were more likely found in totalitarian regimes like the Soviet Union and Nazi Germany. As he did in *Knauff*, Justice Jackson also dissented in *Mezei*. "Because the respondent has no right of entry, does it follow that he has no rights at all," Jackson asked. "Does the power to exclude mean that exclusion may be continued or effectuated by any means which happen to seem appropriate to the authorities?" If so, what would stop the government from ejecting Mezei "bodily into the sea or to set him adrift in a rowboat?"

A defeated Mezei returned to Ellis Island in April 1953. His only hope was that Congress might act on his behalf. He arrived at the ferry slip carrying his clothes, his tools, and a bag of apples. "I feel as if I was walking to death," he said. Mezei still vigorously denied that he was a Communist. "If I were a Communist I would stay in Hungary," he said, "plenty of jobs in Hungary for Communists." The prospect of indefinite detention understandably weighed heavily on Mezei. "You don't do nothing on Ellis Island," he complained, "you go crazy."

Unlike Ellen Knauff, Mezei did not elicit a great deal of sympathy from the public, the press, or Congress. Knauff had seen her family die in the Holocaust, had served in the British military during the war, had worked for the American military after the war, and was married to an American GI. Mezei, on the other hand, had arrived in the United States illegally, had lived in the country for twenty-five years without becoming a citizen, and had married Julia Horvath, an American citizen, while in Hungary under suspicious circumstances, most likely in hopes of easing his entry back into the country. "But when we come to this guy," wrote one of Justice Jackson's clerks and future Supreme Court chief justice, William Rehnquist, "I have some trouble crying."

While there was not a great deal of public sympathy for Mezei, much had changed in the United States by the summer of 1954. The new president, a Republican war hero, Dwight D. Eisenhower, successfully sought an end to the unpopular stalemate in Korea. Though the new president had been cautious in his public comments about Senator Joseph McCarthy, it was clear that Eisenhower wanted to cool the domestic anti-Communist fires of the past few years. His new attorney general, Herbert Brownell, would set a new tone in the Justice Department. Mezei would receive his first hearing in February 1954, nearly four years after he was initially detained.

In an unusual move, Brownell created a three-man board to hear Mezei's case, which consisted not of immigration officials but of outside lawyers, including law professors from Columbia University and New York University.

The government had a strong case against Mezei. The evidence against Ellen Knauff was scant and she could not be directly tied to any espionage. Mezei, however, had been a member of the Hungarian Workers' Sick Benefit and Education Society, which later merged with the International Workers Order, which the government considered a Communist organization. Mezei admitted to being a leader in his local lodge, but denied being a Communist.

Unfortunately for Mezei, the government had a number of witnesses who contradicted his story. Two former Communists testified that they had seen Mezei at Communist Party meetings and one told the hearing that he had personally recruited him for the party. Three other witnesses told officials that they had heard Mezei making pro-Communist statements. In addition to his political problems, Mezei had also been convicted of petty larceny and fined $10 in his earlier stay in Buffalo. While the crime was rather minor, having to do with his possession of bags of stolen flour, this did mean that Mezei could be excluded under the moral turpitude clause.

Whereas Knauff was articulate and made an excellent case for herself, the same could not be said for Mezei. "His testimony was riddled with inconsistencies, and he seemed to have great difficulty understanding and answering many questions," according to one sympathetic account. "Several of his statements lacked credibility." Mezei had also repeatedly lied on government forms about his place of birth.

Not surprisingly, the board unanimously voted to exclude Mezei as a security risk in April 1954. He appealed the decision to Washington, but a board of immigration appeals upheld the decision to exclude him in August. Just two days later, however, the government reversed itself and announced that it had released Mezei on parole.

The special three-man board that had affirmed Mezei's exclusion also recommended in private to Attorney General Brownell that he use his authority to release Mezei, since his role in the Communist Party was minor. That is exactly what he did. Mezei would return to his wife and stepchildren in Buffalo, where he would live an unassuming life until his wife died in 1969. In that year, he sold his house and mysteriously moved back to Communist Hungary, where he lived until his death in 1976.

Mezei's release occurred at the same time that the career of Senator Joe McCarthy was quickly unraveling, thanks to the public humiliation caused by his unwise investigation of alleged Communism in the U.S. Army in the spring of 1954. As Mezei was released from Ellis Island, censure proceedings against McCarthy were about to come to the floor of the Senate. Anti-Communism was not dead, but its rough edges were being sanded down. The Eisenhower administration had no need to burnish its anti-Communist bona fides and could therefore tone down the government's antiradical crusade.

By 1954, Ellis Island had been tainted by its unfortunate connection to the Cold War detention of aliens, which was increasingly becoming a public relations problem. It was being referred to as a concentration camp, and the United States' role as the leader of the free world in opposing Communist tyranny made its detention policies untenable. "Unlike the totalitarians and despots," wrote the *New York Times*, "we Americans abhor imprisonment by administrative officers' fiat."

In this political environment, the Eisenhower administration began to consider closing Ellis Island for good. Publicly, it sold the move as a cost-saving measure. The federal government could move its immigration offices to Manhattan and would no longer have to keep up the many buildings on the twenty-seven-acre compound. But there is little doubt that the public attention of the Knauff and Mezei cases helped seal Ellis Island's fate.

On Veterans Day 1954, Attorney General Brownell spoke before two massive naturalization ceremonies in New York City. He used the occasion to set out a new policy on immigrant detentions. Those whose admissibility to the United States was under question would now no longer be detained while their cases were decided. Only those deemed "likely to abscond" or whose freedom would be "adverse to the national security or the public safety" would be held. The others would be released under conditional paroles or bonds until their cases were cleared. Brownell estimated that authorities had in the past year temporarily detained some 38,000 people, of whom only 1,600 were excluded from entering the United States. Holding so many individuals in detention had become an administrative, civil liberties, and public relations nightmare.

As part of this new policy, Brownell announced the closing of six detention facilities run by the government, including Ellis Island. Washington would save nearly $1 million a year by shuttering it and moving its offices to Manhattan. No longer needed to inspect and process hundreds of thousands of new immigrants, Ellis Island was now no longer wanted as a detention facility.

Ellis Island closed its doors to little fanfare just a few days after Brownell's speech. From now on, those lucky enough to qualify for admission, after filling a quota position and proving they were not subversives, would no longer concern themselves with the little island in New York Harbor. After decades of attention from journalists, politicians, missionaries, and immigrant aid societies, Ellis Island was now drifting off the nation's radar screen. With only 5 percent of Americans claiming foreign birth, the heyday of Ellis Island—with its inspection process, its medical and mental tests, its boards of special inquiry, its hasty wedding ceremonies, its tearful family reunions and even more tearful family separations because of deportation—was over.

Ellis Island's last detainee was Arne Peterssen. The Norwegian seaman was not an immigrant in the traditional sense, but someone who had overstayed his shore leave. Under the newly relaxed immigration rules, officials released Peterssen on parole with a promise that he would rejoin his ship and return home.

"They rewarded with magnificent gifts the country that had received them with such magnificent hospitality," declared a *New York Times* editorial looking back with pride at the achievements of immigrants who had passed through Ellis Island. "Perhaps some day a monument to them will go up on Ellis Island," it continued, admonishing its readers that the "memory of this episode in our national history should never be allowed to fade."

In the glow of postwar prosperity, assimilation, and suburbanization, few cared to keep that memory alive. That would have to wait for another day.

Part V

MEMORY

Chapter 18

Decline

———•◦•———

What the son wishes to forget, the grandson wishes to remember.
—Marcus Lee Hansen, 1938

A BUSINESSMAN READING THE SEPTEMBER 18, 1956, EDI-
tion of the *Wall Street Journal* would have come across an advertisement for an exciting new opportunity. The federal government's
landlord, the General Services Administration (GSA), was soliciting
sealed bids for the purchase of "one of the most famous landmarks
in the world."

The GSA offered to sell the entire twenty-seven-acre Ellis Island
facility, including all thirty-five buildings and the old ferryboat *Ellis
Island*, which had previously carried immigrants from the Manhattan
piers to the island. Ellis Island, the advertisement proclaimed, would
be the perfect location for an oil-storage depot, warehouses, manufac-
turing, or import-export processing.

The sale was made possible by the fact that the facilities at Ellis
Island had been deemed surplus property by the U.S. government since
it had closed its doors in November 1954. The United States had witnessed only about two hundred thousnd immigrants that year, with
fewer than half of them passing through New York. Ellis Island had
served its purpose; its heyday was well in the past. While a *Times* editorial hoped that the memory of Ellis Island's peak years and its role
in American history would not fade away, the GSA had more pressing

matters. No other government agency wanted the vacant island, and Uncle Sam could not hold onto it indefinitely, especially when it was paying $140,000 a year for security and upkeep.

So the GSA opened up bidding for Ellis Island to private individuals and corporations. The idea of selling the historic site did not sit well with everyone. "If you can auction off Ellis Island," a Jersey City congressman wrote President Eisenhower, "perhaps you will be auctioning off the Statue of Liberty next." A Greek American wrote Eisenhower of his arrival at Ellis Island as a child in 1914. "I first sensed the grandeur of this great country," this first-generation immigrant wrote, "when I landed on the Island."

In response, the Eisenhower administration temporarily suspended the sale less than a week after the *Journal* advertisement appeared. Some suggested turning Ellis Island into a national monument that would pay tribute to the contribution of immigrants. That ran into opposition from a group already preparing to open up the American Museum of Immigration at the base of the Statue of Liberty. One of the leaders of that project argued that Ellis Island was the wrong place for a national shrine. "No immigrant was ever attracted to America by Ellis Island," wrote William Baldwin. "Liberty Island is a happy place of continuing inspiration, not a depository of bad memories."

Some of the proposals for Ellis Island included a clinic for alcoholics and drug addicts, a park, a "world trade center," a modern and innovative "college of the future," private apartments, homes for the elderly, and a shelter for juvenile delinquents. Other proposals were less realistic. Bronx congressman Paul Fino suggested a national lottery center would be in keeping with the history of the island, since immigrants "gambled for a new life in this land of ours."

When bidding opened in 1958, the highest offer was just over $200,000 for a property the government considered worth more than $6 million. The high bidder was a New York builder named Sol Atlas, who wanted to turn Ellis Island into Pleasure Island, a high-end resort with a convention center, marina, and recreational and cultural facilities. Though Atlas would later increase his bid, it was still not enough and the island remained surplus government property. Ellis Island had become, in the words of *Business Week*, "Uncle Sam's Red Brick Elephant in New York Harbor."

Ellis Island's future would depend on how Americans viewed what had happened—or what they thought had happened—there. If Americans associated negative memories with Ellis Island, then there was no reason why it should not become an oil storage depot or some other commercial venture. But clearly some Americans were beginning to feel the tug of positive memories. As Harvard's Oscar Handlin put it at the time, the buildings of Ellis Island should "be preserved not simply for their symbolic quality as monuments of an important part of our past but also for the service they can still render."

"This is not just another piece of real estate," Edward Corsi told a congressional committee in 1962. Corsi had come through Ellis Island fifty-six years earlier and later became commissioner there in the 1930s. Now he was arguing, along with historians Handlin and Allan Nevins, that the island's future "should symbolize what it stands for in the history of our nation and in the hearts of countless Americans—the welding of many nationalities, races and religions into a united nation, bound together by freedom and equality of opportunity."

To Corsi and a growing number of first- and second-generation Americans, Ellis Island was no longer just an inspection center created to soothe the concerns of native-born Americans by weeding out undesirable immigrants. Instead those immigrants and their descendants were beginning to shape the historical memory of Ellis Island. In the midst of the Cold War, the island was slowly becoming a symbol of national unity and freedom. During the much bleaker years of the Great Depression, however, Corsi had taken a much different tack. His 1935 history of Ellis Island included a chapter entitled "Who Shall Apologize?" dealing with the "crimes" committed against immigrants there. The passage of twenty-five years had apparently tempered Corsi's views.

During that time, eastern and southern European immigrants and their offspring were now entering the American mainstream, slowly shedding the stigma of being considered undesirable immigrants. The fears of nativists like Francis A. Walker, Prescott Hall, and Madison Grant were in fact realized as the descendants of eastern and southern Europeans took their place in American society. In turn, American culture and society became less Anglo-Saxon.

In the midtwentieth century, Americans enjoyed movies like *It's a Wonderful Life* and *On the Waterfront*, directed by Frank Capra and Elia Kazan. They went to Broadway plays like *Gypsy* and *Funny Girl*, with music by Jule Styne. They laughed at the jokes of Bob Hope, watched

Edward G. Robinson star in movies like *Key Largo* and *Double Indemnity*, and revered the football legend of Notre Dame's Knute Rockne. All arrived as immigrants at Ellis Island. Most poignantly, Americans sang "God Bless America," written by Irving Berlin, who had arrived at Ellis Island in 1893 as Israel Beilin, the Yiddish-speaking son of a Jewish cantor.

While the nativism of the earlier period was dying, the quotas that severely restricted eastern and southern Europeans still remained in place. Not for much longer. The 1964 Civil Rights Act, a landmark piece of legislation that struck a fatal blow against Jim Crow segregation, prohibited discrimination on the grounds of race, color, religion, or national origin. While such legal prohibitions did not extend to immigration, it became politically and morally unacceptable to retain a form of discrimination based on national origins in immigration policy. The days of the quotas were numbered.

In his 1965 State of the Union address, Lyndon Johnson laid out an ambitious legislative plan known as the Great Society. As part of it, he called for an immigration law "based on the work a man can do and not where he was born or how he spells his name." Later that year, Johnson traveled to Liberty Island to sign the bill formally ending forty-four years of immigration quotas biased against eastern and southern Europeans, which he called a "cruel and enduring wrong." The House and Senate overwhelmingly passed the bill.

Although the bill has been widely hailed as a liberal piece of legislation that ended racial and ethnic discrimination in U.S. immigration law, it still kept much of the restrictive apparatus intact. Overall quotas still remained, and restrictions were placed on immigrants from the Western Hemisphere for the first time. In a move that would have a deep impact on the future of U.S. immigration, the bill made family reunification the cornerstone of immigration policy, setting that outside of the overall quota limit.

While legislation to end quotas based on national origin made its way through Congress, Johnson went ahead and settled the question of who should own Ellis Island, if not what the island's future would look like. In May 1965, Johnson signed a proclamation making Ellis Island a part of the National Park Service by adding it to the Statue of Liberty National Monument. The private sale of the island was now off the table.

With full control over the island, the Johnson administration com-missioned architect Philip Johnson to create plans for the development of the island. Frank Lloyd Wright had been drafted a few years earlier to come up with a design for the private development of the island as a self-contained city of the future. His plan went nowhere. Now it was Johnson's turn and he did not disappoint. Rather than renovating and restoring the main buildings of the island, Johnson called for stabiliz-ing them and keeping them as historical ruins. Vines and trees would be allowed to grow untended about the buildings, adding to the feeling of abandoned ruins. "The effect would be a romantic and nostalgic grouping through which the visitor would pass," Johnson said.

The centerpiece of Johnson's plan was a 130-foot-high truncated cone that would be called the Wall of Sixteen Million. Ramps would wind along the cone, allowing visitors to read the names of every immi-grant who had passed through Ellis Island. Some in the press dubbed Johnson's design the "Cult of Instant Ugliness."

There were other problems. A *New York Times* editorial argued that Johnson had gotten it all wrong. Ellis Island was built as a "gate-way," not a wall "built to exclude." Adding some Cold War imagery, the paper saw Johnson's Wall of Sixteen Million as more akin to the Berlin Wall. This interpretation stripped the restrictive function from Ellis Island's past; the gate that barred undesirable immigrants had now evolved into a gateway, a welcoming station rather than an obstacle de-signed to sift out immigrants. The forgetting of the restrictive nature of Ellis Island was not new. In a 1954 article on the "passing of Ellis Island," the *American Mercury* falsely noted that prior to 1921 "there were no restrictions on immigration."

The debate over Ellis Island in the late 1950s and early 1960s took place during an historic lull in U.S. immigration history. The decade following 1955 saw an average of just 288,000 immigrants entering per year. In 1960, just 5.4 percent of all Americans were foreign-born, a historic low, compared to the nearly 15 percent of foreign-born Ameri-cans in 1910.

As immigration slowed to a trickle, the children and grandchildren of those who arrived at Ellis Island were assimilating into American life. In this world, Ellis Island was part of the cultural baggage left behind in the rush toward assimilation, together with tenement apartments, Eu-ropean accents, and unpronounceable names. Despite occasional pleas by people like Oscar Handlin and Edward Corsi, there was little public groundswell for saving Ellis Island.

The island was a mess. One newspaper referred to it as "a seedy ghost town." Though the buildings were structurally sound, vandalism and neglect took their toll. Thieves stole the copper fixtures in the buildings; Mother Nature did the rest. Chunks of plaster and tile had fallen from the ceilings; paint was peeling from the walls; wood was rotting; the roofs leaked. Artifacts of the island's previous life—mattresses, tables, medical equipment—were strewn about. Jungle-like vegetation weaved its way around the island unchecked and unmolested. Combined with the decaying buildings it helped create an eerie and spooky atmosphere on the island.

By the late 1960s, officials in Washington, and the public at large, were distracted by more pressing problems at home and abroad, and Philip Johnson's grand design for restoration was left unfunded. Ellis Island simply sat there, neglected, in New York Harbor amid both the affluence and growing chaos of postwar America.

DURING THE YEARS AFTER the closing of Ellis Island, race, not immigration, came to dominate the national agenda. At the same time that the *Wall Street Journal* ad appeared regarding the possible sale of Ellis Island, blacks in Montgomery, Alabama, were boycotting that city's public transportation system to protest the arrest of Rosa Parks for refusing to give up her seat on a bus to a white man. A young minister named Martin Luther King Jr. became the public face of the bus boycott and the protest against Jim Crow segregation. The modern civil rights movement had begun.

Race and immigration in America have an intertwined and complex relationship. The nation's racial history is a tortured field littered with the tragedy of slavery, discrimination, violence, false promises, and missed opportunities. In contrast, the history of immigration is largely painted in optimistic hues, where plucky immigrants overcome poverty and discrimination to live the American Dream, if not immediately, then over a few generations. Too often, the history of African-Americans is contrasted with that of immigrants, and none too favorably.

For some white European immigrants, their first sight of a black person was on Ellis Island. Austrian immigrant Estelle Miller remembers coming to Ellis Island as a thirteen-year-old and upon seeing a black man there for the first time, she grew so scared that she dropped her family's antique china bowl. But in truth her presence in America was more problematic to the black man. A Norwegian immigrant named Paul Knaplund remembers seeing a "Negro charwoman" during his time at Ellis Island. "Her face expressed utter disdain," he remembered as she watched the streams of immigrants passing before her.

American blacks have had at best an ambivalent attitude towards immigration. Periods of mass immigration have coincided with low points in African-American history. The Progressive Era of the early 1900s, which pushed liberal reform to the forefront of the nation's agenda, was driven largely by fears of mass European immigration and the changes that industrialism had wrought. Though reformist in nature, very little of Progressivism dealt with the rights of blacks. If anything, Jim Crow segregation hardened during this period. The great concern of middle-class, northern, urban reformers was not civil rights for southern blacks but the problems they saw in front of them, which had to do with the massive European immigration.

Meanwhile, black leaders such as Booker T. Washington and A. Philip Randolph were immigration restrictionists, seeing the constant demand for cheap immigrant labor as detrimental to the status and wallets of native-born blacks.

It is no surprise, then, that the civil rights movement of the postwar era took place at the point of lowest sustained immigration in American history. Unencumbered with the problems of immigrants, the nation's attention could focus upon the demands of African-Americans for full political and social rights.

The civil rights movement had some unexpected effects upon Ellis Island immigrants in those postwar years. Despite the rising political power of white ethnic groups, their solid position in the New Deal Democratic coalition, and the rise to power of the first Irish Catholic president, immigration quotas stubbornly remained in place. It was only in light of the Civil Rights Act that Congress and President Johnson could muster enough support to end discrimination against immigrants based on national origins.

The civil rights movement was about more than just changing laws; it was about the expression of racial pride and the inclusion of groups previously left on the margins of the nation's historical narrative. Both

386 / AMERICAN PASSAGE

themes would become tied up with the post-1960s history of Ellis Island. As immigrants took their place in the American mainstream, other groups looked to Ellis Island as they made their pleas for acceptance.

In the early morning hours of March 16, 1970, a small group of American Indians attempted to set off for Ellis Island undetected before daybreak. Their goal was to turn the island into a center for Indian culture, but a gas leak foiled their plans. After that, the Coast Guard stepped up patrols and proclaimed a zone of security around the island.

Perhaps the most bizarre incident occurred later that same year. It was an event that demonstrated what happened when you mixed the machinations of the Nixon administration with Black Power and Black Capitalism.

In 1966, a neurosurgeon named Thomas Matthew formed a group called NEGRO, the National Economic Growth and Reconstruction Organization. Arguing that welfare dependency had harmed blacks, Dr. Matthew called for a program of self-help. To that end, NEGRO would build hospitals, start black-owned businesses, and rebuild the inner city. Matthew planned to fund the organization by selling bonds at a block party and using the money raised to leverage government funds.

But the bond issue didn't quite work, and in a few years Matthew found himself convicted of failing to file his income tax returns since the early 1960s and accruing as much as $150,000 in back taxes and penalties. In late 1969, he began a six-month jail sentence and also agreed to make restitution to the IRS.

While Matthew's rhetoric was out of step with the Great Society and mainstream civil rights movement, his views caught the attention of Richard Nixon and his aides. Once in office, Nixon was stung by criticism that he was insensitive to civil rights. His administration would never win over traditional civil rights groups, so it took another tack by proclaiming its support for black capitalism to help minorities enter the nation's economic mainstream. The Nixon administration made money available to assist blacks with business opportunities. It was the perfect way to mix opposition to welfare with concern for blacks. And Dr. Thomas Matthew seemed made to order for Nixon.

Perhaps that was the reason that Nixon commuted Matthew's sentence for tax evasion, the administration's first executive clemency.

Matthew could be useful to the new administration, a black voice supporting Republican policies. In fact, the move began paying political dividends almost immediately when Matthew came out to support Nixon's embattled Supreme Court nominee, G. Harold Carswell. Matthew's views did not win him friends among other civil rights leaders, but it did give him political access to the Nixon administration, which was eager to have its Commerce Department and Small Business Administration assist black entrepreneurs.

It would prove to be an uneasy relationship, as a 1971 discussion made clear. In a White House meeting discussing the possible pardon of Jimmy Hoffa, Nixon and his aides brought up the case of Matthew in ways that laid bare their mixed feelings about the NEGRO leader and blacks in general. "He stole everybody blind," Nixon said of Matthew, referring to his earlier trouble and somewhat confusing Matthew's actual crime, "after all he was trying to do well by his people so we let him out. . . . They all steal—I mean not all. . . . People do when they are over their heads. He probably didn't know that he was stealing." At that point, one of the aides joked that Matthew "just liberated that money," to which Nixon responded in a more sympathetic vein that Matthew "was a very nice man, very nice. Had wonderful ideas."

Two days after being released from prison, Matthew announced his newest scheme. NEGRO would ask the Nixon administration to turn over control of Ellis Island to the organization under a "lend-lease" agreement. Matthew and his followers would then create an experimental community for one thousand black families.

Six months after Matthew's release from prison, there still was no formal agreement with the federal government on NEGRO's plan for the island. So Matthew and some sixty other members of the group began quietly squatting on the island. Unlike the earlier attempt by American Indians, the Coast Guard did nothing to drive them away. It appeared that the Nixon administration had given tacit approval to the move. Matthew's followers, many of them on welfare or recovering drug addicts, began to clear away the thick brush that had begun to take over the island. They hoped that the government would see this as a good-faith effort and grant them permanent control over the island.

The secret settlement on Ellis Island would soon end when a traffic helicopter for a local television station noticed laundry hanging out to dry at the supposedly deserted island. The press attempted to land on the island to interview the squatters, who were reluctant to cooperate. The unwanted publicity meant the end of the experiment, and after thirteen days, the small band left the island.

This did not deter Matthew, who offered a new and more detailed proposal that the National Park Service approved just a few weeks later. NEGRO received a special five-year permit for the island for no money. In turn, Matthew would turn the deserted island into an Eden of black capitalism. NEGRO would first rehabilitate the island and create "a living memorial to the American immigration experience on Ellis Island." Decaying buildings would be restored, crumbling sea-walls rebuilt, and the grounds cleared. The second, and more impor-tant, goal was the creation of a "rehabilitative community" for drug addicts, alcoholics, welfare recipients, and ex-cons, who would learn skills that would aid their reentry into mainstream society. Ellis Island would become a self-supporting community: NEGRO would build fac-tories that would make shoes, costume jewelry, and metal castings. The money made from these enterprises would allow NEGRO to expand its efforts to help more people. Matthew saw a future island with 1,700 workers, 700 hospital patients, and 100 schoolchildren.

Matthew continually referred to blacks as "new immigrants." If Ellis Island marked the rebirth of European peasants in the New World, Matthew sought to transfer that symbolism to American-born blacks. In his vision, Ellis Island could serve as a gateway for dispossessed and unskilled blacks to reenter American society, essentially turning them into immigrants in their own country. That poor American-born blacks should become like immigrants, new and old, has been a controversial trope in American history, adding further tensions between immigrants and native-born blacks.

Not surprisingly, Matthew's utopian plan never bore fruit, despite support from the Nixon administration. Part of the problem was the disconnect inherent in the idea that blacks were new immigrants. But the real problem was Matthew himself. Part opportunist, part genu-ine humanitarian, and part con artist, the doctor had a vision that far outstripped his managerial abilities and business skills. Matthew was unsuccessful in raising funds to bring his dream to fruition, and few blacks seemed ready to sign up for the arduous work of rehabilitating Ellis Island.

Only a handful of people remained on the island through the winter of 1970–1971. Conditions—a lack of potable water and inadequate heating and plumbing—hampered the efforts, and Matthew's group showed little aptitude for rectifying those problems. By the spring of 1971, a safety engineer found that the "deteriorated, dilapidated, unsanitary" conditions at Ellis Island could cause disease, injury, or even death to NEGRO members. The engineer recommended revoking NEGRO's permit to use Ellis Island.

Since the mid-1960s, Matthew had received at least $11 million in federal loans, grants, and contracts, and the Nixon administration, eager to aid the cause of black capitalism, refused to pull the plug on the Ellis Island operation. In the end, they didn't have to—the ineptitude and grandiose vision of Dr. Matthew did that for them. By the summer of 1971, only five people remained on the island; by the fall that number had dwindled to three. Instead of a vibrant industrial community with schools and hospitals, Ellis Island remained as it had been before: deteriorated and largely abandoned.

As the Ellis Island colony was falling apart, Matthew's Interfaith Hospital in Queens was drawing attention for its filthy conditions and poor treatment of patients. Reports suggested that top Nixon administration officials had refused to cooperate with, and even impeded, investigations into the business practices and contracts of NEGRO. In April 1973, Matthew was arrested on charges of illegally diverting $250,000 in Medicaid payments designated for Interfaith Hospital to other NEGRO projects.

By this time, Richard Nixon had been driven from office, Ellis Island remained a fallow wasteland, and Thomas Matthew's dreams of black capitalism—part scheme and part dream—had long since died.

In the flawed vision of Thomas Matthew, the renewed racial pride of African-Americans could not redeem a decaying and forgotten Ellis Island. Yet black power did bestow a peculiar—and unintended—gift on the descendants of white immigrants. The civil rights and black power movements challenged the concept of the melting pot, noting that black Americans were not so easily melted into the larger American stew. Race was a marker that white Americans did not seem to want to ignore and blacks seemed not to want to forget.

Around the same time, Nathan Glazer and Daniel Patrick Moynihan published a study of New York racial and ethnic groups entitled *Beyond the Melting Pot*. If, as the authors suggested, ethnicity had never completely disappeared in the melting pot, the growth of black power and racial pride among African-Americans helped spur white ethnic groups to more public displays of their own identity. "Kiss Me I'm Irish" and "Kiss Me I'm Polish" buttons appeared. By the 1960s, differences became badges of honor, not shame. Ethnic-themed novels like Philip Roth's *Portnoy's Complaint* and Mario Puzo's *The Godfather* climbed the bestseller list.

Ethnic pride and ethnic defensiveness went hand in hand. A young writer of Slovakian descent named Michael Novak published a jeremiad called *The Rise of the Unmeltable Ethnics*. Defending white ethnics from a variety of charges, Novak also lashed out against "Nordic prejudices" and moralizing, liberal WASPs. It was now the children and grandchildren of the Ellis Island immigrants who found themselves in conflict against "progressive" Nordics and Anglo-Saxons. Outright prejudice and discrimination may have disappeared, but cultural and political conflicts remained.

In the early twentieth century, Americans debated who should or should not be allowed to enter the country at places like Ellis Island. By the second half of the twentieth century, Ellis Island had been forgotten and sat in New York Harbor as a rotting symbol of a bygone era. Before the twentieth century ended, it would be reborn under a different guise—as a museum and a national monument. But the debate over its meaning would continue.

Chapter 19

The New Plymouth Rock

---

*Once I had set foot again on Ellis Island, I knew that I had
come to one of God's places, and that those of us who had
been there were tied to it forever.*
—Mark Helprin, *Ellis Island and Other Stories*

LINO ANTHONY IACOCCA HAD MUCH TO BE PROUD OF
on the night of July 3, 1986. At age sixty-one, he already had a successful career in the auto industry, running Ford Motor Company before
guiding Chrysler out of bankruptcy—with a little help from Uncle
Sam. His recently published autobiography had sold more than 5 million copies. He received as many as five hundred letters a day from
average Americans asking him for advice or thanking him for providing
inspiration in their own lives. Newspapers called him a folk hero for
the 1980s. And he had just overseen a nationwide campaign that raised
almost $300 million for the renovation of the Statue of Liberty and
Ellis Island.

On this patriotic Fourth of July weekend, Iacocca presided over a
glitzy celebration in New York Harbor that featured the relighting of
the newly refurbished Statue of Liberty on its hundredth anniversary.
Politicians, celebrities, and other dignitaries filled the stands to watch
the fireworks display. President Ronald Reagan was on hand to pull
the switch that would light the statue. It arguably could not have been
done without Iacocca, and the shrewd salesman was not shy in letting
everyone know it.

That was not a shabby record for the son of Italian immigrants who had grown up in Allentown, Pennsylvania. Iacocca was an Italian-American mix of Horatio Alger and Dale Carnegie. With his craggy features and gravelly voice, Iacocca was an icon of modern-day America. Had they been alive, immigration restrictionists Francis Walker or Prescott Hall would have been shocked at the presence of an Italian-American head of a major U.S. corporation.

Some wondered how a private businessman ended up in charge of the restoration of public icons like the Statue of Liberty and Ellis Island. It was partly a matter of timing. The federal government had neglected Ellis Island for thirty years. Then Ronald Reagan rode into the White House on a wave of antigovernment sentiment. "Government is the problem, not the solution," he said, tapping into a national mood that had less faith in government after the social, political, and economic turmoil of the 1960s and 1970s. Rather than relying on the public sector, the Reagan administration pushed for what it called "public-private partnerships."

In this vein, the National Park Service began to solicit private assistance to raise funds to restore Ellis Island in 1981. Richard Rovsek, a marketing executive who produced the Easter egg rolls at the Reagan White House, founded the Statue of Liberty–Ellis Island Foundation to raise private money to restore both monuments in New York Harbor. Thus, the private half of the public-private partnership was born.

To oversee the fundraising efforts, Interior Secretary James Watt created the Statue of Liberty–Ellis Island Centennial Commission in 1982. Here was the public half of the equation. As implied by the commission's name, it was hoped that the Statue of Liberty could be restored by its hundredth anniversary in 1986 and Ellis Island by its hundredth anniversary in 1992. Watt named Lee Iacocca to chair the new commission. Not happy with the largely advisory role of the Centennial Commission, Iacocca soon maneuvered to become head of the private foundation as well.

Iacocca also maneuvered to make the Statue of Liberty–Ellis Island Foundation the sole fundraiser for the project, despite the existence of other organizations, such as Philip Lax's Ellis Island Restoration Commission. In the end, Iacocca had become the boss of both the fundraising and the restoration efforts.

Although restoration of the two monuments was linked, it was clear that Ellis Island would play second fiddle. The centennial anniversary of the Statue of Liberty in 1986 made its renovation a more pressing matter, but also it was far better known to the public. "Ellis Island, in the public mind, was a poor cousin to the Statue of Liberty," wrote F. Ross Holland, who was involved in the fundraising and restoration effort. "The foundation had publicized Ellis Island, but it was evident the public was more interested in the Statue of Liberty."

The Statue of Liberty therefore became the center of Iacocca's fundraising. A master salesman, he wasted no time. While individual donations would be important, he knew that if he wanted to raise $200 million he would need to solicit corporate sponsorships—which he did. Coca-Cola, *USA Today*, Stroh's Brewery, Chrysler, Kodak, Nestlé, Oscar Mayer, and U.S. Tobacco were all granted exclusive rights to use the Statue of Liberty in their advertisements. The public seemed to respond to the fundraising effort. When American Express promised to donate a penny from each purchase, AmEx card use jumped by 28 percent.

The Statue of Liberty and Ellis Island were now bound up with the larger political and ideological controversies of the day. It was the height of the Reagan Revolution, whose championing of free-market capitalism and the entrepreneurial spirit did not sit well with everyone.

In November 1985, the left-wing magazine *The Nation* began a series of articles by journalists Roberta Gratz and Eric Fettmann attacking Iacocca and his fundraising campaign. The first article, "The Selling of Miss Liberty," was accompanied by a cover featuring a cartoon of Iacocca dressed as the Statue of Liberty, smoking a cigar and holding a money bag in place of the usual torch. Gratz and Fettmann argued that the fundraising effort was trashing an American icon. "What follows is the story of a corporate takeover of a national shrine at a time when corporate raids are an everyday occurrence," Gratz and Fettmann wrote.

Despite the criticisms, fundraising continued at a record pace, culminating in the unveiling of the Statue of Liberty on the night of July 3, 1986. The event was a huge spectacle. While Iacocca's efforts had made the night a reality, television producer David Wolper was in charge of the entertainment. The producer of *Roots* put together a star-studded lineup for the weekend that included Frank Sinatra, Helen Hayes, Neil

Diamond, Gregory Peck, and José Feliciano. There were song-and-dance numbers as well as historical films, fireworks, tall ships in the harbor, and the release of balloons and doves. Chief Justice Warren Burger swore in 2,000 new citizens—including Mikhail Baryshnikov—at Ellis Island, while 38,000 more participated by video hookup. All 40,000 would simultaneously join in the singing of "America the Beautiful."

For some, it was all too much. Jacob Weisberg, in a dyspeptic anticipatory piece for *The New Republic*, wrote that the celebration was "likely to be remembered as the most revolting display of patriotic glitz and tacky pageantry in this country's history." Despite this, most Americans seemed happy with what they saw of the newly refurbished Statue of Liberty. The criticisms of Iacocca, however, did not end.

Months before Liberty Weekend, Secretary of the Interior Donald Hodel, who replaced Watt, had fired Iacocca from the Statue of Liberty–Ellis Island Centennial Commission. The businessman still remained as head of the Statue of Liberty–Ellis Island Foundation. Some suggested that Republicans feared that the politically ambiguous Iacocca might use his celebrity as a platform to run for office as a Democrat. Others suggested that the administration was not happy with Iacocca's plans for Ellis Island.

Whatever one thought of the Liberty Weekend extravaganza or of Iacocca, there was still more work to do. Ellis Island was still far from being ready for its public unveiling. By March 1987, Iacocca's foundation had raised over $300 million from private sources. By 1991, the figure would reach $350 million.

If the public seemed to be more captivated by the Statue of Liberty, Iacocca made it clear that the driving force behind his work was Ellis Island. For him, the statue was "a beautiful symbol of what it means to be free," but Ellis Island was the "reality." If you want to prosper, Iacocca wrote, "there's a price to pay. . . . Apply yourself. . . . It isn't easy, but if you keep your nose to the grindstone and work at it, it's amazing how in a free society you can become as great as you want to be." For Iacocca, Ellis Island had become a symbol of immigrant success and American greatness.

His father, Nicola Iacocca, had come to America in 1902 at the age of twelve and eventually ended up in Allentown, Pennsylvania. Nineteen years later, Nicola returned to Italy to bring back a wife. When the newlywed couple arrived at Ellis Island, according to Iacocca family lore, the bride was sick with typhus fever and had lost her hair. When inspectors tried to hold her for further examination, Nicola, an "aggressive, fast-talking operator," convinced them she was just suffering from seasickness. It worked and the couple was allowed to land. It is not a terribly plausible story—especially considering the fear that typhus fever had caused in the past—but one that Iacocca often repeated.

Because Ellis Island had great meaning in the Iacocca household, Lee saw his fundraising work as a "labor of love for my mother and father." For him, the Great Hall took on near-religious significance. It was "a cathedral, a churchlike setting, a place to pray. It brings tears to your eyes." Iacocca wrote in his autobiography that Ellis Island "was part of my being, not the place itself, but what it stood for and how tough an experience it was."

"Hard work, the dignity of labor, the fight for what's right—these are the things the Statue of Liberty and Ellis Island stand for," Iacocca argued. Although the Iacocca family's experience at Ellis Island was one of potential pitfalls and tragedy averted, it had now become a symbol of pride and success for the descendants of immigrants who passed through there. For Iacocca and many others with similar backgrounds, Ellis Island was increasingly entwined with their vision of the American Dream.

To others, that vision had distinct political and ideological implications. Some historians did not want the museum's theme to be about the old melting pot, but rather about cultural pluralism. Gratz and Fettmann, who criticized the fundraising for the Statue of Liberty, also took on the restoration of Ellis Island. "Should Ellis . . . portray the history of the great immigration wave, warts and all, or will it become . . . 'an ethnic Disneyland'?" The authors worried that its history might be "prettified" and wondered how "historical appropriateness" would be balanced with "commercial hucksterism." Deeply suspicious of the private sector, Gratz and Fettmann could only see the "logoization" of the Statue of Liberty and Ellis Island. "As often happens when private control is substituted for public accountability, the unifying power of the public good is diminished," they wrote. "A great opportunity was lost to place our common heritage above private gain."

A historian made a similar point, worrying that the new museum would reflect corporate values and become nothing more than "a Disney-like 'Immigrant Land'—with smiling native-garbed workers selling

Coca-Cola to strains of 'It's a Small World After All.'" Even worse, the museum might actually end up glorifying Ellis Island immigrants in a kind of "ethnic populism."

How should the old immigration station be remembered? Two 1984 letters to the *New York Times* symbolized this conflicted memory. The first called Ellis Island a "best forgotten" symbol. "It offered neither welcome nor haven," the writer continued. "Like the Bastille, it has not been missed." The second letter argued that it was the "struggle and eventual triumph" of immigrants "that Ellis Island rightly commemorates." How people interpreted the meaning of Ellis Island was becoming more important than what had actually occurred there.

The former inspection station was well on its way to becoming a national shrine, which meant linking Ellis Island to that original founding place of memory: Plymouth Rock. This formulation not only elevated the dreary former inspection station into the nation's symbolic pantheon; it also resonated with the idea that newer immigrant groups were supplanting the nation's Pilgrim founders. Much as groups like the Society of Mayflower Descendants helped to establish their claim to ownership of America, the descendants of Ellis Island immigrants were now claiming their place. Ellis Island was the new Plymouth Rock and the immigrants who passed through it were the Pilgrims of a modern, multicultural America.

This process began much earlier than most people believe. One can trace Ellis Island's evolution into a national icon as far back as 1903 when Jacob Riis pronounced it "the nation's gateway to the promised land." Two years later, the *Boston Transcript* dubbed it "the Twentieth Century Plymouth Rock," while *The Youth's Companion* wrote about "The New Plymouth Rock."

In 1914, a writer named Mary Antin argued that the "ghost of the Mayflower pilots every immigrant ship, and Ellis Island is another name for Plymouth Rock." For a Russian Jewish immigrant like Antin, linking Plymouth Rock to Ellis Island was a forthright way to express her Americanness and rebuke opponents of immigration.

That an immigrant like Antin would have the temerity to equate Plymouth Rock with Ellis Island was too much for the novelist Agnes Repplier. "Had the Pilgrim Fathers been met on Plymouth Rock by immigration officials, had their children been placed immediately in good free schools, and given the care of doctors, dentists, and nurses," she asked, "what pioneer virtues would they have developed." To equate Plymouth Rock with Ellis Island assumed that modern immigrants were the equal of the original settlers and their descendants, a leap of judgment that was just too far-fetched for Repplier.

Other native-born Americans nervously saw the passing of the baton from Plymouth Rock to Ellis Island as inevitable. A New York City schoolteacher in the early twentieth century was unable to get her largely first- and second-generation pupils to answer basic questions about U.S. history. When all else failed, she asked: Where is Ellis Island? She had finally hit upon the right question, as every hand in the room was raised and "the light of intelligence gleamed from every pair of eyes." While the teacher had always looked with veneration upon Plymouth Rock, the history of these schoolchildren and millions of new Americans now began at Ellis Island.

In the late 1930s and early 1940s, a Slovenian immigrant named Louis Adamic traveled the country giving a speech entitled "Plymouth Rock and Ellis Island."

> The beginning of their vital American background as groups
> is not the glorified Mayflower, but the as yet unglorified
> immigrant steerage; not Plymouth Rock or Jamestown,
> but Castle Garden or Ellis Island or Angel Island or the
> International Bridge or the Mexican and Canadian border,
> not the wilderness of New England, but the social-economic
> jungle of the city slums and the factory system.

With the United States heading toward involvement in another European war, Adamic hoped the inclusion of Ellis Island into America's historic pantheon would help unify the diverse nation. "Let's make America safe for differences," he exhorted his audiences. "Let us work for unity within diversity."

After the war, Ellis Island fell off the nation's collective radar, finding itself uncomfortably in the news with the detentions of enemy aliens during World War II and suspected radicals during the Cold War. However, the revival of white ethnic identity during the late 1960s and 1970s helped bring more attention to Ellis Island.

In deeply nostalgic tones, Leo Rosten wrote an article for *Look* in 1968 entitled: "Not So Long Ago, There Was a Magic Island." Around the same time, Senator Ted Kennedy, a descendant of pre-Ellis Island

Irish immigrants, penned a piece in *Esquire* about Ellis Island and those who passed through it. "They came—creative, industrious, unafraid," Kennedy wrote. "Today Ellis Island stands as a symbol, in new glory, of the oldest theme in our history. It reminds us all that the nobility to which America has risen was born of humble origins." In 1975, National Parks Service historian Thomas Pitkin published the first comprehensive history of Ellis Island. "There is nothing really fanciful in calling Ellis Island, as has been done, the Plymouth Rock of its day," Pitkin wrote in conclusion. "There is no single point in the country where American social history for a generation and more comes to a sharper focus."

In the late 1970s, a group of Armenian-Americans gathered at Ellis Island "to express gratitude to their adopted land of freedom." Set Charles Momjian, one of the event's organizers, captured its meaning for Armenians and other immigrants. "For many, Ellis Island was a sad and disconcerting beginning to life in the United States," Momjian wrote. "It is therefore a measure of our success as Americans that we return to this place, no longer afraid, intimidated, or bewildered, but confident and grateful for the blessings we have experienced in this country." The novelist William Saroyan described how his grandmother was almost excluded upon arrival because of poor eyesight. Though Saroyan was born in the United States, he wrote that Ellis Island was in his "very marrow."

Riding this wave of nostalgia and ethnic pride, Peter Sammartino, a university official and son of Italian immigrants, began the Restore Ellis Island Committee in hopes of eventually opening it to visitors. It succeeded in getting Congress to appropriate $1 million for the effort, as well as $7 million to rebuild the island's seawall. Thanks to Sammartino's efforts, the National Park Service opened the main building to the public for a limited number of guided tours in 1976, but the island was still a mess. Journalist Sydney Schanberg called it "about as romantic as a row of hollow buildings in the South Bronx." Ellis Island was closed again for repairs in 1984.

In that same year, Geraldine Ferraro became not just the first woman from a major American political party to run on a presidential ticket, but also the first Ellis Island descendant. Walter Mondale's selection of Ferraro as a running mate signaled the importance of ethnic identity to the campaign. So did the choice of Mario Cuomo, another New

York Democrat, to give the keynote address at the party's convention.

"The Battle for Ellis Island" is how political writer Michael Barone dubbed the campaign. "What's important in 1984 is not how each ticket appeals to specific ethnic groups," Barone wrote, "but which is more successful in appealing to the Ellis Island tradition generally." Republicans would point to free-market capitalism as instrumental in the success of Ellis Island immigrants, while Democrats would argue for the importance of the New Deal in making second- and third-generation Americans part of the middle class.

In 1988, the Democratic Party nominated another child of Ellis Island as its presidential candidate. Running against blue-blood Republican George H. W. Bush, Michael Dukakis played up his background as the son of Greek immigrants. In his acceptance speech, the Massachusetts governor made prominent mention of his late father, "who arrived at Ellis Island with only $25 in his pocket, but with a deep and abiding faith in the promise of America." During the campaign, Dukakis made a red, white, and blue appearance at Ellis Island, where he discussed his parents' arrival decades earlier. "Their story is your story," he said. "It is our story; it is the story of America."

Ferraro and Dukakis were political losers, but as *Newsweek*'s Meg Greenfield noted, Ellis Island "has become the East Coast equivalent of the log cabin, poor farm-boy upbringing and the rest of that Americana unavailable to so many people with exotic surnames."

Now firmly entrenched in the nation's psyche and historical memory, Ellis Island was once again ready to take its place on the public stage. After years of restoration and fundraising by the Statue of Liberty–Ellis Island Foundation, a renovated Ellis Island was finally reopened on September 9, 1990. An economic recession that year led to a far more restrained event than the glitzy 1986 unveiling of the refurbished Statue of Liberty.

At the cost of over $150 million, the main building on the island's north side was opened to the public as an immigration museum. Visitors disembarking from the ferry would stroll up the path toward the building just as many of their ancestors had. Arriving in the first floor, they would then make their way up a set of stairs, a replica of the original, where inspectors and doctors once closely examined immigrants as they wound their way upstairs. Visitors would then enter the Great Hall.

Though once filled with immigrants marching toward clerks waiting to question them, the renovated Great Hall was starkly empty. Side rooms contained explanations of the immigration inspection process. Tourists could visit a hearing room used by the boards of special inquiry, as well as the detention rooms where immigrants slept in canvas bunk beds stacked three to the ceiling, hanging from wires. The renovation of the main building won rave reviews from *New York Times* architecture critic Paul Goldberger, who called it "skillfully designed, brilliantly executed."

On many levels, the restoration of Ellis Island has been a success. Many visitors come not just to see the renovated main building and museum, but also for something called the American Immigrant Wall of Honor. Iacocca took this idea from the Wall of Sixteen Million that appeared in Philip Johnson's 1960s design, but he added his own salesman's twist. With his philosophy of "Give 'em a piece of it," Iacocca decided to charge people to put their names or the names of their ancestors on the wall. By 1993, the wall had raised more than $42 million, with the potential for more money from more names on expanding walls in the future.

By accident, Iacocca added to the confusion of Ellis Island. Most visitors believe that the Wall of Honor lists the name of every immigrant who passed through the island. Many are upset when they do not find their ancestor's name on the wall. The reality is that an immigrant's name appears on the Wall of Honor only after their descendants donate at least $100. No money, no name.

More confusion followed. Samuel Freedman found that his grandparents are listed twice since both his father and an uncle or aunt had separately given money to list their parents' names. To confuse matters even more, when visitors see the Wall of Honor, they are liable to find scattered among the over 700,000 names (as of 2008) such "immigrants" as Miles Standish, Paul Revere, and Thomas Thayer, whose name was added thanks to the donation of one of his descendants, former First Lady Barbara Bush.

At the other side of the historical spectrum, Ellis Island immigrants share space on the walls with people like Shin Ki Kang, a Korean immigrant who came to America in 1977 and Parvis Mehran, a recent arrival from Iran. The grab-bag nature of the Wall led a reporter to describe it as the "apotheosis of the American dream, allowing anyone and everyone to purchase a place in American history."

What name does one put on the wall? One woman complained that she would have loved to have her grandfather's name on the Wall of Honor, but did not know which name to use? Should it be Nehemiah Nohr, his given name at birth, or "the name assigned by the authorities at Ellis Island and by which he would be known for the next 50 years as a naturalized American, Jacob Friedman?" This woman complained that the Nohrs had vanished into history, "obliterated . . . without known reason, creating a dilemma that those Ellis Island clerks could hardly have foreseen."

The connection between Ellis Island and the issue of names remains tightly drawn in the public mind. As in the mind of the daughter of Jacob Friedman, Ellis Island has become synonymous with the changing of immigrant names.

The most famous story of an Ellis Island name change is that of Sean Ferguson. This Jewish immigrant was reportedly given his Scottish-sounding name by inspectors who, after asking the confused immigrant his name, received a response in Yiddish: *Schoen vergessen,* meaning "I forgot." Thus was baptized Sean Ferguson.

The stories multiply. Immigrants from Berlin received the last name Berliner from officials. Then there is the story of a Jewish orphan who told inspectors he was a *yosem,* an orphan, and found his new name as Josem. Another immigrant was supposedly told by officials to "Put your mark in this space" and found his name had become Yormark. In the HBO series *The Sopranos,* a mobster named Phil Leotardo complains at a family gathering that the family's original name was Leonardo—after Leonardo Da Vinci—but was changed to Leotardo at Ellis Island. When his grandchild asks why, Phil responds: "Because they are stupid, that's why. And jealous. They disrespected a proud Italian heritage and named us after a ballet costume." A popular 1994 children's book is entitled *If Your Name Was Changed at Ellis Island.*

In an interview later in her life, Sophia Kreitzberg retells the story that her stepfather told her about his time at Ellis Island. Officials asked him his name and he replied Kogan. "Kogan Shmogan," the inspector allegedly told Sophia's stepfather, "that's not an American name," and the official renamed him Sam Cohen. "They gave everybody the name of Cohen or Schwartz or something," she said. "That's why you find so many Jewish people with the same ethnic names. They were given those names by the people in Ellis Island."

Then there is the joke about the Chinese laundry owner named Moishe Pipik. When asked how a Chinese man got such a strange name, Pipik explained: "At Ellis Island, I stand in line behind man named Moishe Pipik. When my turn come, man ask my name, I say 'Sam Ting.'" That a Chinese immigrant would easily pass through Ellis Island, despite the Chinese Exclusion Act, signals the apocryphal nature of the story.

Nearly all of these name-change stories are false. Names were not changed at Ellis Island. The proof is found when one considers that inspectors never wrote down the names of incoming immigrants. The only list of names came from the manifests of steamships, filled out by ship officials in Europe. In the era before visas, there was no official record of entering immigrants except those manifests. When immigrants reached the end of the line in the Great Hall, they stood before an immigration clerk with the huge manifest opened in front of him. The clerk then proceeded, usually through interpreters, to ask questions based on those found in the manifests. Their goal was to make sure the answers matched.

The only time immigration officials at Ellis Island wrote down names was when immigrants were held for hearings or medical help. Officials would include aliases and possible permutations of the names of such immigrants on their paperwork. However, these were not official documents, just internal paperwork, and did not have the power to change an immigrant's name officially.

Name changes largely occurred either on the other side of the Atlantic, when steamship officials recorded names in their manifests, or after Ellis Island, when immigrants filled out naturalization papers or other official documents. Often immigrants voluntarily chose to Americanize their names to adapt to their new home.

There is at least one instance of a name change at Ellis Island, however. Frank Woodhull, who had been born a woman named Mary Johnson, but had lived the previous fifteen years of her life passing as a man, arrived at Ellis Island listed as Frank Woodhull on the ship's manifest. After spending one day in detention while authorities figured out whether to admit him, Woodhull was finally allowed to proceed to New Orleans, but not before officials crossed out Woodhull's name on the ship's manifest and penciled "Mary Johnson" in its stead. But this was clearly an exceptional case.

Yet the name change story lives on as urban legend. Many Americans are convinced of its truth because their grandparents told them the story. It is a convenient myth that emphasizes the traumatic nature of Ellis Island and the supposed rough treatment of immigrants, as well as the facility's role in Americanizing immigrants, often against their will. The story serves as a convenient cover for the uncomfortable fact that many immigrants voluntarily discarded their Old World names in an effort to assimilate into American society. Better to blame insensitive immigration officials than Grandpa for the fact that your name is Smith and not Hryczyszyn.

The inclusive nature of the Wall, encompassing Massachusetts Puritans and Korean businessmen, as well as those who actually passed through Ellis Island, brings up larger questions about the memorialization of Ellis Island. Should it be seen as a shrine to celebrate the experiences of those who passed through it? Should it represent immigrants from every era of American history? Or should it commemorate the experiences of everyone who came to America in myriad ways, from eighteenth-century slave ships to early nineteenth-century coffin ships to modern immigrants who arrive at airports?

Even though the restoration of Ellis Island has drawn public acclaim, many scholars have been critical of its evolution into a national icon. Their concerns revolve around three issues. First, the memorializing of Ellis Island should not be used to make negative comparisons with newer immigrant groups. Second, the refurbished Ellis Island should not lead to ideological celebrations of the free-market or "up-by-the-bootstraps" homilies. Last, critics contend that the "nation of immigrants" saga embodied in the Ellis Island story leaves out groups that did not voluntarily emigrate to the United States, namely American Indians and the descendants of African slaves.

Historians are supposed to clear out the fog created by the construction of historical memory, but too often their work betrays an attempt to construct a historical memory that serves an ideological purpose. For example, historian Mike Wallace complained about the lack of "fresh thinking" at the island's museum and helpfully suggested exhibits on "the effect on immigration flows of actions taken by the International Monetary Fund, major multinationals, and the Central Intelligence Agency." He believed that the new immigration museum had nothing that would help people probe contemporary anti-immigrant attitudes. "It would be perfectly possible to leave Ellis," Wallace writes, "with warm feelings toward the old migrants and preexisting resentments of gooks, spicks and towel-heads left intact."

In addition, Wallace and other leftists were concerned that the restoration of Ellis Island abetted the rise of American conservatism. "At the heart of the Reagan/Iacocca reading of the history of immigration was the 'up-from-poverty' saga of the model of white ethnics," wrote Wallace. This amounted to nothing more than an "antigovernment screed" that facilitated the contemporary policies of the Reagan administration.

Art professor Erica Rand sees Ellis Island through the edgier prisms of gender and queer studies. Her 2005 book *Ellis Island Snow Globe* devotes space not only to predictable denunciations of commercialism, but also to more entertaining discussions of same-sex eroticism, with chapters such as "Breeders on a Golf Ball: Normalizing Sex at Ellis Island." Rand is also concerned about the exclusionary nature of the site, which she sees as privileging the historical narrative of one group. She worried that the "claim that the Ellis Island museum honors all immigrants, all migrants, or even all who 'people America' also functions to mask the inequity involved in the concentration of heritage resources at a site that honors and documents primarily white people."

It is hard for some to disentangle the memory of Ellis Island from discussions of race. Many black Americans felt left out of the celebrations of the Statue of Liberty and Ellis Island, even though few Americans seem aware that black nationalist Marcus Garvey, social scientist Kenneth Clark, and Harlem Renaissance writer Claude McKay were among the roughly 143,000 black immigrants—mostly from the Caribbean—who came through Ellis Island between 1899 and 1937. The disconnect was exemplified by black historian John Hope Franklin, who admitted that the renovation of the Statue of Liberty and Ellis Island was "a celebration for immigrants and that has nothing to do with me. I'm interested in it as an event, but I don't feel involved in it."

David Roediger's *Working Toward Whiteness: How America's Immigrants Became White: The Strange Journey from Ellis Island to the Suburbs* exemplifies this unease. What makes the journey from Ellis Island to suburbia so "strange" is not immediately apparent, but it has something to do with the idea that immigrants had to consciously "become white" in order to move into the mainstream of society, and in doing so they bought into ideas of white supremacy, turned their backs on African-Americans, and failed to place themselves in the vanguard of the proletariat for a revolution against capitalism. Apart from the title, Ellis Island barely makes an appearance in the book, but serves as a convenient symbol for Roediger's ideological tract.

For historian Matthew Frye Jacobson, the memorialization of Ellis Island is tied to the troublesome idea of America as a "nation of immigrants." This idea is problematic because it excludes from our national mythology those black Americans and American Indians not descended from immigrants. Just as bad for Jacobson, "the immigrant myth and immigrants' real-life descendants contributed to the swing vote that rendered the Republicans the majority party in the electoral realignment beginning in 1968," an outcome that he abhors. Jacobson implies that the arrival of European immigrants was a bad deal for civil rights. Channeling Malcolm X, he writes: "We didn't land on Ellis Island, my brothers and sisters—Ellis Island landed on us."

Another group was also feeling left out of the whole "nation of immigrants" celebration. To describe the United States in that way, says political scientist Samuel Huntington, "is to stretch a partial truth into a misleading falsehood." Huntington is speaking for white Anglo-Saxon Protestants, whose ancestors, he argues, were settlers, not immigrants. On a similar note, the author of a history of Plymouth Rock argues that, as they visit the Ellis Island museum, "the descendants of Pilgrims do not have to be told that this is a society to which they need not apply."

These criticisms suggest another question about the rehabilitation of Ellis Island. As the federal government originally created the inspection station to exclude undesirable immigrants, is the National Park Service now practicing another kind of exclusion in its celebration of the immigrants who came through there?

In the years since, the National Park Service and the Statue of Liberty–Ellis Island Foundation have made great efforts to be historically inclusive. "It doesn't matter whether your family arrived on the *Mayflower* or recently got off the airplane from Honduras," explained Gary G. Roth, the National Park Service's project manager for the immigration museum. "Ellis Island is a symbol of four hundred years of immigration. The story of it all is told here, including that of Native

Americans and of forced immigrations, the slaves who were brought here against their will."

In 2006, the foundation began fundraising for a project entitled "The Peopling of America Center." The new museum will show "the entire panorama of the American experience," and look beyond the traditional tale of immigration, which excludes those brought over in slave ships and native peoples residing on the continent prior to European colonization. As if to emphasize the inclusive nature of the project, as opposed to the allegedly narrow and exclusionary nature of the current museum, which focuses almost exclusively on Ellis Island immigrants, the center's motto is: "It's About *All* of Us!"

ELLIS ISLAND'S ICONIC STATUS is ever-present. When the online brokerage firm TD Ameritrade launched a new advertising campaign, it chose as its theme the celebration of America's independent spirit. To embody that spirit, it picked Ellis Island immigrants. "When immigrants came to Ellis Island they carried a dream," intoned the company's spokesman, Sam Waterston: "Work hard and opportunity will follow." The company's newspaper ad featured Waterston standing next to a large photo of an immigrant family standing on Ellis Island and looking at the Statue of Liberty, as well as a copy of the famous painting of the signing of the Declaration of Independence. In big letters, the ad stated, "Independence is the spirit that drives America's most successful investors."

In an episode from the fifth season of *The Apprentice*, those vying for the opportunity to work for Donald Trump were given the task of creating a new souvenir booklet for visitors to Ellis Island. "Yes, even Donald Trump seems to appreciate the historic importance and magical allure of this great national monument to freedom and opportunity," proclaimed the newsletter of the Statue of Liberty–Ellis Island Foundation.

The island takes a prominent role in movies like *Godfather II, Hitch, Hester Street,* and *Brother from Another Planet.* The 2006 Italian film *Nuovomondo*, titled *Golden Door* in its American release, deals with Sicilian immigrants who pass through Ellis Island. The film is evocative of the dislocation and confusion of the inspection process; however, it is also historically inaccurate. It shows all immigrants undergoing rigorous physical and mental testing. In reality, the relatively small staff at Ellis Island meant a hasty inspection for most who passed through. Only if immigrants were suspected of having some deficiency did they undergo the full battery of mental and physical testing. William Williams only wished that he could have inspected every immigrant as closely as we see in *Golden Door*.

In the 1990s, New York and New Jersey fought a protracted legal battle for jurisdiction over Ellis Island. In 1998, this battle eventually made its way to the U.S. Supreme Court, which ruled in a 6-3 decision that all but three acres of the site belonged to New Jersey. The Court's majority relied on an 1834 agreement between the two states that granted the then three-acre island to New York while allowing New Jersey to retain the rights to the surrounding waters and submerged land. Some of that land was eventually added to Ellis Island as it expanded.

Despite the rhetoric from both sides, the fight had little to do with lofty issues and more to do with who would control the development of the rest of the island and the taxes it would generate. Beginning in the 1980s, there had been talk of redeveloping the south side of the island, which used to house medical facilities. The new plan included demolishing some of the abandoned buildings and replacing them with a hotel and conference center, with the money from the commercial sites paying for the restoration of the rest of the buildings. New Jersey wanted to build a footbridge from its side of the Hudson to the island. Lee Iacocca had other ideas, including a nebulous plan for an "ethnic Williamsburg," an exhibition center devoted to ethnic arts and crafts and food.

In the end, none of the plans was approved and the southern half of the island remained fallow as preservationists staunchly opposed the idea of commercial development. Unfortunately, they had little money with which to restore the decaying southern section of the island. At this point, the National Park Service stepped in and entered into an agreement with a newly formed nonprofit organization called Save Ellis Island, which was now authorized to raise money for the rehabilitation of the island's southern section.

"Establish the Ellis Island Institute and Conference Center in the thirty unrestored buildings on Ellis Island," its mission statement declares. "The Ellis Island Institute and Conference Center will capture the power of place to become a world class facility for civic engagement and life long learning on the topics of immigration, diversity, human health and wellbeing, the themes of Ellis Island."

To help with fundraising and raise awareness of the project, the clothing maker Arrow launched a nationwide public relations effort. It created a high-production-value advertising campaign with television spots and posters featuring actors Elliot Gould and Christian Slater, pro-football Hall of Famer Joe Montana, *American Idol* finalist Kathryn McPhee, and cast members from *The Sopranos*. Everyone was fashionably dressed—no doubt in Arrow clothing—as they walked through the abandoned buildings of the island's south side to the haunting notes from a string orchestra. To support the effort, the public can buy "Save Ellis Island" T-shirts and leave their family's immigrant stories on a website.

For those wondering what a clothing maker has to do with immigration, Arrow created the slogan: "Ellis Island. Where the World Came Together and American Style Began." Posters reinforced the link between Ellis Island, the American Dream, and the themes of family, opportunity, and freedom. Although Christian Slater's ancestors were decidedly old immigrants from Ireland and England and it is not clear whether they came through Ellis Island, his poster reads: "Ellis Island represents our foundation—a place of possibility and new beginning." To Kathryn McPhee, Ellis Island is about "the collective heritage of the American Dream." For Joe Montana, it is about his Italian immigrant ancestors who worked in the mines of Pennsylvania to create opportunity for their family. "Triumph against the odds," his poster reads. "That's authentic American style."

In a different context, the National Park Service's superintendent of Ellis Island supported the restoration for just the opposite reasons. "It is haunting," Cynthia Garrett said of the island's abandoned south side, whose hospital buildings witnessed many tragedies of disease and death. "It tells us that our history isn't all positive stories and success."

Whether Ellis Island is a story of uplift and success or harrowing tragedies, it has evolved into something akin to a national shrine. In an editorial on the Supreme Court case, the *New York Times* referred to "Ellis Island's sacred history." In 2001, New York City's mayor, Rudy Giuliani, summed up this trend when he said at a naturalization ceremony on the island: "Ellis Island is a wonderful place, it's a sacred place, and it's hallowed ground in American history."

Such talk would no doubt have amused William Williams, Frederic Howe, and so many others who had worked at Ellis Island during its heyday. It would have baffled those immigrants who had to navigate the obstacle course at Ellis Island and probably saw little of the sacred in their experiences.

It was not preordained that Ellis Island should end up as a national shrine. San Francisco's Angel Island holds none of the same allure for descendants of Chinese immigrants, who received a much harsher reception than European immigrants. The Hotel de la Inmigración in Buenos Aires, known as Argentina's Ellis Island, pales in comparison to its northern namesake. Though it has also been turned into an immigration museum where modern Argentines can trace their ancestors who arrived a century earlier, it receives few visitors and is nestled off the city's beaten path.

Having said that, each generation makes its own history and there is nothing wrong with the descendants of Ellis Island reclaiming a historic site that was created, in part, to exclude their ancestors. Too many critics, eager to score political points, ignore the ways that the memorialization of Ellis Island stands as a sharp rebuke to nativists, both past and present.

That said, the historical memory of Ellis Island, like all memory, has been created over time, and that memory will continue to evolve in the future. What exactly this historic site symbolizes can be a matter of debate even with the same family.

After visiting the restored Ellis Island in 2004, a woman sought out her grandmother's name on the Wall of Honor. This successful New York professional called her grandmother, who had passed through the facility many decades earlier, to share the moment. "What are you doing *there*," her grandmother testily responded from the other end of the phone line. The grandmother clearly did not share the same positive thoughts that her granddaughter associated with Ellis Island.

By the dawn of the twenty-first century, Ellis Island's former life as an immigrant inspection station had given way to its latest incarnation as a national shrine and icon, a modern-day Plymouth Rock. As this transformation occurred, America was in the midst of another wave of mass immigration. What kinds of lessons—if any—could Americans learn from how immigrants were treated a century earlier?

Epilogue

"WE SHOULD NOT LET ANYONE IN. WHEN WE CAME, the rules were you could not be a burden to the state," eighty-three-year-old Sophie Wolf told a reporter on a visit to Ellis Island in 1980. "There were no schools where you could learn the language." Wolf had arrived in the United States from Germany in 1923, and for her the new immigrants of the 1980s and beyond were inferior to those of her day. Wolf and many others whose ancestors came through Ellis Island believed that late-twentieth-century immigrants were treated with greater leniency and received more help from the government than the generation that arrived at Ellis Island.

At first glance, Wolf seems to validate the recent criticism of Ellis Island for its ethnic triumphalism. Yet when she continued with her thoughts about Mexican, Vietnamese, and Cuban immigrants, her views seemed to shift. "But you've got to give people a chance," she said. "You can't send them back." Wolf's conflicted response tells us a great deal about American ambivalence toward immigration.

Wolf may have believed that things were tougher for immigrants in the past than they are today, but not everyone agrees. "At the turn of the century [1900]," the National Park Service's Richard Wells told a reporter in 1998, "America treated its immigrants far better than it does today." Americans continue to fight over the memory of Ellis Island in the wake of another era of mass immigration.

In one of history's many ironies, Ellis Island reopened to the public in 1990 just as the United States was witnessing its largest influx of immigrants ever, surpassing the previous record from 1907. The following year would see even greater numbers. During the 1990s, an average of just under 1 million immigrants entered each year, a trend that would continue into the new century.

However much Americans may feel overwhelmed by the sheer volume of this new immigration, the total U.S. population was nearly four times larger in 1990 than it was when Ellis Island opened a century earlier. Therefore, immigration in the 1990s was not as large on a percentage basis as that of the Ellis Island years.

Much like the late nineteenth century, the demographics of immigrants in the late twentieth century were also evolving. The great migration of Europeans has largely run its course. In the 1990s, only 14 percent of immigrants came from Europe, while 22 percent came from just one country: Mexico. Another 22 percent came from the Caribbean and Central and South America, while 29 percent arrived from Asia. By 2004, the percentage of Americans who were foreign-born had risen to nearly 12 percent, from its low of 5 percent in 1960, although still below the almost 15 percent during the heyday of Ellis Island.

However, that does not include illegal or undocumented immigrants. In 2005, the government estimated that there were 10.5 million unauthorized immigrants living in the United States. Some have estimated that the number might be as high as 20 million.

This new wave of immigration has produced its own fearful reaction among some native-born Americans. While some discomfort has stemmed from the sheer number of immigrants as well as the fact that so many are nonwhite, much of the current debate centers on illegal immigration. First there was the 1986 Immigration Reform and Control Act, which allowed a pathway to citizenship for a few million illegal immigrants and penalized businesses for hiring undocumented workers. Then there was the fight in California in the 1990s over Proposition 187, which would have denied government benefits to illegal immigrants. Most recently, Congress failed to pass immigration reform legislation in 2007, which opponents blocked because they argued it would have given amnesty to illegal immigrants.

In this most recent debate, some conservative opponents of the bill have harkened back to the memory of Ellis Island. Former Republican state legislator and columnist Matt Towery called for an "Ellis Island solution" to America's immigration problems. He argued for the creation of modern replicas of Ellis Island, where illegal immigrants would "surrender" to authorities. While officials processed their cases and decided whom to admit, these immigrants would be put to work on public projects like building roads and schools. "Why the heck not," says Towery, ignoring the fact that detained immigrants were rarely put to work while kept at Ellis Island.

"Why was an 'Ellis Island' approach good for the goose of American history, but not for the gander of the present day," Towery asked. Ellis Island immigrants, he continued, had to do more than "put two feet on American soil before earning their status." Why not demand the same for new immigrants? For Towery, Ellis Island regains its former sievelike quality, which weeded out the desirable from the undesirable, and he wonders why we cannot return to such a process in the twenty-first century.

In a similar vein, Congressman Mike Pence, a Republican from Indiana, crafted his own immigration reform plan that included something called "Ellis Island centers." Such centers would be located in NAFTA and CAFTA-DR countries and be managed by "American-owned private employment agencies." The goal would be to screen nonimmigrant temporary workers who could demonstrate that they had employment in the States and no criminal record.

Despite their name, these new centers would be fundamentally different from the historic Ellis Island. Pence's plan deals with nonimmigrants, not immigrants; the centers would not be run by the federal government, but by private companies; they would be located in foreign countries, not on American soil; and they would mandate that immigrants prove they had jobs, while Ellis Island strictly enforced contract-labor laws to prevent the importation of immigrant labor to undercut the wages of native workers.

Most likely, Pence wanted to capitalize on the belief that the words "Ellis Island" harkened back to a supposedly happier and more successful time in American history. Pence no doubt hoped that these centers would ease the concerns of Americans and provide a screening process for those entering the country.

Unhappy with the new immigration, Harvard political scientist Samuel Huntington also looks to Ellis Island. For him, it was a great success, when "control of immigrants coming by ship was fairly easy and a good proportion of those arriving at Ellis Island were denied entry." Huntington inflates the estimate of those denied entry to 15 percent, well over the average of 2 percent excluded. His naïve belief in both the restrictive nature of Ellis Island and its relative ease in regulating immigrants would have amused those who ran the inspection station and believed they were barely holding back the flood of European immigrants seeking entry. For Huntington and the other commentators, Ellis Island stands as an historical rebuke to what they feel is America's current unrestrictive immigration policy.

The use—or misuse—of the memory of Ellis Island masks a common thread in past and present immigration debates. Then as now, Americans are asking themselves: How does the United States decide who gets to enter the country and who does not?

There has never been a time when Americans did not ask themselves that question. It is a myth that before the late 1800s America had an open door to all immigrants. Prior to the rise of federal laws regulating immigration, state governments passed laws banning paupers, criminals, and those with diseases. Those laws were not well enforced, and it was not until the era of Ellis Island that the federal government nationalized and regularized immigration inspection.

It is another myth that debate about immigration divides us between noble liberal ideals that support immigration and ignoble, illiberal ideals that seek to restrict it. In reality, this debate highlights fundamental conflicts and contradictions within the American ideal. Our adherence to the virtues of democratic self-rule collides with the universalist strains of our national creed in which all men and women are created equal. If everyone is equal, then how can the government sift through immigrants and decide who may or may not enter the country? But if we are a representative democracy, then should our immigration laws not reflect the popular will?

To take the belief in the universalist creed to its logical end means to deny the relevance or justice of borders and boundaries, which are by definition designed to exclude. American sovereignty, under which the nation secures our freedoms, liberties, and democracy, sits uneasily with the notion of the United States as a refuge for the world's poor and oppressed. Similarly, the United States has increasingly sought to erase invidious forms of discrimination in its laws, but discrimination is at the heart of any immigration policy. How can these beliefs be reconciled?

As Barbara Jordan, the chair of the U.S. Commission on Immigration Reform, put it in 1995, "immigration to the United States should be understood as a privilege, not a right," and has been so held for most of our history. But as Jordan's own life as a female, lesbian, African-American politician demonstrates, the late twentieth century witnessed the legal and social explosion of a rights revolution that has expanded liberties to women, minorities, and homosexuals. The logical next step was an expansion of rights not just to immigrants legally residing in the country, but also to those who seek entry into the country and who traditionally have possessed fewer constitutional rights.

Much of the discussion of expanding rights has been rooted in the human rights revolution that followed World War II. Universalist notions of human rights and a greater emphasis on international law strike at the heart of the nation-state. There is a contradiction in one of the bedrock documents of international law—the 1948 Universal Declaration of Human Rights. It grants all human beings the right to asylum in other countries, the right to flee persecution, and the right to change their nationality if they please. However, the document does not say that nations have the obligation to accept such refugees and grant them entry and citizenship. It is an unenforceable one-way right. The authors of the Universal Declaration created these rights, but in their silence upheld the sovereignty of nation-states.

As Americans, we believe that all individuals are created equal *and* that this is a democracy where the people are allowed a voice in the writing of their country's laws. We believe in the sovereignty of the United States and the sanctity of borders, *but* we recognize the moral obligation and historical legacy of this country as a sanctuary for those who no longer can abide the conditions in their homeland. That we believe all of these things—often at the same time—comes out most starkly when we talk about immigration. It is this confusion and intellectual schizophrenia that has helped muddle how we think and talk about immigration.

To make matters more complicated, many recent social, economic, political, and ideological trends are helping to weaken the idea of national sovereignty. We live in an era of rapid globalization, which has diluted the idea of national borders. Money, goods, and jobs flow back and forth across borders with little regulation in a world of increasing free trade.

As migrants continually cross borders, ideas of transnationalism are taking root and challenging notions of sovereignty. In an era with relatively easy access to airplanes, telephones, and satellite television, immigrants are able to stay connected with their homelands in ways unimaginable in earlier years. Combine these technological innovations with the ideological ascendancy of multiculturalism over assimilation, and you get the realization of an age of transnationalism whose outlines Randolph Bourne could only have sketched in 1916.

The plenary power doctrine, which has dominated immigration law since the late nineteenth century, is slowly losing its hold. It gives Congress and the executive branch broad authority to regulate immigration with little interference from the courts. The practical result is that would-be immigrants seeking entry to the United States have fewer constitutional guarantees than citizens or nonnaturalized immigrants already in the country. Legal scholars and immigrant activists have been training their collective critical fire against the plenary power doctrine for years. Although never explicitly overturned by the courts, the government has been less and less able to shield the execution of immigration law from court challenges.

In a 2001 Supreme Court case dealing with the rights of a detained immigrant who had been ordered deported, Justice Antonin Scalia repeated words that were familiar for over a hundred years of immigration law. "Insofar as a claimed legal right to release into this country is concerned," Scalia wrote, "an alien under final order of removal stands on an equal footing with an inadmissible alien at the threshold of entry: He has no such right." But in 2001, Scalia's defense of the plenary power doctrine was found in the dissent. He was correct to note that the majority of the Court refused to overturn the precedent of the *Mezei* decision, but instead chose to "obscure it in a legal fog." In the twenty-first century, it appears that much of U.S. immigration law is mired in that same fog.

The terrorist attacks of September 11 and the war on terror have led to more questions about immigrant admissions and the rights of aliens. Could the United States pass new laws—and more strictly enforce the current ones—against immigrants from the Middle East, a form of ethnic and racial discrimination akin to the Chinese Exclusion Act? On the other side of the equation, the status of suspected terrorists detained by the U.S. military has brought about a legal debate as to the constitutional rights of noncitizens. None of these issues have been settled and the future will likely bring only more confusion and conflict over these laws.

In the early twenty-first century, few Americans are satisfied with the nation's immigration law. Some Americans find that the law is broken and unenforced, that millions enter the country illegally and little is done. Others complain about the arcane regulations that govern much of immigration law, for instance, treating refugees from one country differently than those from another. They criticize an immigration law that can be so inflexible that it forces millions to enter the country surreptitiously.

Immigrants are faced with a bifurcated system. Those who seek to enter legally are often faced with a daunting bureaucratic challenge in dealing with an increasingly byzantine system. Those who enter illegally bypass the red tape, but live under the radar and outside the bounds of the national community. Many find the two-tiered society of legal and illegal immigrants troubling and un-American. Wherever you fall on the immigration spectrum, current American immigration law is bound to disappoint, frustrate, and anger.

If the regulation of immigration is somehow tied to the rise of the burgeoning federal government, then our contemporary attitudes toward that government are not helping. Many of those who wish stricter controls and regulation of immigrants are often on the political right and are often the same people who call for limited government intervention in the marketplace. Conservative attacks on government make the call for stronger action against immigrants seem hollow.

Many who call for fewer restrictions on immigration and support open borders are on the political left, but they are often the same people who call for greater government involvement in the economy. They increasingly minimize the right of national sovereignty and dismiss nationalism as an outmoded and reactionary ideal, yet they wish to galvanize the nation for universal health care and other welfare-state programs. They claim to be concerned about rising income inequality, but are relatively uninterested in the idea that cheap immigrant labor keeps wages down for foreign- and native-born alike. These liberals support federal action everywhere but immigration, where they turn as laissez-faire as the editorial page of the *Wall Street Journal*.

To make matters even more confusing, labor unions, which used to be counted among the most ardent supporters of immigration restriction because of their genuine concerns for the effects of cheap labor on wages, have by the dawn of the twenty-first century become supporters of a laissez-faire approach to immigration. This has more to do with the increasingly weak position of unions in this country and the unions' belief that their survival hinges upon support from foreign-born workers in the service economy.

The battle over the status of immigrants in a globalized world where borders are increasingly fuzzy will only grow more heated. Americans are in for a debate that might prove itself even more contentious than the one fought over Ellis Island. It will involve not only the question of American identity but also the relevance of America as a nation. Back in 1908, Henry Cabot Lodge put forth the primacy of national sovereignty when he said: "No one has a right to come into the United States, or become part of its citizenship, except by permission of the people of the United States." Americans are still trying to come to grips with the implications of this idea some one hundred years later.

Can modern Americans learn any lessons from the history of Ellis Island? Historians should be wary of writing history that provides a "usable past." Studying Ellis Island's history provides little ammunition for those who wish either stricter or more lenient immigration laws. Studying this history should lead us neither to the elevation nor the condemnation of Ellis Island immigrants for their successes. History rarely provides neat lessons that can be utilized for present political purposes. If history teaches anything, it is that the past was filled with imperfect people who made imperfect decisions in dealing with an imperfect world. In this, they are very much like ourselves.

Even though very little of what was done at Ellis Island could be replicated today, its history can shed some light on our own times. The dustbin of history is littered with the now-discredited warnings of anti-immigrant writers like Francis Walker, Prescott Hall, and William Williams. Their fears about the quality of immigrants passing through Ellis Island now appear unfounded and mean-spirited. Not only did the United States absorb this wave of immigrants, it thrived in the twentieth century in part due to their contributions. We should be aware of similar fears about the supposed poor quality of today's immigrants and not make the same mistakes of that bygone era.

That is not to say that the problems of today are exactly comparable to those of the Ellis Island era. History does not simply repeat itself in an endless loop. However, the history of Ellis Island should remind us that the problems that the United States is dealing with today are not unique and the questions Americans are asking themselves today are very similar to the ones that bedeviled those who came before us.

During the Ellis Island years, most Americans sought a balance between a completely open door to immigrants and a completely closed door. As modern-day Americans seek ways to deal with immigration, they too will have to find their own balance between the competing ideals of universalism versus national sovereignty, a policy of nondiscrimination versus democratic self-rule, and feelings of generosity versus pragmatism.

Most Americans today believe they live in a great country and understand why people from around the world want to live here. The idea of America as a nation of immigrants is a powerful image that rings true for many across the political spectrum. Yet many of those same Americans are also a little anxious that their country will somehow change in a fundamental way because of mass immigration; they worry that new immigrants are a little too different and that assimilation is not occurring fast enough. Many Americans believe this even if these are the same arguments that were used against their immigrant forebears.

Keeping in mind these deeply conflicted ideas about immigration, modern Americans must find their own compromise, one that takes into account the fears and concerns—legitimate or not—of native-born Americans, while respecting the rights and humanity of those who arrive at our borders. The United States cannot open its doors to the entire world, but it cannot close its borders either. A successful immigration policy will keep the gates open to continue our long history of welcoming strangers who in turn help build this unfinished nation, while reassuring native-born Americans that the laws are being enforced and social dislocations that arise from immigration are minimized.

If Americans are not reassured that immigration is taking place in a legal and orderly manner that is beneficial to the economic well-being, social cohesion, and national security of the nation, then the entire ideal of immigration is at risk. That elusive balance is what Americans have to debate.

Before the rise of national quotas in the 1920s, the debates of the

Ellis Island era—in which restrictionists supported some kind of immigration and their opponents favored some kind of restriction of undesirable immigrants—tried to find that balance. We can see that same dynamic playing out in our own time. Legal scholar Peter Schuck, a pro-immigrant liberal, admits that, "the tension between liberalism's universal aspirations and our need as a society to achieve the degree of solidarity that effective activist government requires must be resolved at *some* level of exclusion." On the other side, restrictionist conservative Mark Krikorian supports a "pro-immigrant policy of low immigration, one that admits fewer immigrants but extends a warmer welcome to those who are admitted." Where to draw that line of exclusion? The devil, of course, is in the details.

As the United States yet again comes to grips with this question, it will have to do it without an active facility like Ellis Island. Rather than passing through something like Ellis Island, today's immigrants enter the country through airports like JFK or LAX, or pass across the Canadian and Mexican borders. The decision to allow immigrants entry into the country is made at American consulates abroad, not at immigration stations at American ports, although increasing numbers of immigrants are bypassing the cumbersome visa process and entering the country illegally.

In this new era of mass immigration, is it not too far-fetched to ask whether Ellis Island will go the way of that other once-totemic symbol of the American founding? One hundred years from now, will Ellis Island seem as quaint, distant, and unrepresentative as Plymouth Rock does now to a lot of Americans? Instead, will the descendants of Hispanic immigrants seek to build a memorial along the Mexican border fence that asserts their entrance into the American mainstream?

The future is notoriously hard to predict, but the fight over the meaning of Ellis Island—and the meaning of immigration in general—will most likely remain a part of our national dialogue for as long as individuals feel the need to pick themselves up from their homelands and make that American passage, whether by boat, plane, or foot.

Acknowledgments

WHEN DISCUSSING THE TITLE OF THE BOOK WITH MY EDITOR, I requested a slight change. Modestly, I asked that the subtitle be changed to "A History of Ellis Island" instead of "The History of Ellis Island." I lost that argument but still think that it would have been a more appropriate subtitle. My modesty stemmed from my belief that it is impossible to write a comprehensive history of Ellis Island. Its history is literally that of millions of stories, of those who arrived and those who processed them, and of the numerous political and legal battles fought over the inspection station. Most readers will not find the names of their ancestors in this book, but I hope that they come away with a better sense of the many meanings of Ellis Island and how this country dealt with immigrants in the late nineteenth century and early twentieth century.

Readers should keep in mind a few things. First, all of the names that appear in these pages are real. Some historians have used pseudonyms when discussing immigrants passing through Ellis Island because of the personal nature of their stories. Because of the passage of time and the public nature of these records, I have chosen to use actual names, though I have tried to tell all the stories with tact and sensitivity.

Second, it is important to keep in mind that many of the stories told about immigrants in these pages come from government records. Many of the immigrants do not "speak" directly to us, but instead "speak" through reports of government officials or transcripts of hearings at Ellis Island. Many of those recording the words of immigrants did not sympathize with those before them. To complicate matters, many of the words had to be translated by other officials before being added to the record. This is not to discount the importance of such historical records (often they are the only ones we have). It just reminds us that all sources have their own set of limitations.

Lastly, terms like "moron," "idiot," "lunatic," "imbecile," "mental defective," "undesirable," and "desirable" appear throughout the text, usually without quotation marks. This is a stylistic decision to make the narrative flow better, but does not imply that the author concurs in the often harsh judgments made against many immigrants by those who used such terms.

RESEARCHING AND WRITING A book is ultimately a solitary endeavor. However, I would be remiss if I didn't acknowledge the help that I received from various people along the way.

Phil Costopoulos, Matt Dallek, Tim Hacsi, Adam Rothman, and Tevi Troy all read parts of the manuscript and provided much-needed feedback. Kevin Swope deserves special mention for reading almost the entire manuscript and giving voluminous comments throughout. Chris Capozzola graciously shared with me his own research on World War I.

Kitty and Ira Carnahan have unfailingly provided support and friendship through the years in ways too numerous to count. Brittany Huckabee has been an invaluable editor and sounding board over the course of two books. Stephen Haas has shared his love of good books and good wine. Steve Thernstrom provided some important help at a crucial time for which I am grateful. Seth Kamil, owner of Big Onion Walking Tours and a good friend, unwittingly helped with the book many years ago when he scheduled me to give tours at Ellis Island while I was working my way through graduate school. Susan Ferber graciously shared her Ellis Island story with me.

Anyone working on immigration history knows Marian Smith. To use a cliché, she is a national treasure. As the senior historian at the U.S. Citizenship and Immigration Services (formerly the INS), Marian generously shared her vast knowledge of the topic and assisted me in navigating some bureaucratic hurdles.

Justin Kehoe, Dalton Little, Dennis Bilger, Amy Lewis, and Ben D'Amore provided research assistance at various stages of the process. Jim Thayer deserves separate thanks. As an undergraduate and graduate student, Jim has been a faithful research assistant and computer guru who has generously offered his help above and beyond what was required under his assistantship. Douglas Baynton, J. T. E. Richardson, and William Forbath graciously shared their research with me. Binkie McSweeney Orthwein and Susan Womack shared material relating to their ancestors who worked at Ellis Island. Robert Murphy of the Knights of Columbus Museum used his detective skills to track down an important photograph.

Besides providing me with a steady paycheck, the University of Massachusetts, Boston, also gave me a Joseph P. Healey Endowment Grant, which allowed a summer of research in Washington, D.C. I want to thank Donna Kuizenga, Roberta Wollons, Spencer Di Scala, and Lester Bartson.

The publication of this book would have been delayed even more had it not been for a Fellowship from the National Endowment for the Humanities. This generous grant allowed me to take a year's leave from teaching to concentrate on reading, research, and writing.

My agent, Rafe Sagalyn, deserves a great deal of credit for helping this book along. After a chance meeting at a Washington party many years ago, Rafe took a chance on a then-unpublished author. I appreciate his patience over these past seven years and the faith he has shown in this book.

My editor, Tim Duggan, has proven to be a wise editor whose comments and edits have pushed me to make this a more readable narrative, while at the same time not losing sight that this is also a serious work of history. This is a better book for his efforts. At HarperCollins, I would also like to thank Tim's assistant, Allison Lorentzen, for her help, and Martha Cameron for her excellent copyediting.

Donna Beath came into my life toward the end of this project. Not quite realizing what was ahead, she threw herself into the role of reader and critic, sometimes going over chapter drafts while sitting on the beach. She has put up with the ups and downs that are an inevitable part of any book project with her warm smile, good cheer, and an always ready cup of tea.

My ties to Ellis Island are not merely professional. It was from my late grandfather that I first learned of Ellis Island, which Pop passed through at least once as a young immigrant from Italy. His wife, my grandmother Antoinette, was born in New York's Little Italy, but her parents, stepmother, and brothers passed through Castle Garden and Ellis Island.

The joy of finishing this book is mixed with a great deal of sadness. Over the course of researching and writing, I have lost two of my aunts. I wish that Marion Marino and Kitty Molinari were still here to see this book.

As I write these words, it has been two months since my father passed away. In addition to the countless hours we spent over the years watching innumerable baseball and football games and boxing matches, it was my father who first encouraged my interest in history and politics. He taught me an important lesson that too few young people learn: Not only does history matter, but it is also endlessly fascinating.

My father suffered from many health problems over the years. He never thought he would see me graduate from college, but he did. He never thought he would see me get my PhD, but he did. He never thought he would see the publication of my first book, but he did. He desperately wanted to see this book published, and although he didn't say so, I know that his nagging over the last year or so to finish it was brought on by the fact that he wasn't sure how much longer he could hold on.

As my father's condition worsened this past summer, I spent a great deal of time driving back and forth to New York to be with him. In my spare time, I was finally able to finish the manuscript. But it was too late. This fall, when we knew that the time was near, I told him I was sorry that he wouldn't see the book. "I tried," he told me with a smile. "I tried." And he did. He fought so hard for so many years, but in the end it was all too much.

Although his suffering has ended and he is now in a better place, that doesn't take away the deep sadness I feel from his absence. It is hard to imagine that I won't hear his voice again or that I won't be able to share the reviews of this book with him. There is still so much more that I want to say to him and so much more that I want to hear from him. However, I was blessed to have him around for as long as I did, and I am grateful for everything that he did for me. I think about him every day, as I will for the rest of my life.

Thankfully, my mother, Maria, is still here to share the joy of this book's publication. She is a "JFK immigrant" who arrived in New York by airplane after the closing of Ellis Island, coming during the quota years of the early 1960s. Words cannot describe how grateful I am for her love and support.

Watertown, Massachusetts
December 2008

Notes

SG   Samuel Gompers Papers, University of Maryland, College Park
SP   *Saturday Evening Post*
TR   Theodore Roosevelt Papers, Library of Congress
TVP   Terence V. Powderly Papers, The Catholic University of America
WC   William E. Chandler Papers, Library of Congress
WGH   Warren G. Harding Papers, Library of Congress
WL   William Langer Papers, University of North Dakota
WSJ   *Wall Street Journal*
WHT   William Howard Taft Papers, Library of Congress
WP   *Washington Post*
WW   Woodrow Wilson Papers, Library of Congress
WW-NYPL William Williams Papers, New York Public Library
WW-Yale   William Williams Family Papers, Yale University

INTRODUCTION

1   **By 1912, thirty-three-year-old**: On the Tyni family, see File 53525-37, INS.
2   **Unlike the Tyni family**: For the story of Anna Segla, see File 52880-77, INS.
3   **Other immigrants**: Letter from Louis K. Pittman, December 3, 1985, Public Health Service Historians Office, Rockville, MD.
3   **Others, luckier than Pittman**: For the story of Frank Woodhull/Mary Johnson, see *NYT*, October 5, 6, 1908; *NYTrib*, October 5, 1908; *New York Herald*, October 5, 1908; and Erica Rand, *The Ellis Island Snow Globe* (Durham, NC: Duke University Press, 2005), Chapter 2.
5   **For these individuals**: Bruce M. Stave, John F. Sutherland, with Aldo Salerno, *From the Old Country: An Oral History of European Migration to America* (New York: Twayne Publishers, 1994), 44–45.
5   **No one story**: James Karavolas, who arrived as a six-year-old in 1915, told of his memories of Ellis Island years later. "Ellis Island didn't impress me at all. The memory is faint," Karavolas admitted. Peter Morton Coan, *Ellis Island Interviews: In Their Own Words* (New York: Checkmark Books, 1997), 279.
7   **The process at**: Edward A. Steiner, *On the Trail of the Immigrant* (New York: Fleming H. Revell Company, 1906), 72; Stephen Graham, *With Poor Immigrants to America* (New York: Macmillan, 1914), 44.
7   **The central sifting**: Allan McLaughlin, "How Immigrants Are Inspected," *PSM*, February 1905; J. G. Wilson, "Some Remarks Concerning Diagnosis by Inspection," *NYM*, July 8, 1911; Alfred C. Reed, "The Medical Side of Immigration," *PSM*, April 1912; E. H. Mullan, "Mental Examination of Immigrants: Administration and Line Inspection at Ellis Island," Public Health Reports, U.S. Public Health Service, May 18, 1917; and Elizabeth Yew, "Medical Inspection of Immigrants at Ellis Island, 1891–1924," *Bulletin of the New York Academy of Medicine* 56, no. 5 (June 1980).
8   **All of these ideas**: Speech by Henry Cabot Lodge before the Boston City Club, March 20, 1908, reprinted, 60th Congress, 1st Session, Senate Document 423.
9   **Traditional histories**: John Higham, *Strangers in the Land: Patterns of American Nativism, 1860–1925* (New Brunswick, NJ: Rutgers University Press, 1955), 4. For a critique of Higham's "psychopathological approach," see Aristide R. Zolberg, *A Nation by Design: Immigration Policy in the Fashioning of America* (New York: Russell Sage Foundation, 2006), 6–8.
9   **The "nativist theme"**: See John Higham, "Another Look at Nativism," *Catholic Historical Review*, July 1958 and John Higham, "Instead of a Sequel, or How I Lost My Subject," *Reviews in American History* 28, no. 2 (2000).

10 **Few Americans argued**: Allan McLaughlin, "Immigration and Public Health," *PSM*, January 1904.
11 **Take the opinions**: Max Kohler, "Immigration and the Jews of America," *AH*, January 27, 1911.
11 **On the other side**: Frank Sargent, "The Need of Closer Inspection and Greater Restriction of Immigrants," *Century Magazine*, January 1904.
11 **"We desire to"**: American Jewish Committee report quoted in Max J. Kohler, *Immigration and Aliens in the United States: Studies of American Immigration Laws and the Legal Status of Aliens in the United States* (New York: Bloch Publishing Company, 1936), 1.
11 **The laws that dealt**: See Erika Lee, "The Chinese Exclusion Example: Race, Immigration, and American Gatekeeping, 1882–1924," *Journal of American Ethnic History*, Spring 2002; Lucy E. Salyer, *Laws Harsh As Tigers: Chinese Immigrants and the Shaping of Modern Immigration Law* (Chapel Hill: University of North Carolina Press, 1995). On the role of Angel Island in historical interpretations of immigration, see Roger Daniels, "No Lamps Were Lit for Them: Angel Island and the Historiography of Asian American Immigration," *Journal of American Ethnic History* 17, no. 1 (Fall 1997).

CHAPTER ONE: ISLAND

19 **Fifty thousand**: Daniel Allen Hearn, *Legal Executions in New York State, 1639–1963* (Jefferson, NC: McFarland & Co., 1997), 40, 299–300.
19 **Pirates bring to**: Rudolph Reimer, "History of Ellis Island," mimeo, 1934, 6–7, NYPL.
20 **When Washington Irving**: Washington Irving, *History, Tales and Sketches* (New York: Library of America, 1983), 628–629.
20 **"Guests from Gibbet Island"**: Washington Irving, "Guests from Gibbet Island," in Charles Neider, ed., *Complete Tales of Washington Irving* (New York: Da Capo Press, 1998). Irving also returns to the theme in his short story "Dolph Heyliger."
20 **Pirate hangings**: "Life and Confession of Thomas Jones," 1824, NYHS.
21 **A similar tale**: "Trial and Confession of William Hill," 1826, NYHS; Frederick Douglass, *Narrative of the Life of Frederick Douglass, an American Slave* (New York: Signet Classics, 1997), 26.
21 **On the night**: *Genius of Universal Emancipation*, January 2, 1827; Ralph Clayton, "Baltimore's Own Version of 'Amistad': Slave Revolt," *Baltimore Chronicle*, January 7, 1998, http://baltimorechronicle.com/slave_ship2.html.
22 **Confusion reigned**: Reimer, "History of Ellis Island," 24; *Commercial Advertiser*, April 23, 1831; *Workingman's Advocate*, April 30, 1831.
22 **Gibbs was a white man**: "Mutiny and Murder: Confession of Charles Gibbs," (Providence, RI: Israel Smith, 1831), NYHS.
23 **Their dead bodies**: *New York Evening Post*, April 23, 1831; *Atkinson's Saturday Evening Post*, April 30, 1831.
23 **The island's last**: Hearn, 46; "The Life of Cornelius Wilhelms: One of the *Braganza* Pirates," 1839, NYHS.
23 **New York City**: For an excellent discussion of New York's waterfront, see Phillip Lopate, *Waterfront: A Journey Around Manhattan* (New York: Crown, 2004).
24 **There are some forty**: See Sharon Seitz and Stuart Miller, *The Other Islands of New York City: A History and Guide*, 2nd ed. (Woodstock, VT: Countryman Press, 2001).
24 **Many of the city's**: Lopate, *Waterfront*, 374.

24 **In upper New York Harbor**: Diana diZerega Wall and Anne-Marie Cantwell, *Touring Gotham's Archaeological Past: 8 Self Guided Walking Tours Through New York City* (New Haven, CT: Yale University Press, 2004), 20–21.

25 **Seals, whales, and porpoises**: Diana diZerega Wall and Anne-Marie Cantwell, *Unearthing Gotham: The Archaeology of New York City* (New Haven, CT: Yale University Press, 2004), 87; John Waldman, *Heartbeats in the Muck: The History, Sea Life, and Environment of New York Harbor* (New York: Lyons Press, 1999); and Mark Kurlansky, *The Big Oyster: History on the Half Shell* (New York: Ballantine Books, 2006).

25 **Little Oyster Island**: Edwin G. Burrows and Mike Wallace, *Gotham: A History of New York City to 1898* (New York: Oxford University Press, 1999), 63.

25 **One of the first orders**: Berthold Fernow, ed., *Records of New Amsterdam*, vol. 1 (Baltimore, MD: Genealogical Publishing Co., 1976), 51, 58–59; Russell Shorto, *The Island at the Center of the World: The Epic Story of Dutch Manhattan and the Forgotten Colony That Shaped America* (New York: Doubleday, 2004), 259; Elva Kathleen Lyon, "Joost Goderis, New Amsterdam Burgher, Weighmaster, and Dutch Master Painter's Son," *New York Genealogical and Biographical Record* 123, no. 4 (October 1992).

26 **Little Oyster Island would also**: Reimer, "History of Ellis Island," 7.

26 **Ellis died in 1794**: I. N. Phelps Stokes, *The Iconography of Manhattan Island*, vol. 5 (New York: Arno Press, 1967), 1198–1199; Thomas M. Pitkin, *Keepers of the Gate: A History of Ellis Island* (New York: New York University Press, 1975), 3.

27 **Over the next few years**: Pitkin, *Keepers of the Gate*, 4–5.

27 **In 1807, Lieutenant Colonel**: Reimer, "History of Ellis Island," 16.

28 **Nature blessed New York's**: Robert Greenhalgh Albion, *The Rise of New York Port, 1815–1860* (New York: Scribner's, 1939), 16–29.

28 **Having such a natural port**: Edward Robb Ellis, *The Epic of New York City: A Narrative History* (New York: Kondasha International, 1997), 223–229.

28 **New York City was**: Burrows and Wallace, *Gotham*, 435–436; Albion, 389; John Gunther, *Inside U.S.A.* (New York: Book of the Month Club, 1997), 555.

29 **For the next few decades**: Reimer, "History of Ellis Island," 17–18.

CHAPTER TWO: CASTLE GARDEN

30 **These men, women**: NYT, August 7, 1855.

31 **The old fort**: On Castle Garden's history, see *Commercial Advertiser*, June 22, 1839; James G. Wilson, ed., *The Memorial History of the City of New York*, vol. 4 (New York: New York History Company, 1893), 441; Phillip Lopate, *Waterfront: A Journey Around Manhattan* (New York: Crown Publishers, 2004), 24; Edwin G. Burrows and Mike Wallace, *Gotham: A History of New York City to 1898* (New York: Oxford University Press, 1999), 815–816; Sharon Seitz and Stuart Miller, *The Other Islands of New York City: A History and Guide*, 2nd ed. (Woodstock, VT: Countryman Press, 2001), 72–74.

31 **The new immigration station**: NYT, August 6, 7, 1855.

31 **The indignation meeting**: NYT, August 7, 10, 1855.

32 **This was an exercise**: Theodore Roosevelt, *New York: A Sketch of the City's Social, Political, and Commercial Progress from the First Dutch Settlement to Recent Times* (New York: Charles Scribner's Sons, 1906), 238, 246.

32 **Born in upstate**: On Rynders, see Tyler Andbinder, *Five Points: The 19th Century New York City Neighborhood That Invented Tap Dance, Stole Elections, and Became the World's Most Notorious Slum* (New York: Free Press, 2001), 141–144, 166–167

and T. J. English, *Paddy Whacked: The Untold Story of the Irish American Gangster* (New York: Regan Books, 2005), 13–15, 26–27.

33 **There were certainly**: Burrows and Wallace, *Gotham*, 736.

33 **Rynders was**: George J. Svejda, "Castle Garden as an Immigrant Depot, 1855–1890," National Park Service, December 2, 1968, 41.

34 **As soon as**: Friedrich Kapp, *Immigration and the Commissioners of Emigration of the State of New York* (New York: Nation Press, 1870), 62; Burrows and Wallace, *Gotham*, 737.

34 **A committee of**: "Report of the Select Committee to Investigate Frauds upon Emigrant Passengers," 1848, excerpted in Edith Abbott, ed., *Immigration: Select Documents and Case Records* (Chicago: University of Chicago Press, 1924), 130–134.

34 **The federal government**: Hans P. Vought, *The Bully Pulpit and the Melting Pot: American Presidents and the Immigrant, 1897–1933* (Macon, GA: Mercer University Press, 2004), 5.

35 **The job of regulating** E. P. Hutchinson, *Legislative History of American Immigration Policy, 1798–1965* (Philadelphia: University of Pennsylvania Press, 1981), 388–404; Daniel J. Tichenor, *Dividing Lines: The Politics of Immigration Control in America* (Princeton, NJ: Princeton University Press, 2002), 58–59; Gerald L. Neuman, *Strangers to the Constitution: Immigrants, Borders, and Fundamental Law* (Princeton, NJ: Princeton University Press, 1996), 19–43. For examples of these state laws, see Abbott, ed., *Immigration*, 102–110.

36 **The Board of Commissioners laid out**: Kapp, *Immigration and the Commissioners*, 109–110. For a history of the Battery, including Castle Garden's many incarnations, see Rodman Gilder, *The Battery* (Boston: Houghton Mifflin, 1936).

36 **Wealthy New Yorkers**: *NYT*, June 15, 1855; Svejda, "Castle Garden," 40.

36 **On Castle Garden's first day**: Svejda, "Castle Garden," 45–46; *NYT*, August 4, 1855.

37 **Having failed**: *NYT*, August 7, 1855. In a letter to the editor the day after the indignation meeting, Rynders clarified his views on the matter. *NYT*, August 8, 1855.

37 **After the final**: *NYT*, August 8, 1855; *New York Daily Tribune*, August 7, 1855.

37 **Throughout the fall**: *NYT*, August 14, 18; December 15, 1855.

37 **The harassment of**: Kapp, *Immigration and the Commissioners*, 108; Svejda, "Castle Garden," 50–57.

38 **Some reports claimed**: Kapp, *Immigration and the Commissioners*, 81.

38 **With the runners**: William Dean Howells, *A Hazard of New Fortunes* (New York: Meridian, 1994), 263; *NYT*, December 23, 1866; Friedrich Kapp, quoted in Charlotte Erickson, ed., *Emigration from Europe, 1815–1914* (London: Adam & Charles Black, 1976), 274; *New York: A Collection from Harper's Magazine* (New York: Gallery Books, 1991), 363.

39 **Between 1860 and**: John Higham, *Strangers in the Land: Patterns of American Nativism, 1860–1925* (New Brunswick, NJ: Rutgers University Press, 1995), 39.

39 **A writer in**: Higham, *Strangers in the Land*, 35; "Dangers of Unrestricted Immigration," *Forum*, July 1887.

40 **Daily newspapers**: Edward Self, "Why They Come," *NAR*, April 1882; Edward Self, "Evils Incident to Immigration," *NAR*, January 1884.

40 **Newspapers throughout**: *Public Opinion*, April 30, May 14, June 30, 1887.

41 **Others used**: "Immigration and Crime," *Forum*, December 1889.

41 **It took Episcopal bishop**: "Government by Aliens," *Forum*, August 1889.

41 **Despite Coxe's florid**: *Public Opinion*, April 30, July 30, 1887, December 28, 1889.

42 **Shortly after the decision**: Hutchinson, *Legislative History*, 65–66.

42 **It was not until**: "An Act to Regulate Immigration," 1882, excerpted in Abbott, ed., *Immigration*, 181–182.

43 **That same year**: Vought, *Bully Pulpit*, 10; Tichenor, *Dividing Lines*, 89–90; Hutchinson, *Legislative History*, 80–83.

44 **The Board of Commissioners**: Document No. 815, Box 4, INS.

44 **Of the estimated**: NYT, January 25, 1883; "Immigration Investigation Report, Testimony and Statistics," House Report 3472, 51st Congress, 2nd Session, Serial 2886.

44 **To many, this cried out**: NYT, February 11, 1883.

45 **In 1880, a twenty-two-year-old**: Robert Watchorn, *The Autobiography of Robert Watchorn* (Oklahoma City, OK: Robert Watchorn Charities, 1959); "Robert Watchorn," *Outlook*, March 4, 1905.

45 **Another sign**: Roll 19, G-7-G20, ANY.

45 **Public concern about**: James B. Bell and Richard I. Abrams, *Liberty: The Story of the Statue of Liberty and Ellis Island* (New York: Doubleday, 1984), 43–45.

46 **In 1887, Pulitzer trained**: NYW, July 27, August 4, 10, 1887; Erickson, ed., *Emigration from Europe, 1815–1914*, 276; NYT, August 31, 1887.

46 **In 1888**: The Ford Report, reprinted in *Congressional Record*, 50th Congress, 2nd Session, 997–999.

48 **Congress never acted**: Vought, *Bully Pulpit*, 12.

49 **As conditions at**: "Immigration and Crime," *Forum*, December 1889.

49 **The decision was inevitable**: John B. Weber, *Autobiography of John B. Weber* (Buffalo, NY: J.W. Clement Company, 1924), 88.

50 **In response, a joint House and Senate**: *Congressional Record*, 51st Congress, 1st Session, Volume 21, 3085–3089.

50 **"Give us a rest"**: Francis A. Walker, "Immigration," *Yale Review*, August 1892; Francis A. Walker, "Immigration and Degradation," *Forum*, August 1891.

51 **Walker also saw**: See Maurice Fishberg, "Ethnic Factors in Immigration—A Critical View," Proceedings of the National Conference of Charities and Correction, May 1906. Australia and New Zealand, largely Anglo-Saxon and with little immigration, saw their birth rates decline during the late nineteenth and early twentieth centuries as well.

51 **Walker's views**: Henry Cabot Lodge, "The Restriction of Immigration," NAR, January 1891.

51 **Lodge used the occasion**: Henry Cabot Lodge, "Lynch Law and Unrestricted Immigration," NAR, May 1891.

52 **Walker and Lodge**: NYT, April 30, 1891; *Boston Traveler*, October 24, 1891; "Report of the Select Committee on Immigration and Naturalization," 51st Congress, 2nd Session, Report No. 3472, January 15, 1891; "Regulation of Immigration and to Amend the Naturalization Laws," House Report, 51st Congress, 2nd Session, Report No. 3808.

52 **The 1891 Immigration Act**: Michael LeMay and Elliot Robert Barkan, *U.S. Immigration and Naturalization Laws and Issues: A Documentary History* (Westport, CT: Greenwood Press, 1999), 66–70; Higham, *Strangers in the Land*, 99–100.

53 **Immigration was now**: Hiroshi Motomura, "Immigration Law After a Century of Plenary Power: Phantom Constitutional Norms and Statutory Interpretation," *Yale Law Journal*, December 1990; Lucy E. Salyer, *Laws as Harsh as Tigers: Chinese Immigrants and the Shaping of Modern Immigration Law* (Chapel Hill, NC: University of North Carolina Press, 1995), 26–28. Salyer claims that the inclusion of this clause that made the executive branch the final arbiter of immigration appeals stemmed from unhappiness over Chinese immigrants using the

courts to challenge the Chinese Exclusion Act. While this could very well be true, it remains speculation.

53 **The new immigration system**: On the rise of the federal government and the administrative state, see Stephen Skowronek, *Building a New American State: The Expansion of National Administrative Capacities, 1877–1920* (Cambridge, UK: Cambridge University Press, 1982); Keith Fitzgerald, *The Face of the Nation: Immigration, the State and the National Identity* (Stanford, CA: Stanford University Press, 1996); Morton Keller, *Affairs of State: Public Life in Late Nineteenth Century America* (Cambridge, MA: Harvard University Press, 1977); Morton Keller, *Regulating a New Society: Public Policy and Social Change in America, 1900–1933* (Cambridge, MA: Harvard University Press, 1994); and Gabriel J. Chin, "Regulating Race: Asian Exclusion and the Administrative States," *Harvard Civil Rights–Civil Liberties Law Review* 37 (2002).

54 **Despite the corruption**: Svejda, "Castle Garden," iii.

CHAPTER THREE: A PROPER SIEVE

57 **As she exited**: The discussion of Annie Moore comes from the NYT, January 2, 1892; *New York Herald*, January 2, 1892; NYW, January 2, 1892; Thomas M. Pitkin, *Keepers of the Gate: A History of Ellis Island* (New York: New York University Press, 1975), 19; and the records of ship manifests found at www.ellisislandrecords.org.

58 **She was soon**: A controversy arose over what happened to Annie Moore. Legend held that she headed out west to Texas, married, and died tragically when she was struck by a streetcar. More recent research found that Annie Moore actually never left New York. Instead, she remained in lower Manhattan, married a German-American named Schayer three years after her arrival, had eleven children of whom only five survived, and died of heart failure at age forty-seven in 1924. "She had the typical hardscrabble immigrant life," said Megan Smolenyak, the genealogist who discovered the story of the real Annie Moore. "She sacrificed herself for future generations." The living descendants of Annie Moore have Irish, German, Italian, Jewish, and Scandinavian surnames, a testament to the American melting pot. NYT, September 14, 16, 2006.

60 **Once on the second floor**: HW, October 24, 1891.

60 **A reporter from**: HW, August 26, 1893.

60 **Politicians, journalists**: NYT, November 7, 1895; Samuel Gompers, *Seventy Years of Life and Labour*, vol. 2 (New York: Augustus M. Kelley Publishers, 1967), 154.

61 **"The existing immigration law"**: "Annual Report of the Superintendent of Immigration to the Secretary of the Treasury for the Fiscal Year Ended June 30, 1892," 11.

61 **In 1875, the Supreme Court**: *Chae Chan Ping v. United States*, 130 U.S. 581 (1889). See also, Hiroshi Motomura, "Immigration Law After a Century of Plenary Power: Phantom Constitutional Norms and Statutory Interpretation," *Yale Law Journal*, December 1990.

61 **Three years later**: *Nishimura Ekiu v. U.S.*, 142 U.S. 651 (1892); Hiroshi Motomura, *Americans in Waiting: The Lost Story of Immigration and Citizenship in the United States* (New York: Oxford University Press, 2006), 33–34; Daniel J. Tichenor, *Dividing Lines: The Politics of Immigration Control in America* (Princeton, NJ: Princeton University Press, 2002), 110; "Developments in the Law: Immigration Policy and the Rights of Aliens," *Harvard Law Review*, April 1983.

63 **The government wanted**: The following discussion is taken from "A Report

of the Commissioners of Immigration Upon the Causes Which Incite Immigration to the United States," 52nd Congress, 1st Session, Executive Document 235, January 1892. See also, John B. Weber, "Our National Dumping-Ground: A Study of Immigration," *NAR*, April 1892.

64 **There was an additional**: John B. Weber, *Autobiography of John B. Weber* (Buffalo, NY: J. W. Clement Company, 1924), 105.

65 **Weber noted that**: A *Harper's Weekly* editorial made the same point, asking, "Who else is there here to do the work which these immigrants are doing for us? We have on former occasions called attention to the important fact that the native American is becoming more and more disinclined to do hard work with his hands. . . . How many native Americans are willing to do the dirt work in railway or canal building or to dig coal or even to serve as farm hands?" *HW*, September 1, 1894.

65 **Following his instructions**: *NYT*, February 15, 1892; Mary Antin, *From Plotzk to Boston* (Boston: W. B. Clarke: 1899), 12.

66 **The emigration of**: Irving Howe, *World of Our Fathers* (New York: Schocken Books, 1976), 5–7.

66 **The two Americans**: Weber, *Autobiography*, 112–128.

67 **By the 1890s**: Howe, *World of Our Fathers* 21; Weber, *Autobiography*, 106.

CHAPTER FOUR: PERIL AT THE PORTALS

70 **Weber was not resentful**: *NYT*, January 31, February 2, 1891.

71 **The *Massilia* had departed**: The discussion of the *Massilia* case comes from "Immigration Investigation, Ellis Island, 1892," 52nd Congress, 1st Session, House Reports, Vol. 12, No. 2090, Series 3053; "Annual Report of the Board of Health of the Health Department of the City of New York for the Year Ending December 31, 1892," City Hall Library, New York City; and Howard Markel, *Quarantine! East European Jewish Immigrants and the New York City Epidemics of 1892* (Baltimore, MD: Johns Hopkins University Press, 1997).

73 **Often confused with**: *NYT*, February 14, 1892.

74 **Within two days**: *NYT*, February 12, 1892.

74 **Edson and his staff**: *NYT*, February 13, 1892.

76 **The actions of Edson**: "Annual Report of the Board of Health of the Health Department of the City of New York for the Year Ending December 31, 1892," 142, City Hall Library, New York City. Howard Markel overemphasizes the role of nativism in explaining the behavior of Edson and other city officials. He complains that the quarantine stigmatized immigrants and that "there was a huge price to pay in the form of violated civil liberties, cultural insensitivities, inadequate financial or physical resources devoted to their medical care." In a more nuanced interpretation, Sherwin Nuland argues that city health officials "did what they believed to be the prudent thing, consistent with measures then current among their colleagues all over the world." The city's response, Nuland continued, mixed anti-immigrant sentiment with "an earnest desire to protect the people for whom they felt primarily responsible: the citizens of their city." Sherwin B. Nuland, "Hate in the Time of Cholera," *New Republic*, May 26, 1997. Markel also misreads an 1895 article by Edson entitled "The Microbe as a Social Leveler." In it, Edson argued that because of contagious diseases, poor and rich, native-born and immigrant, were all tied together. Treating contagious diseases, therefore, called for a more holistic approach. "To the man of wealth, therefore, there is a direct and very great interest in the well-being of the man of poverty," Edson wrote, describing a kind of public health socialism. Edson

did describe Russian Jews as "poor, ignorant, down-trodden" and implied that they could be susceptible to bringing contagious diseases to the United States. However Edson was not scapegoating Russian Jews or calling for their exclusion. If anything, he was saying that native-born Americans had a distinct interest in the well-being of Russian Jews, whether in Russia or in the Lower East Side of Manhattan. See, Cyrus Edson, "The Microbe as a Social Leveller," *NAR*, October 1895.

76 **The sometimes callous treatment**: NYT, February 24, 1892.

77 **New Hampshire senator**: Leon Burr Richardson, *William E. Chandler: Republican* (New York: Dodd, Mead, 1940), 7–11.

77 **As easy as it may be**: Richardson, *William E. Chandler*, 439; Carol L. Thompson, "William E. Chandler: A Radical Republican," *Current History* 23 (November 1952); NYT, March 7, 1892. Historian Morton Keller writes that Chandler "gave voice to a widespread attitude when he warned that trusts . . . tended to destroy competition, crush individualism, and put the control of society into the hands of opulent oligarchs." Morton Keller, *Regulating a New Economy: Public Policy and Economic Change in America, 1900–1933* (Cambridge, MA: Harvard University Press, 1990), 25.

77 **Back in the spring**: NYT, March 6, 1892.

78 **Chandler's investigation**: NYT, June 30, July 29, 1892.

78 **The hearings highlighted**: Transcripts of the Chandler hearings and subsequent report are found in "Immigration Investigation, Ellis Island, 1892," 52nd Congress, 1st Session, House Reports, Vol. 12, No. 2090, Series 3053. For more of Chandler's criticism of Weber, see *Congressional Record*, 52nd Congress, 1st Session, Vol. 23, Part 2, February 15, 1892, 1132.

79 **Weber came across**: John B. Weber, *Autobiography of John B. Weber* (Buffalo, NY: J.W. Clement Company, 1924), 95–96, 99–100.

82 **Then there was**: Markel, *Quarantine!* 49.

82 **Typhus, the *New York Times***: NYT, February 13, 1892. See also Amy L. Fairchild, *Science at the Borders: Immigrant Medical Inspection and the Shaping of the Modern Industrial Labor Force* (Baltimore, MD: Johns Hopkins University Press, 2003), 42–43.

83 **The linkage of**: "Select Committee of the House of Representatives to Inquire into the Alleged Violation of the Laws Prohibiting the Importation of Contract Laborers, Paupers, Convicts, and Other Classes," 1888; Julia H. Twells, "The Burden of Indiscriminate Immigration," *American Journal of Politics*, December 1894.

83 **Cyrus Edson**: Cyrus Edson, "Typhus Fever," *NAR*, April 1892.

84 **Chandler tried to**: William E. Chandler, "Methods of Restricting Immigration," *Forum*, March 1892; Letter from William Chandler to Unknown, 1890, Book 82, WC.

84 **Both extremes**: John Hawks Noble, "The Present State of the Immigration Question," *Political Science Quarterly*, June 1892; HW, September 1, 1894.

85 **Not surprisingly**: AH, March 4, 1892.

85 **Traveling from Turkey**: Howard Markel calls Benjamin Harrison an anti-Semitic restrictionist, claiming that his 1892 reelection platform "contained strong calls for the immigration restriction of Russian Hebrews." The platform calls for no such thing. In fact, the Republican Party platform protested "against the persecution of the Jews in Russia." It did call for "the enactment of more stringent laws and regulations for the restriction of criminal, pauper and contract immigration," a belief in keeping with the general view of regulating against "undesirable" immigrants. Markel also claims that Harrison "was

long a proponent of 'restricting the immigration of Russian Hebrews' and stated so emphatically in his final two annual addresses." That charge is also false. In his 1891 Annual Message to Congress, Harrison discusses the protests made by his government to the Russian czar "because of the harsh measures now being enforced against the Hebrews in Russia." Harrison also sent John Weber on a fact-finding trip to the Pale of Settlement to investigate the rise of anti-Semitism. Harrison was clearly concerned not only about the plight of the Russian Jews, but also about the effect that Jewish emigration might have on America. He wrote: "The immigration of these people to the United States—many other countries being closed to them—is largely increasing and is likely to assume proportions which may make it difficult to find homes and employment for them here and to seriously affect the labor market." Harrison's actual words hardly betray an anti-Semite. "The Hebrew is never a beggar; he has always kept the law—life by toil—often under severe and oppressive civil restrictions. It is also true that no race, sect, or class has more fully cared for its own than the Hebrew race. But the sudden transfer of such a multitude under conditions that tend to strip them of their small accumulations and to depress their energies and courage is neither good for them nor for us." In the wake of the cholera and typhus outbreaks, Harrison's 1892 Annual Message to Congress did argue that the "admission to our country and to the high privileges of its citizenship should be more restricted and more careful. We have, I think, a right and owe a duty to our own people, and especially to our working people, not only to keep out the vicious, the ignorant, the civil disturber, the pauper, and the contract laborer, but to check the too great flow of immigration now coming by further limitations."

85 **What was within:** NYT, September 2, 1892.
86 **Still, the brunt of:** Markel, Quarantine! 120–121, 130.
86 **The quarantine policy:** Richardson, William E. Chandler, 417; Thomas M. Pitkin, Keepers of the Gate: A History of Ellis Island (New York: New York University Press, 1975), 20; John Higham, Strangers in the Land: Patterns of American Nativism, 1860–1925 (New Brunswick, NJ: Rutgers University Press, 1955), 100; NYT, November 7, 1892.
87 **Weber called Chandler's bill:** W. E. Chandler, "Shall Immigration Be Suspended?" NAR, January 1893; Richardson, William E. Chandler, 38; Weber, Autobiography, 133; Arthur Cassot, "Should We Restrict Immigration?" American Journal of Politics, September 1893.
87 **Instead, Congress passed:** Markel, Quarantine! 173–182; Edwin Maxey, "Federal Quarantine Laws," Political Science Quarterly 23, no. 4 (December 1908).
87 **The nation did get:** William C. Van Vleck, The Administrative Control of Aliens: A Study in Administrative Law and Procedure (New York: Da Capo Press, 1971 [1932]), 8–9; Richard H. Sylvester, "The Immigration Question in Congress," American Journal of Politics, June 1893; Pitkin, Keepers of the Gate, 20–21. An article in the Political Science Quarterly agreed, noting that although some had proposed extending "the policy adopted with reference to the Chinese, making race the test of fitness," such a policy would be politically unpopular, cause diplomatic problems, and be "repugnant to the general theory that America is a haven for the oppressed of all mankind." What was needed was a "less clumsy and offensive law." Noble, "The Present State of the Immigration Question."
88 **The new manifests:** Joseph H. Senner, "How We Restrict Immigration," NAR, April 1894; Pitkin, Keepers of the Gate, 20–22.
89 **These boards of special:** Van Vleck, Administrative Control of Aliens, 46–53, 214; Pitkin, Keepers of the Gate, 24.

89 **Health concerns**: Fitzhugh Mullan, *Plagues and Politics: The Story of the United States Public Health Service* (New York: Basic Books, 1989), 40–48.

91 **The epidemic scares**: *NYT*, January 6, July 21, 1894; Joseph Senner, "The Immigration Question," *Annals of the American Academy of Political and Social Science*, July 1897.

92 **The top three**: Daniel J. Tichenor, *Dividing Lines: The Politics of Immigration Control in America* (Princeton, NJ: Princeton University Press, 2002), 79; *HW*, January 8, 1898.

92 **These changes were**: *NYT*, March 6, August 29, 1892; Noble, "The Present State of the Immigration Question"; Henry Cabot Lodge, "Lynch Law and Unrestricted Immigration," *NAR*, May 1891; John Chetwood Jr., "Immigration, Hard Times, and the Veto," *Arena*, December 1897.

92 **Such feelings extended**: James R. O'Beirne, "The Problem of Immigration: Its Dangers to the Future of the United States," *Independent*, November 2, 1893.

92 **While the *Massilia* incident**: Noble, "The Present State of the Immigration Question." On the 1891 lynching of Italians, see Richard Gambino, *Vendetta* (New York: Doubleday, 1977); Jerre Mangione and Ben Morreale, *La Storia: Five Centuries of the Italian American Experience* (New York: HarperPerennial, 1993), 204–213; Lodge, "Lynch Law and Unrestricted Immigration."

93 **If the crimes seemed**: *NYT*, May 18, 1893.

93 **As deportations increased**: *NYT*, May 21, 1894.

93 **The anger of Italians**: *NYT*, April 5, 1896; Pitkin, *Keepers of the Gate*, 24–26. For more on Italian immigrants during this time, see J. H. Senner, "Immigration to Italy," *NAR*, June 1896 and Prescott F. Hall, "Italian Immigration," *NAR*, August 1896.

93 **The fear of Italian**: *BG*, April 26, 1896.

CHAPTER FIVE: BRAHMINS

95 **Boston had long stood**: Barbara Miller Solomon, *Ancestors and Immigrants: A Changing New England Tradition* (New York: Wiley, 1956), 48, 101; Linda Gordon, *Woman's Body, Woman's Right: Birth Control in America*, rev. ed. (New York: Penguin Books, 1990), 135.

96 **It is no surprise**: Francis A. Walker, "Restriction of Immigration," *Atlantic*, June 1896.

96 **Perhaps the best expression**: Solomon, *Ancestors and Immigrants*, 88; Thomas Bailey Aldrich, "The Unguarded Gates," *Atlantic*, March 1895. Today the poem is still popular among supporters of immigration restriction. See, http://www.vdare.com/fulford/unguarded.htm.

97 **Not all of the voices**: *BH*, July 5, 1896.

97 **As the Fitzgeralds**: Solomon, *Ancestors and Immigrants*, 23, 57; Francis Walker, "Immigration," *Yale Review*, August 1892. Charles Francis Adams Jr., brother of Henry, thought the immigration questions was "too big and too intricate . . . to meddle with." Solomon, *Ancestors and Immigrants*, 32.

98 **At just twenty-five years old**: Warren later became a noted constitutional lawyer. The Charles Warren Center for Studies in American History at Harvard University is named after Warren, funded by an endowment from his late wife. It is disappointing, yet unsurprising, that Warren's bio on the Harvard University website makes no mention of his role in the founding of the Immigration Restriction League: http://www.fas.harvard.edu/~cwc/historycwbio.html.

98 **Prescott Hall**: Prescott F. Hall, "The Future of American Ideals," *NAR*, January 1912. For more on the "Anglo-Saxon Complex," and theories of Anglo-Saxon

and Teutonic culture, see Solomon, *Ancestors and Immigrants*, 59–81 and Henry Cabot Lodge, "The Restriction of Immigration," *Our Day*, May 1896.

98 **Hall, who would be**: *Immigration and Other Interests of Prescott Farnsworth Hall*, compiled by Mrs. Prescott F. Hall (New York: Knickerbocker Press, 1922), 119–123.

99 **The deep depressions**: T. J. Jackson Lears, *No Place of Grace: Antimodernism and the Transformation of American Culture, 1880–1920* (New York: Pantheon Books, 1981), 47–58.

99 **In response, the boisterous**: Theodore Roosevelt, "The Strenuous Life," speech delivered to Chicago's Hamilton Club, April 10, 1899; Theodore Roosevelt, "Twisted Eugenics," *Outlook*, January 3, 1914. During his presidency, Roosevelt began speaking of "race suicide," a term coined by Progressive academic Edward A. Ross. The president and father of six famously gave a talk before the National Congress of Mothers arguing against birth control and in favor of larger families. See Theodore Roosevelt, "On American Motherhood," speech delivered to the National Congress of Mothers, March 13, 1905.

100 **In fact, Hall embodied**: Prescott F. Hall, "Representation Without Taxation," unpublished manuscript, in *Immigration and Other Interests of Prescott Farnsworth Hall*.

100 **So it was no surprise**: Morris M. Sherman, "Immigration Restriction, 1890–1921, and the Immigration Restriction League," (Cambridge, MA: Harvard College, 1957).

101 **The IRL's strength**: John Higham, *Strangers in the Land: Patterns of American Nativism, 1860–1925* (New Brunswick, NJ: Rutgers University Press, 1955), 102–103; Solomon, *Ancestors and Immigrants*, 103–104, 123; "Reports of the Industrial Commission on Immigration," vol. 15, 1901, 46.

101 **The IRL worked closely**: Daniel J. Tichenor, *Dividing Lines: The Politics of Immigration Control in America* (Princeton, NJ: Princeton University Press, 2002), 77, 85.

102 **The descendants**: Solomon, *Ancestors and Immigrants*, 73, 107, 114, 120; Julia H. Twells, "The Burden of Indiscriminate Immigration," *American Journal of Politics*, December 1894.

103 **Among its proposals**: "Constitution of the Immigration Restriction League," August 22, 1894, IRL.

103 **For a young man**: *NYT*, December 12, 1894.

103 **Like so many**: *BH*, April 5, 1895.

103 **In mid-December 1895**: "Immigration Restriction League, Annual Report of the Executive Committee for 1895," January 13, 1896, and "IRL Annual Report of the Executive Committee for 1896," January 11, 1897, File 1138, IRL; *Brookline Chronicle*, January 18, 1895; *Boston Journal*, January 25, 1896.

104 **So in April 1896**: *NYT*, April 21, 1896.

104 **In its April 1896 investigation**: "Immigration: Its Effects upon the United States, Reasons for Further Restriction." Publication of the Immigration Restriction League, No. 16, February 13, 1897, IRL. It is certainly true that many Italians were illiterate, due in large part to the poor schools of their native country, but in Italy illiteracy rates went down considerably during the era of peak immigration, from almost 69 percent in 1872 to an estimated 23 percent in 1922. Antonio Stella, *Some Aspects of Italian Immigration to the United States* (New York, Arno Press, 1975), 53.

104 **The IRL members**: "Immigration Restriction League, Annual Report of the Executive Committee for 1895," January 13, 1896, File 1138, IRL; Prescott F. Hall, "Immigration and the Educational Test," *NAR*, October 1897.

104 **Such a test would**: Henry Cabot Lodge, "The Restriction of Immigration," *Our Day*, May 1896.

105 **Not all restrictionists**: Francis A. Walker, "Immigration," *Yale Review*, August 1892.

105 **Writing to the secretary**: Letter from Herman Stump to John Carlisle, February 20, 1897, Grover Cleveland Papers, LOC.

106 **For years, immigration restrictionists**: President Grover Cleveland's Veto Message of the Educational Test Bill, March 2, 1897, reprinted by the National Liberal Immigration League, File 1125, Folder 4, IRL. Many years later, Theodore Roosevelt told Madison Grant that General Leonard Wood had told him that Cleveland had regretted his veto of the literacy test, confirming what many restrictionists had come to believe. There is no definite proof that Cleveland ever expressed regret about his veto. Letter from Madison Grant to Theodore Roosevelt, November 15, 1915, TR.

CHAPTER SIX: FEUD

107 **Just after midnight**: Victor Safford, *Immigration Problems: Personal Experiences of an Official* (New York: Dodd, Mead, and Company, 1925), 199–200.

108 **To some, it was**: Thomas M. Pitkin, *Keepers of the Gate: A History of Ellis Island* (New York: New York University Press, 1975), 26; NYT, June 17, 1897; NYW, June 16, 1897; HW, February 26, 1898.

108 **Officials then moved**: NYT, June 19, 1897.

109 **Victor Safford remembered**: Safford, *Immigration Problems*, 76.

109 **Befitting someone from**: Letter from Edward McSweeney to Archbishop Michael Corrigan, January 12, 1900, ANY; Letter from A. J. You to Terence V. Powderly, June 11, 1900, Box 137, TVP.

110 **McSweeney remained**: Pitkin, *Keepers of the Gate*, 29.

110 **Meanwhile, the McKinley**: Robert E. Weir, *Knights Unhorsed: Internal Conflict in a Gilded Age Social Movement* (Detroit: Wayne State University Press, 2000), 16; Craig Phelan, *Grand Master Workman: Terence Powderly and the Knights of Labor* (Westport, CT: Greenwood Press, 2000), 1–2.

111 **McSweeney seemed**: Phelan, *Grand Master Workman*, 47.

111 **One historian described**: Weir, *Knights Unhorsed*, 15; Vincent J. Falzone, *Terence V. Powderly: Middle-Class Reformer* (Washington, DC: University Press of America, 1978), 174; Terence V. Powderly, *The Path I Trod: The Autobiography of Terence V. Powderly* (New York: AMS Press, 1968; original edition: Columbia University Press, 1940), 287; Letter from Terence V. Powderly to William Scaife, February 6, 1910, Box 153, TVP.

112 **Another critic was**: Falzone, *Terence V. Powderly*, 175.

112 **Powderly fought back**: Letter from Edward F. McSweeney to Samuel Gompers, November 22, 1901, SG.

112 **Powderly's brother**: Letter from T. V. Powderly to Roy W. White, March 1, 1898, Box 128, TVP; T. V. Powderly, "A Menacing Irruption," NAR, August 1888.

112 **Powderly did not stop**: Edward McGlynn, "The New Know-Nothingness and the Old," NAR, August 1887; Powderly, *The Path I Trod*, 5.

113 **That trouble would**: Powderly, *The Path I Trod*, 299; Letter from Edward F. McSweeney to T. V. Powderly, June 6, 1898, Box 133, TVP.

113 **Powderly made**: Memorandum from T. V. Powderly, February 15, 1902, Box 156, TVP; NYT, March 10, 1899.

114 **Such impolitic behavior**: Falzone, *Terence V. Powderly*, 175–182, 188.

114 **The decision on the**: Memorandum from T. V. Powderly, February 15, 1902, Box 156, TVP; Pitkin, *Keepers of the Gate*, 28.

115 **Perhaps that insecurity**: Letter from T. V. Powderly to Thomas Fitchie, August 3, 1898, TR.

115 **Just a few months**: Letter from Sen. T. C. Platt to Thomas Fitchie, February 17, 1898, TR.

115 **Despite Platt's urgings**: Letter from T. V. Powderly to William McKinley, 1901, Series 2, TVP.

116 **More complaints emerged**: Alvan F. Sanborn, "The New York Immigration Service," *Independent*, August 10, 1899; Safford, *Immigration Problems*, 86.

116 **Much as Powderly**: All references to the report come from Report by Campbell and Rodgers, June 2, 1900 to Secretary of the Treasury, Boxes 157–158, TVP.

117 **The most serious charges**: For charges against Lederhilger, see Report by Campbell and Rodgers, June 2, 1900 to Secretary of the Treasury, TVP. See also Letter from Thomas Fitchie to John Lederhilger, September 10, 1900, File 52727-4, INS.

117 **Treasury Department officials**: Letter from Edward F. McSweeney to Archbishop Michael A. Corrigan, September 10, 1900, Roll 19, G-17-G20, ANY.

118 **The report was certainly**: NYT, June 6, 1900.

118 **Edward Steiner**: Edward A. Steiner, *On the Trail of the Immigrant* (New York: Fleming H. Revell Company, 1906), 79–80.

118 **Stewart had been**: Letter from Thomas Fitchie to J. Ross Stewart, September 10, 1900, File 51841/119, INS; NYT, October 5, 1900; Eric Foner, *Freedom's Lawmakers: A Directory of Black Officeholders during Reconstruction* (New York: Oxford University Press, 1993), 204. The *Times* referred to Stewart as "J. Ross Stewart" and claimed he had been a Georgia state legislator. However, it is fairly certain that the man fired at the Barge Office was Jordan R. Stewart and he was from Louisiana. Stewart was also a friend of P. B. S. Pinchback, the first black governor in the nation's history, who had served one month as Louisiana's governor. Pinchback was also living in New York City in the 1890s. George McKenzie, the Republican Colored Leader of the 25th Assembly District in New York, had known Stewart for forty years and wrote to the Treasury Department to protest the charges against his friend, calling him a "brave soldier during the war of the rebellion." McKenzie did not believe the charges against Stewart because they came from "a band of conspirators, trying to reflect discredit on the administration of Commissioner Thomas Fitchie." Letter from George McKenzie to H. A. Taylor, September 19, 1900, File 51841/119, INS.

119 **While men like Stewart**: Letter from Terence V. Powderly to A. J. You, May 16, 1900, Letterbox 73; Letter from Terence V. Powderly to President William McKinley, undated, Box 156, TVP.

119 **Powderly wanted**: Letter from Terence V. Powderly to T. F. Lee, June 19, 1900, Letterboook 73, Box 152, TVP.

119 **Some of Powderly's friends**: See File 51841-97, INS. Fitzharris possesses one of history's all-time great nicknames, which apparently derived from the time he killed a goat he kept in his backyard and skinned it to make some money. James Joyce, no doubt taken by the unusual nickname, immortalized Fitzharris in his novel *Ulysses*.

119 **A Powderly ally**: Letter from A. J. You to Terence V. Powderly, May 24, 1900, Box 137, TVP.

120 **In what was probably**: Letter from Terence V. Powderly to Hon. William McKinley, undated, Series 2, TVP.

120 **In mid-December 1900**: NYW, December 18, 1900; *Leslie's Illustrated Weekly Newspaper*, January 5, 1901.

121 **The centerpiece of the**: Pitkin, *Keepers of the Gate*, 33; *Architectural Record*, December 1902.

121 **Ellis Island now consisted**: *NYT*, December 3, 1900.

122 **Ensconced in Washington**: Letter to T. V. Powderly, September 20, 1900, TVP. For examples of intercepted McSweeney letters, see Box 125, Series 2, TVP.

122 **When not bogged down**: Letter from Terence V. Powderly to Hon. William McKinley, undated, Series 2, TVP.

123 **By the summer of 1901**: Letter from Roman Dobler to T. V. Powderly, August 16, 1901, TVP.

CHAPTER SEVEN: CLEANING HOUSE

127 **It was not a name**: Eric Rauchway, *Murdering McKinley: The Making of Theodore Roosevelt's America* (New York: Hill and Wang, 2003), 60.

128 **Theodore Roosevelt had**: President Theodore Roosevelt, "First Annual Message to Congress," December 3, 1901.

128 **The bullets that**: Theodore Roosevelt, "True Americanism," *Forum*, April 1894.

128 **In the previous decade**: Theodore Roosevelt, *An Autobiography* (New York: Scribner's, 1913), 357; Robert Watchorn, *The Autobiography of Robert Watchorn* (Oklahoma City, OK: Robert Watchorn Charities, 1959), 145.

128 **Roosevelt was no newcomer**: Edmund Morris, *The Rise of Theodore Roosevelt* (New York: Ballantine Books, 1979), 376; Letters from Theodore Roosevelt to Henry Cabot Lodge, January 27, 1897, March 19, 1897, in Henry Cabot Lodge and Charles F. Redmond, eds., *Selections from the Correspondence of Theodore Roosevelt and Henry Cabot Lodge, 1884–1918*, vol. 1 (New York: Da Capo Press, 1971).

129 **Roosevelt worried about**: Theodore Roosevelt, "The Immigration Problem," *Harvard Monthly*, December 1888.

129 **The relationship between**: Roosevelt, "True Americanism."

130 **A young inspector**: Watchorn, *Autobiography*, 145–147.

130 **William McKinley**: Hans Vought, *The Bully Pulpit and the Melting Pot: American Presidents and the Immigrant, 1897–1933* (Macon, GA: Mercer University Press, 2004), 22–23; Daniel J. Tichenor, *Dividing Lines: The Politics of Immigration Control in America* (Princeton, NJ: Princeton University Press, 2002), 73–75.

131 **Even with a new**: Letter from T. V. Powderly to Thomas Fitchie, October 4, 1901, Letter from Acting Secretary O. L. Spaulding to Thomas Fitchie, October 9, 1901, Box 123, TVP.

132 **Edward McSweeney had more reasons**: Letter from Edward McSweeney to Theodore Roosevelt, March 26, 1902, Series 1, TR.

132 **Roosevelt's views**: Tichenor, *Dividing Lines*, 122; Barbara Miller Solomon, *Ancestors and Immigrants: A Changing New England Tradition* (New York: Wiley, 1956), 196; Vought, *Bully Pulpit*, 33.

132 **Yet immigrant defenders**: Roosevelt, *An Autobiography*, 186–187; Tichenor, *Dividing Lines*, 33. On Roosevelt's family background, see Morris, *Rise of Theodore Roosevelt*, 36–37.

133 **Whatever may have been**: Elting E. Morison, ed., *The Letters of Theodore Roosevelt*, vol. 3 (Cambridge, MA: Harvard University Press, 1951), 170–171.

133 **Powderly left the meeting**: Letter from T. V. Powderly to John Parsons, October 25, 1904, Box 139, TVP.

133 **Even with the charges**: Letter from Terence V. Powderly to President William McKinley, undated, TVP; "Reports of the Industrial Commission on Im-

migration," vol. 15, 1901, 72, 170; Letter from Nicholas Butler to Theodore Roosevelt, October 12, 1901, NMB.

134 **"Nicholas Miraculous" Butler**: Letter from Nicholas Murray Butler to Theodore Roosevelt, October 7, 1901, NMB.

134 **Jacob Riis**: Letter from Jacob Riis to TR, March 17, 1902, Series 1, TR.

134 **What Roosevelt really**: Morison, ed., *Letters*, vol. 3, 221, 250.

134 **Finally, in the spring**: NYT, March 24, 1902.

135 **Powderly demanded**: Letter from Terence V. Powderly to Robert Watchorn, March 22, 1902, Letterbook 79, Box 153, TVP.

135 **The sheer number of**: Powderly, 381–382.

136 **The son of a New London**: Robert Williams's grandson John was kidnapped by Indians in the infamous Deerfield Indian raid of 1704 and held for two years. John Williams's book about his experiences inspired James Fenimore Cooper's *Last of the Mohicans*. Other direct descendants of Robert Williams include Louisa May Alcott, the Wright Brothers, Ephraim Williams, founder of Williams College, General George B. McClellan, and Eli Whitney.

137 **At lunch, he sat Williams**: Letter from Edward Van Ingen to Theodore Roosevelt, March 27, 1902, Series 1, Reel 25, TR.

137 **Roosevelt always had**: John Morton Blum, *The Republican Roosevelt* (Cambridge, MA: Harvard University Press, 1954), 12–13.

137 **Roosevelt felt**: Roosevelt, 57–63.

138 **Williams informed Roosevelt**: Letter from William Williams to Theodore Roosevelt, August 8, 1902, Series 1, TR.

138 **The cases of Murray and Braun**: Letter from James Sheffield to William Williams, April 29, 1915, Williams Papers, WW-NYPL.

139 **Murray replaced McSweeney**: Letter from Terence V. Powderly to Robert Watchorn, March 22, 1902, Letterbook 79, Box 153, TVP.

140 **Williams let nothing**: "Annual Report of the Commissioner-General of Immigration," 1902, 56.

140 **Others also felt**: Letter from William Williams to Senator Thomas Platt, May 26, 1902, WW-NYPL.

140 **Nor would the abusive**: Letter from William Williams to N. J. Sparkling, May 26, 1903; Letter from Williams to John Bell, gateman at Ellis Island, November 3, 1903, WW-NYPL.

141 **To protect immigrants**: Letter from Herbert Parsons to William Williams, April 3, 1902, WW-NYPL; "Annual Report of the Commissioner-General of Immigration," 1902, 56.

142 **New bids were put out**: Letter from William Williams to Theodore Roosevelt, June 24, 1902, WW-NYPL.

142 **Williams even tackled**: NYT, July 12, 1903.

143 **In addition to**: Letter from William Williams to Theodore Roosevelt, September 17, 1902, Series 1, TR.

143 **Roosevelt then ordered**: Letter from William Williams to Theodore Roosevelt, February 4, 1903, Series 1, TR.

144 **Because of an electoral**: BG, June 27, 1903.

144 **The case took on**: Letter from Theodore Roosevelt to Elihu Root, October 3, 1903; Letter from Theodore Roosevelt to Curtis Guild, Junior, October 20, 1903, Morison, ed., *Letters*, vol. 610–611, 633–634.

144 **The case remained**: NYT, December 10, 1903.

144 **Though McSweeney tried**: Letter from William Williams to Theodore Roosevelt, February 4, 1903, Series 1, TR; BG, June 15, 1904; BH, July 11, September 18, 1903.

145 **As McSweeney was**: Francis E. Leupp, *The Man Roosevelt* (New York: Apple-
ton, 1904), 136.

145 **Shortly after his dismissal**: Watchorn, *Autobiography*, 92.

145 **These were difficult**: Letter from Terence V. Powderly to Robert Watchorn,
July 4, 1902, Letterbook 79, Box 153; Terence V. Powderly to T. L. Lee, July 7,
1902, Letterbook 80, Box 153, TVP.

146 **Powderly's depression**: Letter from Robert Watchorn to George R. Cullen,
May 18, 1903, TVP.

146 **But Roosevelt had not**: Letter from Robert Watchorn to Terence V. Powderly,
September 5, 1903, Box 128, TVP; Letter from Theodore Roosevelt to Philander Chase Knox, August 1, 1903, Morison, ed., *Letters*, vol. 3, 538–539.

146 **Nothing came of**: Letter from Terence V. Powderly to John N. Parsons, October
25, 1904, TVP.

146 **On October 23**: NYT, October 24, 1903; March 14, 1904.

147 **Goldman called**: Candace Falk, ed., *Emma Goldman: A Documentary History of
the American Years*, vol. 2: *Making Speech Free, 1902–1909* (Berkeley: University
of California Press, 2005), 121–123.

147 **Writing from his**: John Turner, "The Protest of an Anarchist," *Independent*, December 24, 1903.

148 **Turner certainly had**: *U.S. Ex Rel. Turner v. Williams*, U.S. 279 (1904). For more
on the Turner case, see Daniel Kanstroom, *Deportation Nation: Outsiders in
American History* (Cambridge, MA: Harvard University Press, 2007), 136–138
and David Cole, *Enemy Aliens: Double Standards and Constitutional Freedoms in
the War on Terrorism* (New York: New Press, 2003), 108–109.

CHAPTER EIGHT: FIGHTING BACK

150 **Williams's appointment**: Letter from William Williams to Prescott F. Hall, December 27, 1902, File 999, IRL.

150 **Immigrants were on notice**: Letter from William Williams to Bolognesi, Hart-
field & Co., June 10, 1902, WW-NYPL.

150 **Williams believed that**: "United States Immigration Laws with Annotations
for Guidance of Immigrant Inspectors at the Ellis Island Station," November
1902, TVP; Letter from William Williams to Theodore Roosevelt, November
25, 1902, WW-NYPL.

151 **Compare Williams's 1902 edict**: *Reports of the Industrial Commission on Immigration, 1901* (New York: Arno Press, 1970, reprint), 81.

151 **In his first Annual Message**: President Theodore Roosevelt, "First Annual
Message to Congress," December 3, 1901.

151 **Roosevelt warned Williams**: Letter from Theodore Roosevelt to William Williams, January 21, 1903, Series 2, TR; Letter from Theodore Roosevelt to William Williams, January 23, 1903, in Elting E. Morison, ed., *The Letters of Theodore
Roosevelt*, vol. 3 (Cambridge, MA: Harvard University Press, 1951), 411–412.

152 **Williams at first responded**: Letter from William Williams to Theodore
Roosevelt, January 24, 1903, Series 1, TR.

152 **Williams then shot back**: Letter from William Williams to Theodore Roosevelt,
January 29, 1903, WW-NYPL.

152 **Williams continued**: Letter from William Williams to Theodore Roosevelt,
February 8, 1903, Series 1, TR.

152 **A man like William Williams**: NYT, May 24, 1903.

153 **Williams ended his**: Edited version of Williams's Annual Report for 1903 with
Roosevelt's edits is found in the WW-NYPL.

153 **Over 857,000 immigrants**: Kate Holladay Claghorn, "Immigration in Its Relation to Pauperism," *Annals of the American Academy of Political and Social Science*, July 1904.

154 **In a 1906 book sympathetic**: Edward A. Steiner, *On the Trail of the Immigrant* (New York: Fleming H. Revell Company, 1906), 75.

154 **Not everyone agreed**: Wallace Irwin, "Ellis Island's Problems," *New York Globe and Commercial Advertiser*, June 14, 1904.

154 **It was a sentiment**: "Annual Report of the Commissioner-General of Immigration," 1903, 70.

154 **It was a cold**: "Annual Report of the Commissioner-General of Immigration," 1904, 106.

155 **To his supporters**: Letter from Prescott Hall to William Williams, December 24, 1902, WW-NYPL.

156 **Even the *American Hebrew***: Quoted in Williams Memo, "Comments on Certain Articles Which Appeared in the 'Staats Zeitung' Between December 1902 and October 1903," undated, WW-NYPL; *AH*, January 30, 1903.

156 **Despite the support**: "Hell on Earth," *New Yorker Staats-Zeitung*, September 4, 1903. A translated copy appears in the William Williams Papers at the New York Public Library. Williams, who was fluent in German, either translated the articles himself or had them translated.

156 **Williams may have**: Letter from Frank Sargent to William Williams, April 14, 1903, WW-NYPL; Letter from Robert Watchorn to Terence V. Powderly, September 5, 1903, Box 128, TVP.

157 **Roosevelt's trip began**: The following account of Roosevelt's visit is taken from *NYT*, September 17, 1903, and *BG*, September 17, 1903.

157 **After a quick lunch**: For a slightly different version of the story, see Henry Pratt Fairchild, *Immigration: A World Movement and Its American Significance* (New York: Macmillan, 1913), 188.

158 **Little escaped**: Letter from Theodore Roosevelt to Victor Howard Metcalf, February 22, 1906, in Morison, ed., *Letters*, vol. 5, 162–163; *Morgen Journal*, July 10, 1912. For more on trachoma, see Howard Markel, "'The Eyes Have It': Trachoma, the Perception of Disease, the United States Public Health Service, and the American Jewish Immigration Experience, 1897–1924," *Bulletin of the History of Medicine* 74 (2000).

158 **Among those invited**: Letter from Theodore Roosevelt to Ralph Trautman, November 28, 1903, in Morison, ed., *Letters*, vol. 3, 659–660. In addition to von Briesen, the commission included former district attorney Eugene Philbin; Thomas Hynes, New York commissioner of corrections; Ralph Trautman, Treasurer, New York Palisades Interstate Park Commission; and Lee Frankel, of the United Hebrew Charities.

159 **The Von Briesen Commission**: See File 52727-2, INS.

160 **Roosevelt was happy**: Letter from Theodore Roosevelt to Ralph Trautman, November 28, 1903, in Morison, ed., *Letters*, vol. 3, 659–660.

161 **The final report**: Letter from Eugene A. Philbin to Theodore Roosevelt, December 1, 1903, Series 1; Letter from Theodore Roosevelt to Eugene Philbin, December 2, 1903; Letter from Arthur von Briesen to Theodore Roosevelt, December 4, 1903, TR.

162 **Though Roosevelt said**: Hans Vought, *The Bully Pulpit and the Melting Pot: American Presidents and the Immigrant, 1897–1933* (Macon, GA: Mercer University Press, 2004), 42–43; Letter from Theodore Roosevelt to Henry Cabot Lodge, May 23, 1904 in Henry Cabot Lodge and Charles F. Redmond, eds., *Selections from the Correspondence of Theodore Roosevelt and Henry Cabot Lodge, 1884–1918*, vol. 2 (New York: Da Capo Press, 1971).

162 **Roosevelt's campaign manager:** Letter from George B. Cortelyou to William Williams, September 24, 1904, WW-NYPL; Letter from William Williams to Theodore Roosevelt, October 15, 1904, Series 1, TR.

163 **Apparently, Williams's problems:** Letter from Robert Watchorn to Terence V. Powderly, December 21, 1904, TVP.

163 **In December 1904:** Letter from Theodore Roosevelt to Gifford Pinchot, January 19, 1905, Series 2, TR; Letter from Theodore Roosevelt to Northrop Stranahan, December 24, 1904, in Morison, ed., *Letters*, vol. 3, 1077–1078.

164 **Some immigration defenders:** *AH*, January 20, 1905.

CHAPTER NINE: THE ROOSEVELT STRADDLE

165 **Leaning over the second-story:** H. G. Wells, *The Future in America* (New York: Arno Press, 1974, orig. pub. 1906), 140.

165 **Wells had cemented:** Robert Watchorn, *The Autobiography of Robert Watchorn* (Oklahoma City, OK: Robert Watchorn Charities, 1959), 127–128.

166 **Once there, Watchorn ended up:** In her 1925 autobiography, famed union organizer Mother Jones wrote: "I remember John Siney, a miner. Holloran, a miner. James, a miner. Robert Watchorn, the first and most able secretary that the miners of this country ever had. These men gave their lives that others might live. They died in want." Though she was correct about Watchorn's position, he was still very much alive at the time of the publication of Mother Jones's 1925 memoir. In fact, by the time the autobiography was published, not only was Watchorn alive, he had become a millionaire oilman. Mary Field Parton, ed., *The Autobiography of Mother Jones* (Chicago: Charles H. Kerr, 1925), 240.

167 **On the issue of Joe Murray:** Letter from Theodore Roosevelt to Gifford Pinchot, January 19, 1905, Series 2; Letter from Robert Watchorn to Theodore Roosevelt, January 21, 1905, Series 1, TR; Letter from Robert Watchorn to Oscar Straus, May 4, 1907, Box 6, OS.

167 **Roosevelt was adept:** President Theodore Roosevelt, "Fifth Annual Message to Congress," December 5, 1905.

168 **It was a fine statement:** President Theodore Roosevelt, "Fifth Annual Message to Congress," December 5, 1905.

168 **If Roosevelt wanted:** Henry James, *The American Scene*, republished in Henry James, *Collected Travel Writings: Great Britain and America* (New York: Library of America, 1993) 425–426.

168 **With each passing week:** *NYT*, April 17, 1906.

169 **If Americans thought:** Philip Cowen, *Memories of an American Jew* (New York: International Press, 1932), 185–186; *NYT*, January 7, 1907.

169 **Robert Watchorn, who oversaw:** *NYT*, March 11, 1906.

169 **Watchorn told a Jewish audience:** *NYT*, November 19, 1906; Sheldon Morris Neuringer, *American Jewry and United States Immigration Policy, 1881–1953* (New York: Arno Press, 1980), 60.

169 **College professor Edward Steiner:** Edward A. Steiner, *On the Trail of the Immigrant* (New York: Fleming H. Revell Company, 1906), 91–92.

170 **Watchorn had a chance:** Robert Watchorn, "The Gateway of the Nation," *Outlook*, December 28, 1907.

171 **At a dinner celebrating:** John Morton Blum, *The Republican Roosevelt* (Cambridge, MA: Harvard University Press, 1954), 37. See also, Letter from Theodore Roosevelt to Lyman Abbott, May 29, 1908, in Elting E. Morison, ed., *The Letters of Theodore Roosevelt*, vol. 6 (Cambridge, MA: Harvard University Press, 1951), 1042.

171 **Straus, along with Schiff**: On Straus's background, see Naomi W. Cohen, *A Dual Heritage: The Public Career of Oscar S. Straus* (Philadelphia: Jewish Publication Society of America, 1969).

171 **As part of his**: David Nasaw, *The Chief: The Life of William Randolph Hearst* (Boston: Mariner Books, 2000), 207–209.

171 **The Bureau of**: Oscar Straus, *Under Four Administrations: From Cleveland to Taft* (Boston: Houghton Mifflin, 1922), 216; Letter from Oscar Straus to Robert Watchorn, December 30, 1907, OS.

171 **On the morning of**: Oscar Straus Diary, 3, Box 22, OS.

172 **Some cases were**: Straus, *Under Four Administrations*, 216–217.

172 **"I would be less than human"**: Cohen, *A Dual Heritage*, 154–155.

172 **Straus made his**: Oscar Straus Diary, 67–68, Box 22, OS; NYT, May 22, 1907. Robert Watchorn discusses the same story in his autobiography, but some of the details are different. Watchorn, *Autobiography*, 132–135.

172 **Straus made yet**: "Report of Conference held at the Ellis Island Immigration Station," June 15, 1908, File 51831-101, INS.

174 **The case hinged**: The Department of Commerce and Labor debated this issue in 1909 and 1914. See File 52745-4, INS.

174 **With this in mind**: Letter from Oscar Straus to Robert Watchorn, June 21, 1907, Letterbook 8, Box 20, OS.

175 **"Not only must we treat"**: President Theodore Roosevelt, "Sixth Annual Message to Congress," December 3, 1906.

175 **While many worried**: Letter from Theodore Roosevelt to Lyman Abbott, May 29, 1908, in Morison, ed., *Letters*, vol. 6, 1042.

175 **Throughout the first decade**: "National Liberal Immigration League," File 1125, Folder 1, IRL.

175 **The pro-immigrant group**: Rivka Shpak Lissak, "The National Liberal Immigration League and Immigration Restriction, 1906–1917," *American Jewish Archives*, Fall/Winter 1994; Neuringer, *American Jewry*, 53–54.

176 **The public debate**: *Charities*, December 16, 1905. While Secretary of Commerce and Labor, Straus told a reporter: "The restriction for the purpose of excluding the diseased, the criminal and other undesirable classes that have been incorporated in our laws, are salutary and wise." NYT, November 17, 1907.

176 **Closer to the**: NYT, January 7, 1907.

176 **As an official**: Steiner, *On the Trail*, 93.

176 **Not only did Watchorn**: NYT, May 12, 14, August 12, 15, 1905, February 9, March 17, November 9, 1906; Letter from Theodore Roosevelt to James S. Clarkson, October 3, 1905, in Morison, ed., *Letters*, vol. 5, 43–44; Marcus Braun, *Immigration Abuses: Glimpses of Hungary and Hungarians* (New York: Pearson Advertising Co., 1906); Gunther Peck, *Reinventing Free Labor: Padrones and Immigrant Workers in the North American West, 1880–1930* (Cambridge, UK: Cambridge University Press, 2000), 92–93.

177 **Theodore Roosevelt showed**: Cowen, *Memories*, 187–188. According to Cowen's translation of the article, Hitler argued that Jews were behind America's restrictive immigration quotas in force at the time, believing they wanted to keep out Gentile immigrants while "the Jews are always coming in new swarms." Nothing could be further from the truth, since the American Jewish community was a loud opponent of immigration quotas and Jewish immigrants were severely affected by them.

178 **La Guardia was clearly**: Letter from Louis K. Pittman, December 3, 1985, Public Health Service Archives, Rockville, MD.

178 **La Guardia found**: Fiorello H. La Guardia, *The Making of an Insurgent: An Auto-*

biography, 1882–1919 (Philadelphia: Lippincott, 1948), 62–75; "Efficiency Report for Fiorello H. La Guardia," June 12, 1909, Folder 8, Box 26C7, FLG.

178 **An acquaintance of:** Thomas Kessner, *Fiorello H. La Guardia and the Making of Modern New York* (New York: Penguin, 1989), 24–26; Arthur Mann, *La Guardia: A Fighter Against His Times, 1882–1933* (Philadelphia: Lippincott, 1959), 44–49.

179 **In the early years:** For examples of photographs of immigrants, see *The World's Work*, February 1901; *Outlook*, December 28, 1907; *NYT*, March 11, 1906.

180 **Lewis Hine was one:** On Lewis Hine, see Karl Steinorth, ed., *Lewis Hine: Passionate Journey* (Zurich: Edition Stemmle, 1996); *America & Lewis Hine: Photographs, 1904–1940* (New York: Aperture, 1977); and Maren Stange, *Symbols of Ideal Life: Social Documentary Photography in America, 1890–1950* (Cambridge, UK: Cambridge University Press, 1989), 47–87. Hine's Ellis Island photos can be viewed online at the George Eastman House website: http://www.eastman.org/fm/lwhprints/htmlsrc/ellis-island_idx00001.html.

181 **More photographs made:** See Peter Mesenhöller, *Augustus F. Sherman: Ellis Island Portraits, 1905–1920* (New York: Aperture, 2005). Some of Sherman's more exotic subjects were most likely foreign-born circus performers brought over to perform in the United States by Barnum and Bailey. See Letter from William Williams to Daniel Keefe, March 24, 1910, File 52880-171, INS.

181 **Labor leader Samuel Gompers:** Samuel Gompers, *Seventy Years of Life and Labour*, vol. 2 (New York: Augustus M. Kelley, 1967), 154, 160; Letter from Charles Eliot to Edward Lauterbach, February 1, 1907, File 1125, Folder 1, IRL.

182 **The test for both sides:** On the 1907 Immigration Act, see John Higham, *Strangers in the Land: Patterns of American Nativism, 1860–1925* (New Brunswick, NJ: Rutgers University Press, 1955), 128–130; Hans Vought, *The Bully Pulpit and the Melting Pot: American Presidents and the Immigrant, 1897–1933* (Macon, GA: Mercer University Press, 2004), 54–57; Daniel J. Tichenor, *Dividing Lines: The Politics of Immigration Control in America* (Princeton, NJ: Princeton University Press, 2002), 124–128; and William C. Van Vleck, *The Administrative Control of Aliens: A Study in Administrative Law and Procedure* (New York: Da Capo Press, 1971), 10–12.

183 **Writing to Speaker Cannon:** Letter from Theodore Roosevelt to Joseph Cannon, January 12, 1907, in Morison, ed., *Letters*, vol. 5, 550.

183 **The defeat of:** Cohen, *A Dual Heritage*, 155; Letter from Robert Watchorn to Oscar Straus, February 29, 1908, OS.

184 **Hall took his case:** Letter from Theodore Roosevelt to Prescott Farnsworth Hall, June 24, 1908, in Morison, ed., *Letters*, vol. 6, 1096–1097.

184 **Lodge had been:** Letter from Henry Cabot Lodge to Theodore Roosevelt, July 26, 1908, in Henry Cabot Lodge and Charles F. Redmond, eds., *Selections from the Correspondence of Theodore Roosevelt and Henry Cabot Lodge, 1884–1918*, vol. 2 (New York: Da Capo Press, 1971).

184 **Lodge, however, was not:** These figures on appeals come from the Annual Reports of the Commissioner of Immigration. I could not find figures predating 1906. Straus's tenure covered 1908, and parts of 1907 and 1909. An examination of the data from 1906 to 1915 shows that the deportation figures on appeal fell within the range of 44 to 69 percent. The data under Straus's administration fell on the low end of the range, but are hardly aberrant compared to the policies of his predecessor and successor.

184 **That even Henry Cabot Lodge:** Letter from Prescott Hall to Theodore Roosevelt, February 24, 1909, File 801, IRL.

185 **Roosevelt had sought:** For Reynolds's report, see File 51467-1, INS.

185 **In July 1905**: Letter from Robert Watchorn to Robert DeC. Ward, July 22, 1905, File 916, Folder 1, IRL.

185 **Keeping up a correspondence**: Letter from Robert Watchorn to Prescott Hall, June 5, 1906, File 958; Letter from Prescott Hall to Robert Watchorn, June 7, 1906, File 958, IRL.

185 **That was before**: Letter from William Loeb, Jr. to Rev. Dr. Judson Swift, Field Secretary, American Tract Society, February 1, 1908, Box 9, OS: Grose quoted in Mesenhöller, *Augustus F. Sherman*, 12.

186 **Jewish leaders**: Letter from Robert Watchorn to Oscar Straus, February 3, 1908, Box 9; Letter from Oscar Straus to Robert Watchorn, February 1, 1908, Letterbox 3, Box 20, OS.

186 **Roosevelt had little**: Letter from William Loeb, Jr. to Rev. Dr. Judson Swift, Field Secretary, American Tract Society, February 1, 1908, Box 9, OS.

186 **At the same time**: Thomas Pitkin and Francesco Cordasco, *The Black Hand: A Chapter in Ethnic Crime* (Totowa, NJ: Littlefield, Adams, 1977), 85.

186 **Watchorn noted that**: Letter from Oscar Straus to Robert Watchorn, March 2, 1908; Letter from Oscar Straus to Robert Watchorn, March 19, 1908, OS; Thomas Pitkin, *Keepers of the Gate: A History of Ellis Island* (New York: New York University Press, 1975), 97–100. Other discussions of Black Hand violence can be found in "How the United States Fosters the Black Hand," *The Outlook*, October 30, 1909, and "Imported Crime: The Story of the Camorra in America," *McClure's Magazine*, May 1912.

186 **Immigration restrictionists**: Victor Safford, *Immigration Problems: Personal Experiences of an Official* (New York: Dodd, Mead, and Company, 1925), 88–90.

187 **Samuel Gompers, another friend**: Gompers, *Seventy Years*, vol. 2, 164.

187 **By the summer of 1908**: BG, August 9, 1908, September 5, 1908; Gompers, *Seventy Years*, vol. 2, 164; Oscar Straus Diary, 214, OS.

188 **It is no surprise**: John Lombardi, *Labor's Voice in the Cabinet: A History of the Department of Labor from its Origin to 1921* (New York: AMS Press, 1968), 144–145.

188 **Gompers, who never had** : Gompers, *Seventy Years*, vol. 2, 168; Lombardi, *Labor's Voice*, 147–148.

188 **The in-house journal**: *Journal of The Knights of Labor*, January 1909, quoted in "What of the Future?" Publication of the Immigration Regulation League, No. 5, File 1144, IRL. Powderly even wrote a letter to President-elect Taft urging him to keep Straus as secretary of Commerce and Labor. Showing how out of touch he had become with the labor movement, Powderly claimed that American workers would second his support for Straus. "Talk to labor men anywhere, as I have done, and you will find that what I state is correct and moderate." Letter from Terence V. Powderly to William Howard Taft, January 7, 1909, Series 3, WHT.

189 **Americans tried to**: Allan McLaughlin, "Immigration and Public Health," PSM, January 1904; Frank Sargent, "The Need of Closer Inspection and Greater Restriction of Immigrants," *Century Magazine*, January 1904.

189 **"The advocates of absolutely unrestricted"**: *Outlook*, February 22, 1913; "Reports of the Industrial Commission on Immigration," vol. 15, 1901; NYT, April 14, 1911.

190 **Dr. Victor Safford struck**: Safford, *Immigration Problems*, 88.

CHAPTER TEN: LIKELY TO BECOME A PUBLIC CHARGE

191 **One day, the former president**: Robert Watchorn, *The Autobiography of Robert Watchorn* (Oklahoma City, OK: Robert Watchorn Charities, 1959), 140–141.

192 **While he had expressed**: Letter from Theodore Roosevelt to Herbert Knox Smith, January 18, 1909, TR; Watchorn, *Autobiography*, 149–152.

192 **Prescott Hall had been**: Letter from Prescott Hall to Hon. William Howard Taft, December 8, 1908, File 801, IRL.

192 **Despite the criticism**: NYT, April 25, May 19, 1909.

193 **The personal attacks**: NYT, July 17, 1909; Letter from Robert Watchorn to Charles D. Hilles, January 20, 1913, Series 6, Reel 451, WHT. After leaving Ellis Island, Watchorn took a job with Union Oil Company, having befriended its owner Lyman Stewart. Despite Watchorn's lack of experience in the oil industry or in business in general, he was named treasurer of the company. Many board members opposed him, believing him to be incompetent.

The former coal miner, union leader, and government bureaucrat was soon traveling to New York and London to raise capital among the world's savviest financiers. Watchorn was out of his element, and accusations of ethical impropriety followed him in his new career. He soon managed to upset Stewart and cast doubt on his own honesty and competence when he became involved in a controversy over a million dollars' worth of stock options given to him by Stewart. Details of the deal remain murky, but it led to Watchorn's resignation under a cloud of suspicion. Then, having entered the world of oil wildcatting in Oklahoma and Texas, Watchorn became a millionaire and by the 1930s had turned his attention to philanthropy. He endowed a church in his hometown of Alfreton, England, and a music hall at the University of Redlands. In 1932, Watchorn presented his greatest piece of philanthropy—the Lincoln Memorial Shrine—to his adopted hometown of Redlands, California. See Frank J. Taylor and Earl M. Welty, *Black Bonanza: How an Oil Hunt Grew into the Union Oil Company of California* (New York: McGraw-Hill, 1950) 165–166; Watchorn, *Autobiography*, 154–162, 185–211.

193 **Just as Roosevelt**: Letter from William Howard Taft to Herbert Parsons, May 17, 1909, Series 8, WHT; NYT, May 20, 1909.

193 **Even after leaving**: William Williams, "The Sifting of Immigrants," *Journal of Social Science*, September 1906.

194 **Although he questioned**: Williams, "The Sifting of Immigrants," NYT, July 18, 1909.

194 **The letter of the law**: "Annual Report of the Commissioner-General of Immigration," 1909, 132; NYT, June 5, 1909.

195 **It also possessed**: William C. Van Vleck, *The Administrative Control of Aliens; A Study in Administrative Law and Procedure* (New York: Commonwealth Fund, 1932), 54.

195 **Realizing this**: "Annual Report of the Commissioner-General of Immigration," 1909, 133; President Theodore Roosevelt, "First Annual Message to Congress," December 3, 1901.

196 **Now Williams was**: He called the $25 test "nothing more than a timely warning to immigrants that they cannot land without funds adequate for their support until such times as they are likely to obtain profitable employment." Letter from William Williams to A.J. Sabath, July 15, 1909, File 52531-12, INS.

196 **Williams's edict had**: "Annual Report of the Commissioner of Ellis Island to Commissioner-General of Immigration," August 16, 1909; NYT, June 30, 1909.

196 **Conditions worsened**: NYT, July 14, 1909.

197 **On July 4, Rudniew**: Isaac Metzker, ed., *A Bintel Brief: Sixty Years of Letters from the Lower East Side to the Jewish Daily Forward* (Garden City, NY: Doubleday, 1971), 98–100; AH, July 16 1909, 278.

197 **Williams was unmoved**: NYT, July 10, 1909.

197 **Many Americans were**: Letter from Russell Bellamy to William Williams, July 12, 1909; Letter from Prescott Hall to William Williams, July 14, 1909, WW-NYPL.

197 **Eighty-two-year-old Orville Victor**: Letter from Orville Victor to William Williams, July 17, 1909; Letter from William Patterson to William Williams, July 8, 1909, WW-NYPL.

198 **Not all of Williams's**: Letter from an anonymous pupil at PS 62 in Manhattan to William Williams, undated, WW-NYPL.

198 **The child who wrote**: On the history of the Hebrew Immigrant Aid Society, see, Mark Wischnitzer, *Visas to Freedom: The History of HIAS* (Cleveland, OH: World Publishing Company, 1956).

198 **The HIAS took on**: "Brief for the Petitioner in the Matter of Hersch Skuratowski," 1909, File 52530-12, INS; Esther Panitz, "In Defense of the Jewish Immigrant, 1891–1924," in Abraham Karp, ed., *The Jewish Experience in America*, vol. 5 (Waltham, MA: American Jewish Historical Society, 1969).

199 **The lawyers were not**: NYT, July 16, 1909; Max J. Kohler, *Immigration and Aliens in the United States: Studies of American Immigration Laws and the Legal Status of Aliens in the United States* (New York: Bloch, 1936), 54–55.

199 **There was something**: "Brief for the Petitioner in the Matter of Hersch Skuratowski," 1909, 46–61, File 52530-12, INS.

200 **The controversy over**: For an overview of the issue of racial classifications, see Marian L. Smith, "INS Administration of Racial Provisions in U.S. Immigration and Nationality Law Since 1898," *Prologue*, Summer 2002.

200 **Powderly and his colleagues**: See File 52729/9, INS; Joel Perlmann, " 'Race or People': Federal Race Classifications for Europeans in America, 1898–1913," Jerome Levy Economics Institute Working Paper No. 320, January 2001; "Reports of the Industrial Commission on Immigration," vol. 15, 1901, 132–133; "Annual Report of the Commissioner-General of Immigration," 1898, 33–34; "Annual Report of the Commissioner-General of Immigration," 1899, 5. Patrick Weil claims that "immigration officials continued to use the statistics provided by the list to deny admission to immigrants of certain ethnic backgrounds, even when their exclusion was not specifically provided for by law." Weil provides no support for his hypothesis. Patrick Weil, "Races at the Gate: A Century of Racial Distinctions in American Immigration Policy, 1865–1965," *Georgetown Immigration Law Journal* 15 (2001).

201 **For Jews, this new classification**: Panitz, "In Defense of the Jewish Immigrant, 1891–1924," 55–57; Nathan Goldberg, Jacob Lestchinsky, and Max Weinreich, *The Classification of Jewish Immigrants and its Implications* (New York: YIVO Institute for Jewish Research, 1945); Perlmann, " 'Race or People': Federal Race Classifications for Europeans in America, 1898–1913." The Hebrew classification was eliminated in 1943.

201 **Decades later**: Kohler, *Immigration and Aliens in the United States*, 400–401.

201 **Now it was William Williams's**: Memo from William Williams to Commissioner-General of Immigration, September 8, 1909, File 52531-12A, INS.

201 **Williams was not happy**: NYT, July 16, 1909.

201 **Williams assumed**: Letter from Frank Larned to Williams Williams, July 23, 1909, File 52531-12A, INS; Letter from William Williams to Frank Larned, July 20, 1909, File 52531-12, INS.

202 **After the resolution**: NYT, July 27, 1909; Letter from Charles Nagel to William Williams, July 16, 1909, CN.

202 **"There is no more need"**: NYT, July 27, 1909; Letter from Charles Nagel to William Williams, July 31, 1909, CN.

203 **The following year**: *Canfora v. Williams*, 1911, reprinted in Edith Abbott, ed., *Immigration: Select Documents and Case Records* (Chicago: University of Chicago Press, 1924), 256–258; File 53139-7, INS.

203 **These cases show**: *U.S. v. Ju Toy*, 198 U.S. 253 (1905).

204 **However, the Department of Commerce**: File 53438-11, INS.

205 **These supposedly weak**: Amy Fairchild argues that the immigration inspection process was part of the shaping of a modern, industrial workforce. See Amy L. Fairchild, *Science at the Borders: Immigrant Medical Inspection and the Shaping of the Modern Industrial Labor Force* (Baltimore, MD: Johns Hopkins University Press, 2003). For a discussion of the exclusion of immigrants with physical deficiencies, see Douglas C. Baynton, "Defectives in the Land: Disability and American Immigration Policy, 1882–1924," *Journal of American Ethnic History*, Spring 2005.

205 **In 1902, commissioner-general**: Letter from Frank Sargent to William Williams, October 6, 1902, WW-NYPL.

205 **Medical officials**: For more on the designation of "poor physique," see Fairchild, *Science at the Borders*, 165–169.

205 **Sargent defined**: Letter to all Commissioners of Immigration and inspectors from Frank Sargent, Commissioner General, Bureau of Immigration, April 17, 1905, File 916, Folder 1, IRL.

205 **William Williams agreed**: Letter from William Williams to Prescott Hall, April 10, 1904, File 916, Folder 1, IRL; Williams, "The Sifting of Immigrants."

205 **He had been**: Allan McLaughlin, "Immigration and the Public Health" PSM, January 1904.

206 **Doctors with the**: Fairchild, *Science at the Borders*, 166–167; Elizabeth Yew, "Medical Inspection of Immigrants at Ellis Island, 1891–1924," *Bulletin of the New York Academy of Medicine* 56, no. 5 (June 1980).

206 **This did not mean**: "Book of Instructions for the Medical Inspection of Aliens, Bureau of Public Health and Marine-Hospital Service," January 18, 1910.

206 **In the first**: Letter from Robert Watchorn to Prescott Hall, May 12, 1908, File 958, IRL.

207 **When Williams took**: William Williams, "Notice Concerning Detention and Deportation of Immigrant," March 18, 1910, Folder 10, Box 13, MK. On the connection between deafness and the "likely to become a public charge" clause, see Douglas C. Baynton, " 'The Undesirability of Admitting Deaf Mutes': U.S. Immigration Policy and Deaf Immigrants, 1882–1924," *Sign Language Studies* 6, no. 4 (Summer 2006).

207 **In his first**: "Annual Report of William Williams, Ellis Island Commissioner," September 19, 1910, Folder 5, File 1061, IRL. Also found in "Annual Report of the Commissioner-General of Immigration," 1910, 134–135.

207 **The amount of work**: "Annual Report of the Commissioner-General of Immigration," 1911, 147.

207 **Williams worried**: Letter from William Williams to Commissioner-General of Immigration, June 24, 1910, WW-Yale.

208 **To the thousands**: On the case of Wolf Konig, see File 53452-973, INS.

208 **Michele Sica was**: On the case of Michele Sica, see File 53305-74, INS.

209 **Although much younger**: On the case of Bartolomeo Stallone, see File 53370-234, INS.

210 **Williams himself**: On the case of Jacob Duck, see File 52880-127, INS.

210 **Though Williams may have**: On the Kaganowitz family, see File 53390-146, INS.

211 **Meier Salamy Yacoub**: File 53257-34, INS.

211 **Jewish groups were**: "Extracts from Minutes of Second Annual Meeting of National Jewish Immigration Council Held February 18, 1912," File 53173, INS: NYT, November 14, 1909.

212 **Whether Uhl was**: Edward Alsworth Ross, *The Old World in the New: The Significance of Past and Present Immigration to the American People* (New York: Century, 1914), 289–290.

212 **Meter was detained**: File 53370-699, INS.

212 **While HIAS continued**: Max J. Kohler, "Immigration and the Jews of America," *AH*, January 27, February 3, 1911; *NYT*, January 19, 1911. For a response to Kohler's charges, see Memorandum for the Secretary from Commissioner-General of Immigration Daniel Keefe, February 16, 1911, File 53173-12, INS.

213 **Not all Jewish leaders**: Panitz, "In Defense of the Jewish Immigrant, 1891–1924."

213 **Simon Wolf**: *NYT*, July 18, 1909; Letter from Lipsitch to Kohler, March 7, 1911, Folder 11, Box 11, MK.

213 **HIAS President**: Letter from Leon Sanders to Max J. Kohler, July 29, 1910, Folder 11, Box 11, MK.

213 **Secretary Nagel**: Kohler, 198–199. See also, Otto Heller, ed., *Charles Nagel: Speeches and Writings, 1900–1928*, vol. 1 (New York: Putnam's, 1931), 151, 157.

214 **Jewish groups attempted**: *AH*, January 28, 1910.

214 **After that, Williams's**: "Annual Report of the Commissioner General of Immigration," 1911, 152.

215 **In response, members**: Letter from Moe Lenkowsky and Anton Kaufman, Chairman and Secretary of the Citizens Committee of Orchard, Rivington and East Houston Streets, to William Howard Taft, April 9, 1912, WW-NYPL; Letter from William Williams to Theodore Roosevelt, January 31, 1912, Series 1, Reel 126, TR; Letter from William Williams to Daniel Keefe, September 13, 1912, Series 6, Number 1579, WHT.

CHAPTER ELEVEN: "CZAR WILLIAMS"

217 **Taft listened to a number**: *NYT*, October 19, 1910; Letter from William Williams to Charles Nagel, October 19, 1910, Folder 64, Box 4, Series I, WW-Yale.

217 **But President Taft's**: *NYTrib*, December 16, 1910; Letters from Charles Nagel to Charles D. Norton, December 10, 13, 1910, WHT; Letter from Charles Nagel to William Howard Taft, January 7, 1911, WHT.

217 **These were hard**: "Remarks of President Taft to the Board of Directors of the American Association of Foreign Newspapers at the Executive Office, Washington, DC," January 4, 1911, No. 77, Reel 364, Series 6, WHT; *New York Evening Sun*, January 4, 1911.

218 **"Away with Czarism"**: *Morgen Journal*, April 17, 1911; *New York Evening Journal*, May 24, 1911; *Morgen Journal*, June 23, 1911.

219 **The *Morgen Journal* listed**: *Morgen Journal*, April 17, 1911; *Szabadsag*, October 11, 1910.

219 **O. J. Miller**: Memorandum from William Williams to Daniel Keefe, October 14, 1910, Folder 63, Box 4, CN; File 53139-7, INS. Nagel sent a detective to investigate Miller and his organization. The investigation discovered that the German Liberal Immigration Bureau was only a paper organization and Miller a reporter for the *New Yorker Staats-Zeitung*. Nevertheless, Miller's agitation caught the attention of government officials, congressmen, and German-American organizations.

219 **Groups such as**: File 53139-7, INS.

219 **At first, Williams was**: Letter from William Williams to Charles Nagel, April 5, 1911; Letter from William Williams to Charles Nagel, April 7, 1911, File 53139-7, INS.

220 **Nor could Charles**: Letter from Charles Nagel to Charles Norton, October 21, 1910, Folder 65, Box 4, WW-Yale.

220 *Harper's Weekly* **asked**: HW, July 7, 1911.

222 **It is hard**: Broughton Brandenburg, "The Tragedy of the Rejected Immigrant," *Outlook*, October 13, 1906; Philip Taylor, *The Distant Magnet: European Emigration to the USA* (New York: Harper & Row, 1971), 123; "Report of the Dillingham Immigration Commission," undated, File 1060, Folder 9, IRL; "Annual Report of the Commissioner General of Immigration," 1907, 83. *Harper's Weekly* estimated that some eight thousand potential immigrants were refused passage by steamship companies at Bremen in 1905. HW, April 14, 1906.

222 **For some immigrants**: Letter from J. M. Jenks to Oscar Straus, March 12, 1907, Box 6, OS.

222 **An American congressional**: "Report of the Sub-Committee of the Immigration Commission," 1907; Senator A. C. Latimer and Rep. John L. Burnett, File 1060, Folder 8, IRL; Taylor, 123.

222 **William Williams was**: Letter from Charles Nagel to William Williams, April 6, 1911, File 53139-7, INS.

223 **Nagel won no friends**: "Hearings on House Resolution No. 166," House Committee on Rules, United States House of Representatives, May 29, 1911, 107; Max J. Kohler, *Immigration and Aliens in the United States: Studies of American Immigration Laws and the Legal Status of Aliens in the United States* (New York: Bloch, 1936), 46.

223 **This was not**: Otto Heller, ed., *Charles Nagel: Speeches and Writings, 1900–1928*, vol. 1 (New York: Putnam's, 1931), xviii, 146; Letter from Charles Nagel to William Howard Taft, April 16, 1912, Number 3D, Series 6, WHT.

223 **The agitation among**: "Hearings on House Resolution No. 166," House Committee on Rules, United States House of Representatives, May 29, 1911, 3–6.

224 **Before the hearings**: Letter from William Williams to Prescott Hall, May 12, 1911, File 916, Folder 2, IRL.

224 **Still, while the earlier**: Letter from William Williams to Charles Nagel, June 5, 1911, Folder 81, Box 5, WW-Yale.

225 **He had arrived**: On Bass, see "An English Pastor's Experience on Ellis Island: The Abuse of the USA Immigration Laws," undated, Reel 409; Letter from William Williams to Commissioner-General of Immigration, January 30, 1911; Letter from Charles Nagel to Charles D. Norton, Secretary to the President, February 25, 1911, Series 6, Reel 409, WHT; "Hearings on House Resolution No. 166," House Committee on Rules, United States House of Representatives, May 29, 1911, 130–135; *New York Evening Journal*, June 21, 1911; Letter from William Williams to Commissioner General of Immigration, March 9, 1911, Box 13, Folder 10, MK.

227 **With the failure**: NYT, October 8, 1911; Charles Thomas Johnson, *Culture at Twilight: The National German-American Alliance, 1901–1918* (New York: Peter Lang, 1999), 76; *Morgen Journal*, January 4, 17, February 7, 1912.

227 **Williams had his defenders**: HW, June 10, 1911.

228 **Arthur von Briesen**: Letter from Arthur von Briesen to William Howard Taft, June 29, 1911, Folder 82, Box 5, Series I, WW-Yale.

228 **Williams's most steadfast**: Letter from William Howard Taft to William Williams, November 25, 1911, Number 90, Reel 509, Series 8, WHT.

228 **In his own way**: Letter from William Howard Taft to William Williams, May 2, 1913, Folder 9, Box 1, WW-Yale.

228 **Williams continued with**: Thomas Pitkin, *Keepers of the Gate: A History of Ellis Island* (New York: New York University Press, 1975), 109.

229 **The economic effects**: NYT, September 28, 1912, September 21, 1913; Philip Cowen, *Memories of an American Jew* (New York: International Press, 1932), 184.

229 **When the U.S. Commission**: On the Dillingham Commission, see Robert F. Zeidel, *Immigrants, Progressives and Exclusion Politics: The Dillingham Commissioner, 1900–1927* (DeKalb, IL: Northern Illinois University Press, 2004); Desmond King, *Making Americans: Immigration, Race, and the Origins of the Diverse Democracy* (Cambridge, MA: Harvard University Press, 2000), 50–81; Daniel J. Tichenor, *Dividing Lines: The Politics of Immigration Control in America* (Princeton, NJ: Princeton University Press, 2002), 128–132; Oscar Handlin, *Race and Nationality in American Life* (Boston: Little, Brown, 1948), 93–138; Jeremiah Jenks and W. Jett Lauck, *The Immigration Problem: A Study of American Immigration Conditions and Needs* (New York: Funk & Wagnalls, 1913); and *Survey*, January 7, 1911. Tichenor found that the Dillingham Commission's "expert findings offered a portrait of southern and eastern European newcomers that legitimized the xenophobic narrative and policy agenda of Progressive Era restrictionists." In response, Zeidel notes that the Dillingham Commission was deeply rooted in the reform movements of the early twentieth century, a fact many historians have ignored "because they have not wanted to equate any form of xenophobia with progress." Its conclusions and recommendations can be found in U.S. Immigration Commission, "Abstracts of Reports of the Immigration Commission with Conclusions and Recommendations and Views of the Minority, Volume One," 61st Congress, 3rd Session, Document 747, 1911.

230 **His new party's platform**: Rivka Shpak Lissak, "Liberal Progressives and Immigration Restriction, 1896–1917," Annual Lecture, American Jewish Archives, 1991; Tichenor, *Dividing Lines*, 135–136; Hans Vought, *The Bully Pulpit and the Melting Pot: American Presidents and the Immigrant, 1897–1933* (Macon, GA: Mercer University Press, 2004), 86–87.

230 **The candidate who**: Woodrow Wilson, *A History of the American People, Volume 5* (New York: Harper & Brothers, 1901), 212–214.

231 **Thanks to newspaper**: Arthur S. Link, *Wilson: The Road to the White House* (Princeton, NJ: Princeton University Press, 1947), 381–387, 499–500; James Chace, *1912: Wilson, Roosevelt, Taft, and Debs—and the Election That Changed the Country* (New York: Simon & Schuster, 2004), 135–137; Wilson quoted in the *Jewish Immigration Bulletin*, November 1916, 8.

231 **Despite the controversy**: Letter from William Howard Taft to A. Lawrence Lowell, November 6, 1910, File 860, IRL.

231 **For two decades**: *Survey*, February 8, 1913.

232 **The numbers support**: LD, May 25, 1912; *Outlook*, February 22, 1913.

232 **With only a few**: Morris M. Sherman, "Immigration Restriction, 1890–1921, and the Immigration Restriction League," (Cambridge, MA: Harvard College, 1957), 33.

232 **A few weeks before**: NYT, January 26, 1913.

233 **Ethnic groups were**: *New Yorker Staats-Zeitung*, May 7, 14, 1913, translation found in File 53139-7C, INS.

233 **Throughout the 1912 campaign**: *Warheit*, July 14, 1912, in "Instances of Continued Abuse of the Ellis Island Authorities by Certain Newspapers Printed in

Foreign Languages in the City of New York," undated, Folder 32, Box 3, WW-NYPL.

233 **The *Deutsches Journal***: *Deutsches Journal*, April 28 1913; "Comments on Annexed Report of Case of Aron Mosberg," April 18, 1913, File 53139-7C, INS.

233 **"Sir, You are the murderer"**: Letter from John Czurylo to William Williams, May 3, 1913; Letter from William Williams to the Commissioner-General of Immigration, May 9, 1913, WW-NYPL.

234 **In an April 1913 letter**: Letter from William Williams to the Commissioner-General of Immigration, April 21, 1913, File 53139-7C, INS.

234 **The uncertainty**: Letter from Prescott Hall to William Williams, November 22, 1912, Box 3, WW-NYPL.

235 **Others remembered Williams**: Letter to William Williams, June 18, 1913, Box 3, WW-NYPL.

235 **Others took issue**: *Morgen Journal*, June 20, 1913.

235 **After the war**: NYT, February 9, 1947; Frederic R. Coudert, "In Memoriam: William Williams," *American Journal of International Law* 41, no. 3 (July 1947).

236 **Two months before**: Case of Lipe Pocziwa, No. 667, Series 6, Reel 404, WHT.

237 **William Williams**: "Annual Report of the Commissioner General of Immigration," 1911, 147; "Annual Report of the Commissioner General of Immigration," 1912, 23.

CHAPTER TWELVE: INTELLIGENCE

238 **During the depths**: On the Zitello family, see File 54050-240, INS.

241 **When Dr. Thomas Salmon**: On the life and career of Thomas W. Salmon, see Earl D. Bond, *Thomas W. Salmon: Psychiatrist* (New York: W.W. Norton, 1950) and Manon Parry, "Thomas W. Salmon: Advocate of Mental Hygiene," *American Journal of Public Health* 96, no. 10 (October 2006).

241 **Salmon saw the chance**: For a description of the work of a psychologist on line examination at Ellis Island, see Thaddeus S. Dayton, "Importing Our Insane," *HW*, October 19, 1912.

242 **Salmon was on the**: Ian Robert Dowbiggin, *Keeping America Sane: Psychiatry and Eugenics in the United States and Canada, 1880–1940* (Ithaca, NY: Cornell University Press, 1997), 203.

242 **The results of Salmon's work**: Salmon would later become the first medical director of the National Committee for Mental Hygiene. During World War I, he served as a consultant for the U.S. Army and worked with returning soldiers suffering from shell shock and other psychological disorders. In 1923, he was elected president of the American Psychiatric Association. Despite having no formal background in psychiatry, Salmon had reached the pinnacle of his profession.

242 **At the time**: On the Binet tests, see Stephen Jay Gould, *The Mismeasure of Man* (New York: W.W. Norton, 1996), 176–188.

243 **There was also**: Leila Zenderland, *Measuring Minds: Henry Herbert Goddard and the Origins of American Intelligence Testing* (Cambridge, UK: Cambridge University Press, 1998), 102–103.

243 **If there was some**: C. B. Davenport, *Eugenics: The Science of Human Improvement by Better Breeding* (New York: Henry Holt, 1910). For more on Davenport, see Daniel J. Kelves, *In the Name of Eugenics: Genetics and the Uses of Human Heredity* (Cambridge, MA: Harvard University Press, 1985), 41–56.

244 **In 1911, Davenport recommended**: Letter from C. B. Davenport, Secretary of the America Breeders Association, Eugenics Section, to Prescott Hall, May 20,

1911, File 342, IRL; Report of the Immigration Committee of the Eugenics Section, American Breeders Association, December 30, 1911, File 1064, Folder 1, IRL. Interestingly, one member of the committee was Columbia anthropologist Franz Boas, who achieved fame for his criticism of eugenics.

244 **Now many IRL members**: Robert DeC. Ward, "National Eugenics in Relation to Immigration," NAR, July 1910; Robert DeC. Ward, "The Crisis in Our Immigration Policy," File 1063, Folder 9; Robert DeC. Ward, "Our Immigration Laws from the Viewpoint of National Eugenics," *National Geographic*, January 1912. "The need is imperative for applying eugenic principles in much of our legislation. But the greatest, the most logical, the most effective step that we can take is to begin with a proper eugenic selection of the incoming alien millions. If we, in our generation take these steps, we shall earn the gratitude of millions of those who will come after us for we shall have begun the real conservation of the American race."

244 **For Prescott Hall**: "Eugenics and Immigration," Prescott Hall, undated, File 1061, Folder 1, IRL; *Immigration and Other Interests of Prescott Farnsworth Hall*, Compiled by Mrs. Prescott F. Hall (New York: Knickerbocker Press, 1922), 53.

245 **One answer for Hall**: Prescott Hall, "Birth Control and World Eugencis," unpublished manuscript, in *Immigration and Other Interests of Prescott Farnsworth Hall*.

245 **As to whether humans**: *Immigration and Other Interests of Prescott Farnsworth Hall*, 33, 83. Interestingly, anthropologist Franz Boas had recently completed his study, published by the Dillingham Commission, which showed a divergence in head size between foreign-born Hebrews and Sicilians and American-born Hebrews and Sicilian Americans. The American environment, Boas concluded, was having some effect on the "race characteristics" that many believed immutable. The irony is that Boas used the discredited theory of craniometry to prove his anti-eugenic, anti-racist theory. See "Changes in Bodily Form of Descendants of Immigrants," Reports of the Immigration Commission, Volume 38, 61st Congress, 2nd Session.

245 **At the intersection**: "Is it any wonder that serious students contemplate the racial future of the Anglo-Saxon American with some concern? They have seen the passing of the American Indian and the buffalo; and now they query as to how long the Anglo-Saxon may be able to survive." William Z. Ripley, "Races in the United States," *Atlantic*, December 1908. See also Robert DeC. Ward, "National Eugenics in Relation to Immigration," NAR, July 1910.

245 **Progressive sociologist**: Ross quoted in M. Victor Safford, "The Business Side of Immigration," speech delivered at Old South Club, October 20, 1913, File 1064, Folder 8, IRL.

246 **Ross proudly noted**: Edward Alsworth Ross, *The Old World in the New: The Significance of Past and Present Immigration to the American People* (New York: Century, 1914), 285–286.

246 **A leading academic**: Ross, *The Old World in the New*, 289–293.

246 **Ross predicted that**: Ross, *The Old World in the New*, 228, 254–256.

246 **These descriptions placed**: NYT, June 20, 1914.

247 **Amidst such pressing**: "Immigration and Insanity," address of William Williams, U.S. Commissioner of Immigration, before the Mental Hygiene Conference at New York City, November 17, 1912, File 53139-13, INS; "The Crisis in Our Immigration Policy," Robert DeC. Ward, File 1063, Folder 9, IRL.

247 **Williams complained**: See File 53139-13A, INS.

247 **Neither Congress**: H. H. Goddard, "The Binet Tests in Relation to Immigration," *Journal of Psycho-Asthenics* 18 (1913); Henry H. Goddard, "The Feeble

Minded Immigrant," *The Training School*, November/December 1912; and
Steven A. Gelb, "Henry H. Goddard and the Immigrants, 1910–1917: The Studies and Their Social Context," *Journal of the History of the Behavioral Sciences* 22
(October 1986). For more general background on Goddard and intelligence testing, see Zenderland, *Measuring Minds*; Franz Samelson, "Putting Psychology on
the Map: Ideology and Intelligence Testing," in Allan R. Buss, ed., *Psychology in
Social Context* (New York: Irvington Publishers, 1979); and Gould, *The Mismeasure of Man*, 188–204.

248 **Believing this was proof**: Goddard's own mathematical abilities were less than
stellar. He translated his assistants' success rate of nine out of eleven into a rate
of "seven-eighths." Goddard, "The Feeble Minded Immigrant."

249 **Goddard magnanimously said**: Goddard, "The Feeble Minded Immigrant";
Goddard, "The Binet Test in Relation to Immigration."

249 **Goddard's test did not go**: Goddard, "The Binet Test in Relation to Immigration."

249 **Goddard's staff chose**: Henry H. Goddard, "Mental Tests and the Immigrant,"
Journal of Delinquency, September 1917. For some unknown reason, perhaps
owing to his sloppiness as a researcher, Goddard claims to have tested "about
165 immigrants." Other scholars have used that figure as well, but a count of the
figures from Goddard's own article comes up with 191: 54 Jews, 70 Italians, 45
Russians, and 22 Hungarians. Even the numbers on Goddard's chart (252) don't
add up to 191, and there is an error of arithmetic in one of the columns.

250 **The results, wrote Goddard**: Gelb, "Henry H. Goddard and the Immigrants,
1910–1917: The Studies and their Social Context"; Gould, *The Mismeasure of
Man*, 194–198.

251 **As for whether**: Zenderland, *Measuring Minds*, 274.

251 **Even a nonscientist**: The debate over Goddard's legacy is contentious. On one
side, Leon Kamin and Stephen Jay Gould have been harshly critical of Goddard's work, methods, and intentions. On the other side, Franz Samelson, Leila
Zenderland, and Steven Gelb have been more measured in their interpretations,
placing the psychologist within the context of his times. Gelb's description
is the most helpful: "Goddard's writings about Ellis Island immigrants, when
placed in their proper context, do not provide evidence of the virulent type
of racism with which his name has become associated. Goddard is more accurately described as a 'decent' man, pursuing questions and conclusions—in the
name of disinterested 'science'—that were, in fact, driven by the engines of an
institutionalized, pernicious social ideology." Gelb, "Henry H. Goddard and the
Immigrants, 1910–1917: The Studies and Their Social Context." For a harsher
view of Goddard, see Leon Kamin, "The Science and Politics of IQ," *Social
Research* 41 (1974).

251 ***The Survey*, the nation's leading**: *Survey*, September 15, 1917.

252 **Goddard had been**: C. P. Knight, "The Detection of the Mentally Defective
Among Immigrants," *JAMA*, January 11, 1913.

252 **For immigrants suffering**: E. H. Mullan, "Mental Examination of Immigrants:
Administration and Line Inspection at Ellis Island," *Public Health Reports*, U.S.
Public Health Service, May 18, 1917, 737, 746.

253 **Ellis Island doctors**: Knight, "The Detection of the Mentally Defective Among
Immigrants"; E. H. Mullan, "Mental Examination of Immigrants: Administration and Line Inspection at Ellis Island," 738.

253 **Howard Knox**: For background on Knox, see John T. E. Richardson, "Howard
Andrew Knox and the Origins of Performance Testing on Ellis Island, 1912–
1916," *History of Psychology* 6, no. 2 (May 2003); John T. E. Richardson, "A

Physician with the Coast Artillery Corps: The Military Career of Dr. Howard Andrew Knox, Pioneer of Psychological Testing," *Coast Defense Journal* 15, no. 4, November 2001.

253 **Knox shared many**: Howard A. Knox, "The Moron and the Study of Alien Defectives," JAMA, January 11, 1913.

253 **Knox was also sensitive**: Howard A. Knox, "Psychogenetic Disorders: Cases Seen in Detained Immigrants," *Medical Record*, July 12, 1913; Howard A. Knox, "The Difference Between Moronism and Ignorance," NYM, September 20, 1913; E.K. Sprague, "Mental Examination of Immigrants," *Survey*, January 17, 1914. "Does the Binet-Simon measuring scale of intelligence or its American modification . . . represent the average normal intelligence of practically the entire human race," asked Ellis Island doctor Bernard Glueck. "Assuredly not. We are convinced of this both from experience with the immigrant and actual experimental investigation of the subject and were it considered necessary to adduce facts to prove the fallacy of such a contention, these could easily be gotten from the hundreds of case histories on file at Ellis Island." Bernard Glueck, "The Mentally Defective Immigrant," NYM, October 18, 1913.

254 **Knox noted one case**: Howard A. Knox, "Psychological Pitfalls," NYM, March 14, 1914; Howard A. Knox, "Diagnostic Study of the Face," NYM, June 14, 1913.

254 **Another Ellis Island doctor**: Glueck, "The Mentally Defective Immigrant."

255 **Ignoring Goddard's work**: Knox, "The Moron and the Study of Alien Defectives."

255 **The testing room**: Howard A. Knox, "Measuring Human Intelligence," *Scientific American*, January 19, 1915; Howard A. Knox, "Tests for Mental Defects," *Journal of Heredity* 5 (1914).

255 **Once the conditions**: Glueck, "The Mentally Defective Immigrant."

256 **This battery of questions**: NYT, November 1, 7, 1912.

256 **The questions that**: Howard A. Knox, "A Comparative Study of the Imaginative Powers in Mental Defectives," *Medical Record*, April 25, 1914.

257 **Immigrants were also**: E. H. Mullan, "The Mentality of the Arriving Immigrant," *Public Health Bulletin* 90 (October 1917): 118–124.

257 **Ellis Island doctors were increasingly bothered**: Bernard Glueck, "The Mentally Defective Immigrant"; Zenderland, *Measuring Minds*, 276–277.

257 **Howard Knox created**: For examples of the various tests, see Howard A. Knox, "Mentally Defective Aliens: A Medical Problem," *Lancet-Clinic*, May 1, 1915; Howard A. Knox, "A Scale Based on the Work at Ellis Island for Estimating Mental Defect," JAMA, March 7, 1914; Mullan, "The Mentality of the Arriving Immigrant." Mullan's report contains detailed results from the whole array of tests used at Ellis Island on a sample of literate and illiterate immigrants in 1914.

258 **These tests were about**: Mullan, "The Mentality of the Arriving Immigrant," 42–43; T. E. John, "Knox's Cube Imitation Test: A Historical Overview and an Experimental Analysis," *Brain and Cognition* 59 (2005).

258 **In 1913, the number of**: NYT, September 16, 1913; Berth Boody, A Psychological Study of Immigrant Children at Ellis Island, reprint (New York: Arno Press, 1970), 65; Knox, "Mentally Defective Aliens: A Medical Problem," 495.

259 **Like others**: Howard Knox, "Mental Defectives," NYM, January 31, 1914.

CHAPTER THIRTEEN: MORAL TURPITUDE

260 **Dressed in a large green**: *Time*, March 1, 1926; Edward Corsi, *In the Shadow of Liberty* (New York: Macmillan, 1935), 201–210.

260 **Vera's problems began**: Vera married for a third time in 1930 to seventy-five-year-old millionaire Sir Rowland Hodge. In 1934, she asked for a divorce. The Earl of Craven died in 1932 in France at the age of thirty-five.

261 **Immigration officials declared**: Quoted in *Black's Law Dictionary*, 7th ed. (St. Paul, MN: West Group, 1999), 1026.

261 **The term entered American**: Jane Perry Clark, *Deportation of Aliens from the United States to Europe* (New York: Columbia University Press, 1931), 164, 171; Brian C. Harms, "Redefining 'Crimes of Moral Turpitude': A Proposal to Congress," *Georgetown Immigration Law Journal* 15 (2001).

263 **Vera could now attend to**: NYT, March 16, 1926.

264 **Angry at the reception**: The Vera Cathcart story was prominent enough to warrant a mention in Frederick Lewis Allen's popular history of the 1920s, where it was listed as a notable event of early 1926, along with Byrd's flight over the North Pole and the disappearance of evangelist Aimee Semple McPherson. Frederick Lewis Allen, *Only Yesterday: An Informal History of the 1920s* (New York: Perennial Classics, 1931; reissued 1990), 181.

264 **Women of all nationalities**: Deirdre M. Moloney, "Women, Sexual Morality, and Economic Dependency in Early U.S. Deportation Policy," *Journal of Women's History* 18, no. 2 (Summer 2006). Moloney claims that the "enforcement of immigration policies concerning women's sexuality differed according to their race and ethnicity." She offers only anecdotal, but not statistical, evidence for the claim.

264 **Giulia Del Favero**: Document No. 16129, Box 23, Entry 7, INS.

265 **Sometimes, though, those vultures**: Campbell and Rodgers Report, June 2, 1900, to Secretary of the Treasury, Boxes 157–158, TVP.

266 **Immigration officials continued**: File 52388-59, INS.

266 **A young Serbian woman** : File 52388-77, INS.

267 **Young women who transgressed**: File 53155-125, INS.

268 **Immigration officials also**: Some scholars have seen the imposition of morality tests as specifically targeted against women. One historian, discussing the exclusion of a pregnant, unmarried woman named Dolan, argued that, "it was highly unlikely that the man who impregnated her would have been similarly excluded. Dolan's story painfully illustrates how the incorporation of patriarchal heterosexual imperatives into immigration policy resulted in the exclusion of women who violated its order." Of course, for practical reasons, had the father of the child entered alone, there would have been no way for inspectors to tell that he had fathered an illegitimate child. Had the father of the child entered with his pregnant girlfriend, however, both man and woman would have been excluded or forced to marry before entering the country. Eithne Luibheid, *Entry Denied: Controlling Sexuality at the Border* (Minneapolis: University of Minnesota Press, 2002), 3–5.

268 **"I had approved exclusion"**: Oscar Straus Diary, Box 22, OS.

268 **In another case**: File 52279-14, INS.

268 **Sometimes women could use**: File 53257-34, INS.

270 **Oftentimes, the moral turpitude**: William M. Sullivan, "The Harassed Exile: General Cipriano Castro, 1908–1924," *Americas* 33, no. 2 (October 1976); J. Fred Rippy and Clyde E. Hewitt, "Cipriano Castro: 'Man Without a Country,'" *American Historical Review* 55, no. 1 (October 1949).

270 **In December 1912**: NYT, December 31, 1912.

271 **He arrived on**: *New York Herald Tribune*, August 18, 1942.

271 **At his hearing**: File 53166-8, INS.

271 **Castro had a number:** *WP*, January 3, 1913.

272 **One month after:** Memorandum in the case of Cipriano Castro, January 30, 1913, Folder 39, Box 59, CN.

272 **Meanwhile, New York Democrats:** *NYT*, February 16, 1913.

272 **Castro returned to America:** On Castro's 1916 visit, see File 53166-8C, INS.

272 **This time, however, officials:** *NYT*, December 8, 1924.

273 **The solicitor of the Department:** File 53371-25, INS.

273 **The case of Marya Kocik:** File 53148-19, INS.

274 **Officials became:** File 53986-67, INS.

274 **Eva Ranc provided officials:** File 54050-228, INS.

277 **Eva Ranc's case shows:** Quoted in Francesco Cordasco and Thomas Monroe Pitkin, *The White Slave Trade and the Immigrants: A Chapter in American Social History* (Detroit: Blaine Ethridge Books, 1981), 26.

277 **There was a term for this:** *Outlook*, November 6, 1909.

277 **The imagery implied:** Jane Addams, "A New Conscience and an Ancient Evil," *McClure's Magazine*, November 1911.

277 **Reports began to filter:** Edwin Sims, "The White Slave Trade," *Woman's World*, September 1908.

278 **Ellis Island inspector Marcus Braun:** File 52484-1-F, 1-G, INS.

278 **French authorities complained:** Letter from Marcus Braun to Commissioner General of Immigration, September 16, 1909, File 52484/1-F, INS.

279 **McSweeney focused on:** Letter from Edward F. McSweeney to Terence V. Powderly, July 27, 1898, Box 125, Series 2, TVP.

279 **In 1908, the case:** Mark Thomas Connelly, *The Response to Prostitution in the Progressive Era* (Chapel Hill: University of North Carolina Press, 1980), 114–115; *U.S. v. Bitty*, 208 U.S. 393 (1908).

280 **Former New York police commissioner:** Gen. Theodore A. Bingham, *The Girl That Disappears: The Real Fact About the White Slave Traffic* (Boston: Richard G. Badger, Gorham Press, 1911), 15.

280 **He found that talent:** George Kibbe Turner, "The Daughters of the Poor," *McClure's Magazine*, November 1909. For more on the Independent Benevolent Association, see Timothy J. Gilfoyle, *City of Eros: New York City, Prostitution, and the Commercialization of Sex, 1790–1920* (New York: W.W. Norton, 1992), 261–262.

280 **The fight against:** Mara L. Keire, "The Vice Trust: A Reinterpretation of the White Slavery Scare in the United States, 1907–1917," *Journal of Social History* 35, no. 1 (2001).

280 **Some, like Theodore Bingham:** Quoted in Cordasco and Pitkin, 22. For more on Bingham, see James Lardner and Thomas Reppetto, *NYPD: A City and its Police* (New York: Henry Holt, 2000), 141–142.

281 **Despite the increased:** File 51777-303, INS.

281 **The 1911 Dillingham Commission:** "Importing Women for Immoral Purposes: A Partial Report from the Immigration Commission on the Importation and Harboring of Women for Immoral Purposes," 61st Congress, 2nd Session, Document No. 196, 1909, 68.

282 **On the other hand:** "Importing Women for Immoral Purposes," 58–59.

282 **Single French women:** Edward J. Bristow, *Prostitution and Prejudice: The Jewish Fight Against White Slavery, 1870–1939* (New York: Schocken Books, 1983), 166.

282 **The charge of:** On the relationship between Jews and prostitution, see Lloyd Gartner, "Anglo-Jewry and the Jewish International Traffic in Prostitution, 1885–

1914," *AJS Review* 7 (1982); Egal Feldman, "Prostitution, the Alien Woman and the Progressive Imagination, 1910–1915," *American Quarterly*, Summer 1967; and Bristow, *Prostitution and Prejudice*.

282 **The link between:** Bristow, *Prostitution and Prejudice*, 156–157, 160.

283 **Were most prostitutes:** "Importing Women for Immoral Purposes," 60; Ruth Rosen, *The Lost Sisterhood: Prostitution in America, 1900–1918* (Baltimore, MD: Johns Hopkins University, 1982), 139–140; Gilfoyle, *City of Eros*, 292.

283 **Were large numbers:** Rosen, *The Lost Sisterhood*, 118; Bristow, *Prostitution and Prejudice*, 156–157.

283 **The Dillingham Commission:** "Importing Women for Immoral Purposes," 51, 54–55.

283 **William Williams also believed:** Letter from William Williams to Commissioner-General of Immigration, December 18, 1912, File 52809-7E, INS.

284 **Williams was probably:** Rosen, *the Lost Sisterhood*, 118, 133–134, 137. On the debate over whether white slavery was myth or reality, see Connelly, *The Response to Prostitution in the Progressive Era*, Chapter 6, and Rosen, *The Lost Sisterhood*, Chapter 7. Connelly argues that white slavery was largely a myth that scapegoated immigrants for the problems in American cities. Rosen argues that "a careful review of the evidence documents a real traffic in women, a historical fact and experience that must be integrated into the record." Rosen writes that various contemporary investigations showed that "the sale of some women into sexual slavery is an inescapable fact of the American past." Another historian agrees with Rosen. "Even a superficial sampling of contemporary evidence leaves no doubt that a white-slave traffic existed in the United States." But while the prostitution business was a reality, "no nationally organized white slave syndicate existed." Roy Lubove, "The Progressives and the Prostitute," *Historian*, May 1962.

284 **The public may have:** File 53155-144, INS.

285 **On June 9, 1914:** File 53986-43, INS.

286 **The Supreme Court failed:** "Redefining 'Crimes of Moral Turpitude': A Proposal to Congress."

286 **The reach of:** INS: I-94W Nonimmigrant Visa Waiver Arrival/Departure Form.

CHAPTER FOURTEEN: WAR

289 **At a few minutes:** On the Black Tom explosion, see Jules Witcover, *Sabotage at Black Tom: Imperial Germany's Secret War in America, 1914–1917* (Chapel Hill, NC: Algonquin Books, 1989); Tracie Lynn Provost, "The Great Game: Imperial German Sabotage and Espionage against the United States, 1914–1917," PhD dissertation, University of Toledo, 2003; NYT, July 31, August 1, 1916; NYW, July 31, August 1, 1916. Witcover called the Black Tom explosion "the centerpiece of one of the greatest and most cunning deceptions ever perpetrated on the United States by a foreign power."

290 **On Manhattan's Lower East Side:** "Why Dveire Kept Her Head," *Jewish Immigration Bulletin*, November 1916.

291 **The few barges:** *Survey*, August 5, 1916. An explosion on the Jersey piers in 1911 also caused damage at Ellis Island. The cause of that explosion was either the careless handling of explosives being loaded onto ships at the Jersey pier or an explosion in a ship's boiler, which set off ten thousand pounds of black powder. See Files 53173-26 and 53173-26B, NA and NYT, February 2, 1911.

292 **The road to**: Quoted in Witcover, *Sabotage*, 310–311.

293 **Any male over**: "President's Proclamation of a State of War, and Regulation Governing Alien Enemies," *NYT*, April 7, 1917. For more on the implications of the detention of German alien enemies in World War I, see Christopher Capozzola, *Uncle Sam Wants You: World War I and the Making of the Modern American Citizen* (New York: Oxford University Press, 2008).

294 **The German officers**: Frederic C. Howe, *The Confessions of a Reformer* (Chicago: Quadrangle Books, 1967), 272.

294 **One exception was**: *NYT*, June 20, 1917.

295 **Another detainee**: File 54188-473E, INS.

295 **Not everyone felt**: File 54188-468M, INS.

295 **Most were not**: "Annual Report of the Commissioner General of Immigration," 1918, 14.

296 **Other cases were**: File 54188-468H, INS.

296 **The militarization of**: "U.S. Immigration Service Bulletin," April 1, 1918, Folder 6, File 1133, IRL; Thomas Pitkin, *Keepers of the Gate: A History of Ellis Island* (New York: New York University Press, 1975), 120; *NYT*, September 23, 1918.

297 **The man in charge**: On Howe's pre–Ellis Island career, see Kevin Mattson, *Creating a Democratic Public: The Struggle for Urban Participatory Democracy During the Progressive Era* (University Park, PA: Pennsylvania State University Press, 1998) and Howe, *Confessions of a Reformer*, 240–251.

298 **Howe sought to humanize**: Howe, *Confessions of a Reformer*, 256–257; *Survey*, October 17, 1914; *Outlook*, October 21, 1914; Pitkin, *Keepers of the Gate*, 113–114.

298 **"Aliens traveling in the cabin"**: Memo from William Williams to Inspectors, Jan. 22, 1912, and Letter from William Williams to Commissioner-General of Immigration, Jan. 22, 1912, File 53438-15, INS.

299 **News of this inspection**: File 53438-15, INS; *NYS*, January 22, 1912.

300 **With Ellis Island overflowing**: File 53139-13B, INS; Frederic C. Howe, "Turned Back in Time of War," *Survey*, May 6, 1916.

301 **At Bennet's urging**: "Ellis Island Immigration Station, Hearings Before the Committee on Immigration and Naturalization, House of Representatives, 64th Congress, First Session, July 28, 1916."

301 **One case that aroused**: "Ellis Island Immigration Station, Hearings," 54.

301 **At the hearing**: "Ellis Island Immigration Station, Hearings," 53.

301 **Not only was Howe**: Howe, *Confessions of a Reformer*, 270–271.

302 **Alice Gouree**: File 54188-482, INS.

303 **Then there was**: "Ellis Island Immigration Station, Hearings," 42–43; Howe, *Confessions of a Reformer*, 270. In the book, Howe does not refer to Lamarca by name, but the reference is clear. The unnamed woman was "an Italian girl, had been married in Algeria and brought to this country. Her husband had taken her clothes away from her and had kept her in confinement. She had been forced by him to receive men. She was arrested and brought to the island. The husband had not been arrested."

303 **Giulietta seemed**: On the case of Giulietta Lamarca, see File 53986-43, INS.

304 **Bennet charged**: *NYT*, July 19, September 6, 1916.

304 **Howe described**: "Ellis Island Immigration Station, Hearings," 55–56.

304 **Howe's inattention**: Letter from Frederic C. Howe to Woodrow Wilson, December 8, 1914, Series 2, and Letter from Frederic C. Howe to Woodrow Wilson, December 31, 1917, Series 4, WW. For Howe's outside interests, see *NYT*, April 28, 1915.

305 **Howe spoke out**: NYT, June 11, 1915.

305 **Even the *Times***: NYT, June 21, 1916.

306 **Even Howe's choice**: Sandra Adickes, *To Be Young Was Very Heaven: Women in New York Before the First World War* (New York: St. Martin's Press, 1997), 59–61, 151.

306 **Randolph Bourne believed**: Randolph S. Bourne, "Trans-national America," *Atlantic*, July 1916. See also, David A. Hollinger, *Postethnic America: Beyond Multiculturalism* (New York: Basic Books, 1995).

307 **For Bourne**: "Americanization," *New Republic*, January 29, 1916.

307 **The president had**: Arthur S. Link, *Wilson: Campaigns for Progressivism and Peace, 1916–1917* (Princeton, NJ: Princeton University Press, 1965), 327–328.

308 **The literacy test**: Letter from Byron Uhl to Fiorello La Guardia, June 16, 1917, Folder 8, Box 26C7, FLG; NYT, March 28, 1917.

308 **Instead of rejoicing**: Barbara Miller Solomon, *Ancestors and Immigrants: A Changing New England Tradition* (New York: Wiley, 1956), 202.

309 **The targeting of Germans**: NYT, April 7, 1917, August 2, 1918; Witcover, *Sabotage*, 66–67.

309 **Then there was**: NYT, December 10, 1915, September 23, 24, 1917, September 2, 3, 1918; "Brewing and Liquor Interests and German Propaganda," Hearings before a Subcommittee of the Committee on the Judiciary, United States Senate, Sixty-Fifth Congress, Second Session.

CHAPTER FIFTEEN: REVOLUTION

311 **As the train approached**: NYT, February 10, 1919; Letter from A. D. H. Jackson to Anthony Caminetti, February 13, 1919, File 54235-36C, INS. There are differing accounts of the number of radicals sent from Seattle to New York. One account lists forty-five radicals, while another counts thirty-six with two more joining the group along the train route. The Jackson letter and a letter from immigration officials in Seattle corroborate that the number of radicals leaving Seattle was forty-seven. Letter from John H. Sargent, acting commissioner of immigration in Seattle to Commissioner General of Immigration Anthony Caminetti, February 7, 1919 in "I.W.W. Deportation Cases," Hearings before a House Subcommittee of the Committee on Immigration and Naturalization, 66th Congress, Second Session, April 27–30, 1920. For the other numbers, see Robert K. Murray, *Red Scare: A Study in National Hysteria, 1919–1920* (New York: McGraw-Hill, 1955), 194–195, and William Preston Jr., *Aliens and Dissenters: Federal Suppression of Radicals, 1903–1933* (New York: Harper Torchbooks, 1963), 198–201.

312 **The train arrived**: Letter from A. D. H. Jackson to Anthony Caminetti, February 13, 1919, File 54235-36C, INS; NYT, February 10, 1919.

312 **When the Red Special**: *New York Call*, February 18, 1919.

312 **Attorneys Caroline Lowe**: Charles Recht, unpublished autobiography, Chapter 10, Folder 18, Box 1, Collection 176, CR; Preston, *Aliens and Dissenters*, 200.

312 **In contrast to**: *New York Call*, February 20, 1919.

312 **The detainees were**: *NYTrib*, February 21, 1919.

313 **McDonald and the other**: Frederic C. Howe, *The Confessions of a Reformer* (Chicago: Quadrangle Books, 1967), 274–275.

313 **Howe was swimming**: Preston, *Aliens and Dissenters*, 182–183; "The Deportations," *Survey*, February 22, 1919.

313 **This expansion of the law**: Memo from Thomas Fisher, Immigration Inspector, to Henry W. White, Commissioner of Immigration, Seattle, Washington,

August 24, 1918; Letter from Henry W. White to Anthony Caminetti, August 28, 1918, File 54235-36B, INS.

314 **Though he was out**: Memo from John M. Abercrombie to All Commissioners of Immigration and Inspectors in Charge, March 14, 1919, File No. 54235-36B, INS.

314 **For those Red Special**: Preston, *Aliens and Dissenters*, 204–205.

314 **Martin de Wal**: *Survey*, May 17, June 14, 1919.

315 **In the middle of this**: *NYT*, June 3, 4, 5, 1919.

315 **Howe was not**: Memo from A. Warner Parker to Anthony Caminetti, April 17, 1919, File 54235-85B, INS.

316 **The attacks on**: Congressional Record, 66th Congress, 1st session, 1522–1524; Arthur Mann, *La Guardia: A Fighter Against His Times, 1882–1933* (Philadelphia: Lippincott, 1959), 101.

316 **During this second hearing**: "Conditions at Ellis Island," Hearing before the Committee on Immigration and Naturalization, 66th Congress, 1st Session, November 24, 26, 28, 1919, 21, 76.

317 **Back inside**: "Conditions at Ellis Island," 29–30.

317 **The press had a field day**: *LD*, December 13, 1919; *NYW*, November 25, 1919.

318 **With Secretary Wilson**: Kenneth D. Ackerman, *Young J. Edgar: Hoover, the Red Scare, and the Assault on Civil Liberties* (New York: Carroll & Graf Publishers, 2007), 50–59, 112. William N. Vayle [*sic*], "Before the Buford Sailed," *NYT*, January 11, 1920.

318 **If the earlier roundups**: Letter from Francis G. Caffey to Frederic C. Howe, July 12, 1917, Folder R57, EG.

318 **Beginning in 1907**: Oscar Straus Diary, March 6, 1908, 165–166, Box 22, OS.

318 **For two years**: Candace Falk (ed.), *Emma Goldman: A Documentary History of the American Years*, vol. 2, *Making Speech Free, 1902–1909* (Berkeley: University of California Press, 2005), 66–68, 254–257.

319 **As Julius Goldman**: File 54235-30, INS.

320 **When released from jail**: "Deportation Hearing of Emma Goldman," Ellis Island, NY, October 27, and November 12, 1919, Folder 63R, EG.

320 **Detained at Ellis Island**: File 54709-449, INS; Constantine Panunzio, *The Deportation Cases of 1919–1920* (New York: Da Capo Press, 1970), 60–62.

321 **Apart from the**: "Deportation: Its Meaning and Menace, Last Message to the People of America by Alexander Berkman and Emma Goldman," Ellis Island, New York, U.S.A., December 1919, LOC.

322 **In his waning**: John Lombardi, *Labor's Voice in the Cabinet: A History of the Department of Labor from its Origin to 1921* (New York: AMS Press, 1968), 132; Louis F. Post, "Living a Long Life Over Again," 309, 322, unpublished manuscript, LOC.

322 **Post complained**: Louis F. Post, "Administrative Decisions in Connection with Immigration," *American Political Science Review* 10 (May 1916).

323 **Still in office**: Louis F. Post, *The Deportations Delirium of Nineteen-Twenty* (Chicago: Charles H. Kerr, 1923), 1–27.

323 **Post found that**: Emma Goldman, *Living My Life*, vol. 2 (New York: Alfred A. Knopf, 1931), chapter 51.

323 **Collecting their things**: Alice Wexler, *Emma Goldman in Exile: From the Russian Revolution to the Spanish Civil War* (Boston: Beacon Press, 1989), 13–15. For more on Goldman's deportation, see Candace Serena Falk, *Love, Anarchy and Emma Goldman* (New Brunswick: Rutgers University Press, 1984), 181–182, and Alice Wexler, *Emma Goldman: An Intimate Life*, (New York: Pantheon Books, 1984), 271–276; Ackerman, *Young J. Edgar*, 160.

324 **Colorado congressman:** Vayle, "Before the Buford Sailed." A slightly different version of this account appears in *Congressional Record*, January 5, 1920.

324 **It must have been:** Ackerman, *Young J. Edgar*, 160.

325 **Upon arrival at:** Post, *The Deportations Delirium*, 27.

325 **The press was quick:** Letter from F. W. Berkshire, Supervising Inspector to Anthony Caminetti, Commissioner General of Immigration, February 11, 1920, File 54235-36G, INS; *LD*, January 3, 1920; Post, *The Deportations Delirium*, 7.

325 **"One could not imagine":** *NYT*, December 22, 1919.

325 **A few years before:** *Bugajewitz v. Adams*, 228 U.S. 585 (1913).

326 **Post made enemies:** Ackerman, *Young J. Edgar*, 274–276.

326 **At the height:** Panunzio, *The Deportation Cases of 1919–1920*, 16; Jane Perry Clark, *Deportation of Aliens from the United States to Europe* (New York: Columbia University Press, 1931), 225.

328 **When the war ended:** Fred Howe believed that big business was behind the war and repeatedly tried to convince Wilson of his theory that "it was not the Kaiser, nor the Czar, but the imperialistic adventurers who had driven their countries into conflict. Secret diplomacy, the conflict of bankers, the activity of munition-makers, exploiters, and concessionaires in the Mediterranean, in Morocco, in south and central Africa, had brought on the cataclysm; glacial-like aggregations of capital and credit were responsible for the war." Howe, *Confessions of a Reformer*, 287.

328 **No one felt:** Howe, *Confessions of a Reformer*, 279–282.

328 **To Howe, the brutality:** Frederic C. Howe, "Lynch Law and the Immigrant Alien," *Nation*, February 14, 1920.

329 **Before leaving Ellis Island:** Howe, *Confessions of a Reformer*, 327–328.

CHAPTER SIXTEEN: QUOTAS

330 **Immigration officials stationed:** *NYT*, July 2, 1923; Henry H. Curran, *Pillar to Post* (New York: Scribner's, 1941), 287–288.

331 **Restrictionists had long:** "Plain Remarks on Immigration for Plain Americans," *SP*, February 12, 1921.

331 **Americans feared that:** *LD*, December 18, 1920; Lothrop Stoddard, "The Permanent Menace from Europe," in Madison Grant and Charles Steward Davison, eds., *The Alien in Our Midst or Selling Our Birthright for a Mess of Pottage* (New York: Galton, 1930), 226.

332 **"The influx of aliens":** *NYT*, November 27, 1920.

332 **This was all too:** *NYT*, November 17, 1920.

333 **As Congress moved:** "The League's Numerical Limitation Bill," Publications of the Immigration Restriction League, No. 69, IRL.

333 **Hall lived long:** *Immigration and Other Interests of Prescott Farnsworth Hall*, compiled by Mrs. Prescott F. Hall, (New York: Knickerbocker Press, 1922).

334 **If one of those ships:** *NYT*, August 1, September 2, 1923.

334 **A major backbone:** Desmond King, *Making Americans: Immigration, Race, and the Origins of the Diverse Democracy* (Cambridge, MA: Harvard University Press, 2000), 112.

334 **The National German-American:** Charles Thomas Johnson, *Culture at Twilight: The National German-American Alliance, 1901–1918* (New York: Peter Lang, 1999), 102, 104–107, 118; *NYT*, March 8, 1916.

335 **The growing popularity:** Prescott F. Hall, "Immigration and World Eugenics," Publications of the Immigration Restriction League, No. 71, IRL. Mark Snyderman and R. J. Herrnstein argue that intelligence testing had little effect on the

passage of immigration quotas, while Leon Kamin argues the opposite. Mark Snyderman and R. J. Herrnstein, "Intelligence Tests and the Immigration Act of 1924," *American Psychologist*, September 1983; Leon Kamin, *The Science and Politics of I.Q.* (Potomac, MD: Lawrence Erlbaum, 1974). Those taking a more nuanced view include Steven A. Gelb, Garland E. Allen, Andrew Futterman, and Barry A. Mehler, "Rewriting Mental Testing History: The View from the American Psychologist," *Sage Race Relations Abstracts*, May 1986; and Franz Samelson, "Putting Psychology on the Map: Ideology and Intelligence Testing," in Allan R. Buss, ed., *Psychology in Social Context* (New York: Irvington, 1979), 135–136. Stephen Jay Gould seems to want to have it both ways, arguing that immigration restriction was inevitable in the 1920s even without eugenics, but that "the timing, and especially the peculiar character, of the 1924 Restriction Act [sic] clearly reflected the lobbying of scientists and eugenicists." Stephen Jay Gould, *Hen's Teeth and Horse's Toes* (New York: W.W. Norton, 1983), 301, and Stephen Jay Gould, *The Mismeasure of Man* (New York: W.W. Norton, 1996), 261–262.

335 **Madison Grant's**: Madison Grant, "The Racial Transformation of America," NAR, March 1924; Madison Grant, "America for the Americans," *Forum*, September 1925.

335 **"These immigrants adopt"**: SP, May 7, 1921.

336 **Such views were**: SP, February 28, 1920; February 12, May 7, November 26, 1921.

336 **America's postwar**: File 53986-43, INS.

338 **His new job**: Curran, *Pillar to Post*, 285–286.

338 **"It was a poor place"**: Curran, *Pillar to Post*, 291–296.

338 **There was little that**: *Outlook*, November 2, 1921; *Delineator*, March 1921.

338 **Complaints by the British**: Von Briesen Commission Report, 1903, File 52727/2, INS; Curran, *Pillar to Post*, 309.

339 **Even Fred Howe**: Frederic C. Howe, *Confessions of a Reformer* (Chicago: Quadrangle Books, 1967), 257–258.

339 **The British seemed**: NYT, July 29, 1923.

339 **A female British journalist**: LD, August 4, 1923.

339 **There had been**: NYT, July 2, 1923; LD, September 22, 1923; Rex Hunter, "Eight Days on Ellis Island," *Nation*, October 28, 1925.

340 **What the British**: NYT, December 19, 1922.

340 **Yet this was not**: "Despatch [sic] from H.M. Ambassador at Washington reporting on Conditions at Ellis Island Immigration Station," 1923, NYPL.

341 **Curran dismissed**: Henry H. Curran, "Fewer and Better," SP, Nov. 15, 1924; Curran, 298–299.

341 **Curran admitted**: Henry H. Curran, "Fewer and Better, or None," SP, April 26, 1924.

341 **Though this made**: Curran, *Pillar to Post*, 296–297.

342 **The shifting of inspection**: William E. Chandler, "Consular Certificates for Intending Immigrants," *Independent*, October 1, 1891.

342 **Fiorello La Guardia**: Letter from Fiorello La Guardia to Anthony Caminetti, September 9, 1916, Folder 8, Box 26C7, FLG.

342 **Though La Guardia**: Thomas Kessner, *Fiorello H. La Guardia and the Making of Modern New York* (New York: Penguin Books, 1989), 120–124.

343 **These new quotas**: Roger Daniels, *Guarding the Golden Door: American Immigration Policy and Immigrants Since 1882* (New York: Hill and Wang, 2004), 56–57.

343 **Stricter quotas led to**: Letter from James J. Davis, Secretary of Labor to President Warren G. Harding, April 16, 1923, Folder 5, File 75, WGH; "President

Calvin Coolidge's Remarks at Governor's Conference at the White House," October 20, 1923, Series 1, File 52, CC.

343 **Deportations also increased**: Jane Perry Clark, *Deportation of Aliens from the United States to Europe* (New York: Columbia University Press, 1931), 29.

344 **To rectify the situation**: For a discussion of the national origins plan, see King, *Making Americans*, 204–228 and *NYT*, June 30, 1929; Harry H. Laughlin, "The Control of Trends in the Racial Composition of the American People," in Grant and Davison, eds., *The Alien in Our Midst or Selling Our Birthright for a Mess of Pottage.*

344 **The commission calculated**: *NYT*, August 7, 1925.

344 **Edward F. McSweeney**: For more on the "history wars" of the 1920s and Mc-Sweeney's role, see Jonathan Zimmerman, "Each 'Race' Could Have Its Heroes Sung: Ethnicity and the History Wars in the 1920s," *Journal of American History* 87, no, 1 (June 2000), and Christopher J. Kauffman, "Edward McSweeney, the Knights of Columbus, and the Irish-American Response to Anglo-Saxonism, 1900–1925," *American Catholic Studies* 114, no. 4 (Winter 2003). See also, *NYT*, September 8, 1921, June 9, 1923, and *BG*, July 10, 1921.

345 **More substantively**: W. E. Burghardt Du Bois, *The Gift of Black Folk: The Negroes in the Making of America* (Boston: Stratford, 1924) and David Levering Lewis, *W. E. B. Du Bois: The Fight for Equality and the American Century, 1919–1963* (New York: Henry Holt, 2000), 95–96.

345 **To a pro-immigration**: *NYT*, August 7, 1925. See also Edward F. McSweeney, "The Immigration Act of 1924: Fallaciousness of the 'National Origins' Theory," *Journal of the American-Irish Historical Society* 223 (1926).

345 **In the late afternoon**: *Framingham News*, November 17, 19, 1928.

346 **Powderly had once**: T. V. Powderly, "Immigration's Menace to the National Health," *NAR*, July 1902; Letter from Terence V. Powderly to Frederick Wallis, September 9, 1920, Box 139, TVP.

346 **Freed from the burdens**: Vincent J. Falzone, *Terence V. Powderly: Middle-Class Reformer* (Washington, DC: University Press of America, 1978), 191–193.

346 **Powderly was not**: Henry H. Goddard, "Feeblemindedness: A Question of Definition," *American Association for the Study of the Feeble Minded: Proceedings and Addresses* 33 (1928); Leila Zenderland, *Measuring Minds: Henry Herbert Goddard and the Origins of American Intelligence Testing* (Cambridge, UK: Cambridge University Press, 1998), 325–327; Gould, *The Mismeasure of Man*, 202–204.

347 **By the time**: Edward Alsworth Ross, *Seventy Years of It: An Autobiography* (New York: Appleton-Century, 1936), 275–277.

347 **Nine-year-old Edoardo Corsi**: Edward Corsi, *In the Shadow of Liberty* (New York: Arno Press, 1969), 3–7, 22. As proof that the memory of immigrants, like all memory, is usually fuzzy around the edges, Corsi's account of his family's arrival is slightly off. The Corsi family arrived in November 1906, not October 1907 as Corsi notes, meaning that Edward was nine years old, not ten. Also, young Edward is listed as "Nerino Corsi" on the steamship list.

348 **Those days were**: *LD*, February 24, 1934; "Report of the Ellis Island Committee," March 1934.

348 **The 1930s would**: *LD*, February 24, 1934.

349 **The combination of**: Corsi, *In the Shadow of Liberty*, 95.

CHAPTER SEVENTEEN: PRISON

350 **"Herzlich Willkommen!"**: Arnold Krammer, *Undue Process: The Untold Story of America's German Alien Internees* (Lanham, MD: Rowan & Littlefield, 1997),

10–11, 25–26, 30. For more on the issue, see John Christgau, *"Enemies": World War II Alien Internment* (Ames, IA: Iowa State University Press, 1985).

351 **On December 8, 1941**: Memo from Major Lemuel B. Schofield to J. Edgar Hoover, December 8, 1941, File 56125-29, INS.

351 **Some of the internees**: Jerre Mangione, *An Ethnic At Large: A Memoir of America in the Thirties and Forties* (New York: Putnam's, 1978), 321.

352 **A large number of enemy**: File 56125-29, INS; "Harbor Camp for Enemy Aliens," *NYTM*, January 25, 1942. See also "The Detention of Krauss," *New Yorker*, March 6, 1943.

352 **The Office of Strategic Services**: The OSS report and other related documents can be found in File 56125-86, INS.

353 **Hoover was right**: Not to be outdone, Hoover later placed his own FBI agents among the detainees at Ellis Island. According to a German who was temporarily detained at Ellis Island: "You see, there were FBI men scattered among us as observers. You don't know them, and once a roommate I'd had for a month or more left, and one of the guards told me that fellow had been an FBI man on duty." "The Detention of Krauss," *New Yorker*, March 6, 1943.

353 **One Justice Department official**: File 56125-86, INS.

354 **Although Bishop was taken**: *NYT*, January 15, 1940. On the Christian Front, see Theodore Irwin, "Inside the Christian Front," *Forum*, March 1940, and Ronald H. Bayor, *Neighbors in Conflict: The Irish, Germans, Jews, and Italians of New York City, 1929–1941* (Baltimore, MD: The Johns Hopkins University Press, 1978), 97–104.

355 **One of them was**: On the Pinza case, see *NYT*, March 13, 1942; Ezio Pinza, *An Autobiography* (New York: Rinehart, 1958), 202–228; Sarah Goodyear, "When Being Italian Was a Crime," *Village Voice*, April 11, 2000; "Statement of Doris L. Pinza," Subcommittee on the Constitution, Committee on the Judiciary, U.S. House of Representatives, October 26, 1999.

356 **The other half**: Rose Marie Neupert, "The Neupert Family Story," http://www.gaic.info/real_neupert.html.

356 **Most of the detainees**: *NYT*, September 23, 1942. For more on these camps, see Mangione, *An Ethnic At Large*, 319–352.

357 **By March 1946**: Stephen Fox, *Fear Itself: Inside the FBI Roundup of German Americans During World War II* (New York: iUniverse, Inc., 2005), 327–328.

357 **One of those not holding up**: On the story of the Hackenbergs, see Fox, *Fear Itself*, 325–332.

357 **Hundreds of these enemy aliens**: *NYT*, January 3, 1947; Letter from Rosina and Max Rapp to Senator William Langer, July 23, 1947, Folder 12, Box 214, WL.

358 **The Fuhr family**: On the Fuhr family, see Fox, *Fear Itself*, 109–126.

358 **While in custody**: Fox, *Fear Itself*, 114, 122.

359 **Langer introduced a bill**: Senate Bill 1749, July 26, 1947, 80th Congress, 1st Session; Fox, *Fear Itself*, 124–126; Eberhard E. Fuhr, "My Internment by the U.S. Government," http://www.gaic.info/real_fuhr.html.

359 **One of those not on Langer's list**: Sworn Statement of William Langer, August 1947, Folder 9, Box 214, WL; Senate Bill 1083, April 10, 1947, 80th Congress, 1st Session; *NYT*, September 11, 1947.

359 **At the end of June 1948**: *Ahrens v. Clark*, 335 U.S. 188 (1948); *NYT*, July 7, 8, 1948; Fox, *Fear Itself*, 140.

359 **In the following weeks**: Fox, *Fear Itself*, 329–333; *NYT*, November 17, 1945. For lists of German detainees and the disposition of their cases, see Folder 1, Box 257, WL.

360 **Although exact numbers**: On the number of enemy alien detainees, see Kram-

mer, *Undue Process*, 171. Two websites document the experience of German internment during World War II: the German American Internee Coalition, http://www.gaic.info/index.html, and http://www.foitimes.com/.

360 **The bill also granted**: On the McCarran Internal Security Act of 1950, see Michael J. Ybarra, *Washington Gone Crazy: Senator Pat McCarran and the Great American Communist Hunt* (Hanover, NH: Steerforth Press, 2004) 509–534.

361 **President Truman came out**: "Text of President's Message Vetoing the Communist-Control Bill," *NYT*, September 23, 1950.

361 **Embarrassed at having**: W. L. White, "The Isle of Detention," *American Mercury*, May 1951.

361 **Gulda arrived at**: *NYT*, October 9, 1950; *Time*, October 23, 1950; *Newsweek*, Oct. 23, 1950; A. H. Raskin, "New Role for Ellis Island," *NYTM*, November 12, 1950.

362 **The law also affected**: Letter from Arthur A. Sweberg to President Harry Truman, March 22, 1951; Letter from Mrs. Josephine Mazzeo to President Harry Truman, March 28, 1951, Folder 2750-C Misc, Box 1717, HST.

362 **George Voskovec**: *New Yorker*, May 12, 1951; *NYT*, December 4, 1950.

362 **As Truman predicted**: *NYT*, March 22, 1951.

363 **Upon his release**: *NYT*, April 3, 1951; *New Yorker*, May 12, 1951.

363 **Voskovec would later**: *NYT*, November 13, 1955.

363 **When Ellen arrived**: Ellen Raphael Knauff, *The Ellen Knauff Story* (New York: W.W. Norton, 1952), 8.

364 **The government's case**: The case against Ellen Knauff is summarized in Memorandum for the President from J. Howard McGrath, Attorney General, received July 14, 1950, Justice Department Folder, Box 22, HST.

364 **It would be more**: Knauff, *The Ellen Knauff Story*, 29.

365 **The Court relied on**: *Knauff v. Shaughnessy*, 338 U.S. 537 (1950); David Cole, *Enemy Aliens: Double Standards and Constitutional Freedoms in the War on Terrorism* (New York: New Press, 2003), 136–137; Charles D. Weisselberg, "The Exclusion and Detention of Aliens: Lessons from the Lives of Ellen Knauff and Ignatz Mezei," *University of Pennsylvania Law Review* 143, no. 4 (April 1995).

366 **Having lost**: *NYT*, May 18, 1950.

366 **In the meantime**: *Time*, April 17, 1950.

366 **In the spring of 1950**: Knauff, *The Ellen Knauff Story*, 138.

366 **The press attention**: Memo for Charles Rose from Edward A. Harris, June 15, 1950; Memo for Steve Spingarn from Harry S. Truman, June 17, 1950, Justice Department Folder, Box 22, HST.

367 **The Justice Department stalled**: Memorandum for Peyton Ford, Deputy Attorney General from Steve Spingarn, August 2, 1950; Memorandum for Peyton Ford, Deputy Attorney General from Steve Spingarn, September 25, 1950, Justice Department Folder, Box 22, HST.

367 **However, the Justice Department**: *NYT*, February 28, 1950; Knauff, *The Ellen Knauff Story*, 81.

368 **It took the board members**: *NYT*, March 27, 1951.

368 **By the end of August**: U.S. Department of Justice, Board of Immigration Appeals, File A-6937471, reprinted in Knauff, *The Ellen Knauff Story*.

368 **McGrath released**: *NYT*, November 3, 1951.

369 **However, there were**: *NYT*, July 3, 1953; Knauff, *The Ellen Knauff Story*, 54.

369 **Though she ultimately**: Anthony Lewis, "Security and Liberty: Preserving the Values of Freedom," in Richard C. Leone and Greg Anrig Jr., *The War on Our Freedoms: Civil Liberties in an Age of Terrorism* (New York: Public Affairs, 2003), 72.

369 **The widespread sympathy**: For the story of C. L. R. James and his detention at Ellis Island, see C. L. R. James, *Mariners, Renegades, and Castaways: The Story of Herman Melville and the World We Live In*, reprint (London: Allison & Busby, 1985), 132–173; Emily Eakin, "Embracing the Wisdom of a Castaway," *NYT*, August 4, 2001; Farrukh Dhondy, *C. L. R. James: A Life* (New York: Pantheon, 2001), 107–111.

371 **Mezei was not**: On the Mezei case, see Cole, *Enemy Aliens*, 138–139; Weisselberg, "The Exclusion and Detention of Aliens"; Richard A. Serrano, "Detained, Without Details," *Los Angeles Times*, December 21, 2005.

372 **The next step**: *Shaughnessy v. Mezei*, 345 U.S. 206 (1953).

373 **A defeated Mezei**: *NYT*, April 23, 1953.

374 **The government had a strong**: Weisselberg, "The Exclusion and Detention of Aliens," 975–978. Many years later, Mezei's stepdaugher remembered how Ignatz would enlist his stepchildren to hand out Communist leaflets on election day. Richard A. Serrano, "Detained, Without Details," *Los Angeles Times*, December 21, 2005.

374 **Whereas Knauff was**: Weisselberg, "The Exclusion and Detention of Aliens," 979.

374 **The special three-man board**: Weisselberg, "The Exclusion and Detention of Aliens," 983–984.

375 **By 1954, Ellis Island**: *NYT*, December 6, 1954.

375 **On Veterans Day**: *NYT*, November 12, 13, 1954.

376 **"They rewarded with"**: *NYT*, November 14, 1954.

CHAPTER EIGHTEEN: DECLINE

379 **A businessman reading**: *WSJ*, September 18, 1956.

379 **The sale was made**: *NYT*, November 14, 1954.

380 **So the GSA opened**: Barbara Blumberg, "Celebrating the Immigrant: An Administrative History of the Statue of Liberty National Monument, 1952–1982," National Park Service, 1985, Chapter 5.

380 **In response**: Blumberg, "Celebrating the Immigrant."

380 **Some of the proposals**: *NYT*, February 3, 1958.

380 **When bidding opened**: *Business Week*, September 29, 1956.

380 **Ellis Island's future**: *NYTM*, May 25, 1958.

381 **"This is not just"**: *NYT*, December 20, 1960; December 8, 1962.

381 **To Corsi**: Edward Corsi, *In the Shadow of Liberty: The Chronicle of Ellis Island* (New York: Macmillan, 1935), 281–295.

383 **With full control**: Blumberg, "Celebrating the Immigrant," chapter 6; *Time*, March 4, 1966; *NYT*, February 25, 1966.

383 **The centerpiece of**: *New York World-Telegram and Sun*, March 7, 1966.

383 **There were other**: *NYT*, February 26, 1966; Harry T. Brundidge, "The Passing of Ellis Island," *American Mercury*, December 1954.

384 **The island was a mess**: *NYT*, July 16, 1964, March 5, 1968.

384 **For some white**: Peter Morton Coan, *Ellis Island Interviews: In Their Own Words* (New York: Checkmark Books, 1997), 220; Paul Knaplund, *Moorings Old and New: Entries in an Immigrant's Log* (Madison, WI: State Historical Society of Wisconsin, 1963), 148. See also David R. Roediger, *Working Toward Whiteness: How America's Immigrants Became White: The Strange Journey from Ellis Island to the Suburbs* (New York: Basic Books, 2005), 118–119.

385 **Meanwhile, black leaders**: On African-American attitudes toward immigration, see Daryl Scott, " 'Immigrant Indigestion': A. Philip Randolph: Radical

and Restrictionist," Center for Immigration Studies, *Backgrounder*, June 1999, http://www.cis.org/articles/1999/back699.html, and "'Cast Down Your Bucket Where You Are': Black Americans on Immigration," Center for Immigration Studies, Paper 10, June 1996, http://www.cis.org/articles/1996/paper10.html.

386 **In the early morning hours**: NYT, March 17, 1970.

387 **It would prove**: Nixon Tapes, Conversation No. 610-1, Nov. 1, 1971, RMN. The conversation is not transcribed and the audio quality of the recording is poor. This is the author's rough transcription of the account. The aides at the meeting included John Mitchell, George Schultz, and H. R. Haldeman.

387 **Two days after**: On the NEGRO takeover of Ellis Island, see NYT, January 8, July 25, 26, August 2, 19, 20, 21, 1970; *Newsweek*, September 28, 1970; and Blumberg, "Celebrating the Immigrant," chapter 6.

388 **This did not deter**: NEGRO brochure, WHCF, SMOF, Leonard Garment, Box 138, RMN.

388 **Matthew continually referred**: In fact, a few years earlier, Irving Kristol wrote a long piece arguing the same idea. See Irving Kristol, "The Negro Today is Like the Immigrant Yesterday," NYTM, September 11, 1966.

388 **Not surprisingly**: Blumberg, "Celebrating the Immigrant," 6.

389 **Since the mid-1960s**: NYT, April 24, 1973.

389 **As the Ellis Island colony**: NYT, November 29, December 11, 1973. In November 1973, Matthew was convicted and sentenced to three years in prison. In March 1975, an appeals court struck down the conviction, arguing that errors by the judge merited a dismissal.

389 **Around the same time**: Nathan Glazer and Daniel Moynihan, *Beyond the Melting Pot: The Negroes, Puerto Ricans, Jews, Italians, and Irish of New York City*, 2nd ed. (Cambridge, MA: MIT Press, 1970).

390 **Ethnic pride**: Michael Novak, *The Rise of the Unmeltable Ethnics* (New York: Macmillan, 1971). On the phenomenon of white ethnicity, see Vincent J. Cannato, *The Ungovernable City: John Lindsay and his Struggle to Save New York* (New York: Basic Books, 2001), 389–441; and Mathew Frye Jacobson, *Roots, Too: White Ethnic Revival in Post Civil-Rights America* (Cambridge, MA: Harvard University Press, 2006).

CHAPTER NINETEEN: THE NEW PLYMOUTH ROCK

391 **On this patriotic**: NYT, July 3, 1986.

392 **Although resoration of the**: F. Ross Holland, *Idealists, Scoundrels, and the Lady: An Insider's View of the Statue of Liberty–Ellis Island Project* (Urbana: University of Illinois Press, 1993), 80, 205.

393 **The Statue of Liberty**: NYT, November 4, 1985.

393 **In November 1985**: Roberta Gratz and Eric Fettmann, "The Selling of Miss Liberty," *Nation*, November 9, 1985. For other articles by Gratz and Fettmann on the topic, see "Mr. Iacocca Meets the Press," *Nation*, March 8, 1986; "Post-Iacocca" *Nation*, April 19, 1986; and "Whitewashing the Statue of Liberty," *Nation*, June 7, 1986. F. Ross Holland dismisses the complaints of Gratz and Fettmann as "scurrilous" and "liberally sprinkled with untruths, half-truths, misinformation, and distorted facts." *Idealists, Scoundrels, and the Lady*, 180–181.

394 **For some, it was all**: Jacob Weisberg, "Gross National Production," *New Republic*, June 23, 1986.

394 **If the public**: Lee Iacocca with William Novak, *Iacocca: An Autobiography* (New York: Bantam Books, 1984), 339–441.

394 **His father, Nicola:** Iacocca with Novak, *Iacocca*, 5; Peter Wyden, *The Unknown Iacocca* (New York: William Morrow, 1987), 260.

395 **"Hard work":** Iacocca with Novak, *Iacocca*, 339.

395 **To others, that vision:** Holland, *Idealists, Scoundrels, and the Lady*, 158–159; Roberta Gratz and Eric Fettmann, "The Battle for Ellis Island," *Nation*, November 30, 1985.

395 **A historian made:** Lynn Johnson, "Ellis Island: Historic Preservation from the Supply Side," *Radical History Review*, September 1984.

396 **How should the:** NYT, January 14, 21, 1984.

396 **The former inspection:** For more on the evolution of the historical memory of Plymouth Rock, see John Seelye, *Memory's Nation: The Place of Plymouth Rock* (Chapel Hill: University of North Carolina Press, 1998).

396 **This process began:** Jacob A. Riis, "In the Gateway of Nations," *Century Magazine*, March 1903; "The New Plymouth Rock," *Youth's Companion*, December 14, 1905.

396 **In 1914, a writer:** Mary Antin, *They Who Knock at Our Gates: A Complete Gospel of Immigration* (Boston and New York: Houghton, Mifflin, 1914), 98.

396 **That an immigrant:** Werner Sollors, "National Identity and Ethnic Diversity: 'Of Plymouth Rock and Jamestown and Ellis Island' or Ethnic Literature and Some Redefinitions of 'America,'" in Genevieve Fabre and Robert O'Meally, eds., *History and Memory in African-American Culture* (New York: Oxford University Press, 1994), 103–105; Agnes Repplier, "The Modest Immigrant," *Atlantic Monthly*, September 1915.

397 **Other native-born Americans:** Thomas Darlington, "The Medic-Economic Aspect of the Immigration Problem," *North American Review*, December 21, 1906.

397 **In the late 1930s:** Sollors, "National Identity and Ethnic Diversity," 108–109; Dan Shiffman, *Rooting Multiculturalism: The Work of Louis Adamic*; Louis Adamic, *From Many Lands* (New York: Harper & Brothers, 1939), 296–299.

397 **In deeply nostalgic:** Leo Rosten, "Not So Long Ago, There Was a Magic Island," *Look*, December 24, 1968; Edward M. Kennedy, "Ellis Island," *Esquire*, April 1967; Thomas M. Pitkin, *Keepers of the Gate: A History of Ellis Island* (New York: New York University Press, 1975), 177.

398 **In the late 1970s:** "Ellis Island Remembered," September 23, 1978, NYPL.

398 **Riding this wave:** F. Ross Holland, *Idealists, Scoundrels, and the Lady: An Insider's View of the Statue of Liberty–Ellis Island Project* (Urbana: University of Illinois Press, 1993), 5–6; NYT, July 25, 1981.

399 **"The Battle for":** Michael Barone, "The Battle for Ellis Island," *Washington Post*, August 14, 1984, and Matthew Frye Jacobson, *Roots Too: White Ethnic Revival in Post-Civil Rights America* (Cambridge, MA: Harvard University Press, 2006), 320–322.

399 **In 1988:** NYT, September 4, 1988; Michael Dukakis, "A New Era of Greatness for America": Address Accepting the Presidential Nomination at the Democratic National Convention in Atlanta, July 21, 1988; and Jacobson, *Roots Too*, 327–331.

399 **Ferraro and Dukakis:** Meg Greenfield, "The Immigrant Mystique," *Newsweek*, August 8, 1988.

400 **At the other side:** NYT, January 14, 1993, August 11, 2000.

401 **What name does:** NYT, September 21, 1990.

401 **The most famous:** The Sean Ferguson story also appears in Alan M. Kraut, *The Huddled Masses: The Immigrant in American Society, 1880–1921* (Wheeling, IL: Harlan Davidson, 1982), 56–57. While Kraut calls the story possibly apocryphal, he uses it to illustrate the changing of names by officials at Ellis

Island. One possible explanation of the story has the original Sean Ferguson as a Yiddish-speaking actor named Berel Bienstock. When Bienstock came to the United States to seek a career in the movies, his agent suggested that he Americanize his name. When he finally got to California and met with a movie producer who asked him his name, the nervous Bienstock replied in Yiddish "Schoen fergessen" and the producer wrote down his name as Sean Ferguson. But that story might also be apocryphal. See Stephen J. Sass, "In the Name of Sean Ferguson," JewishJournal.com, June 21, 2002, http://www.jewishjournal.com/home/preview.php?id=8761.

401 **The stories multiply**: Ellen Levine, illustrated by Wayne Parmenter, *If Your Name Was Changed at Ellis Island* (New York: Scholastic, 1993); Ellen Levine claims that her grandfather's original name was Louis Nachinovsky, but immigration inspectors at Ellis Island "changed many Jewish names to Levine or Cohen." And so her grandfather had become Louis Levine. Another book on Ellis Island, under the header "There's a Man Goin' Round Changing Names," discusses how "tens of thousands" of names were changed at Ellis Island. More discussion of name changes can be found in David M. Brownstone, Irene M. Frank, and Douglass Brownstone, *Island of Hope, Island of Tears* (New York: MetroBooks, 2002), 177–179.

401 **In an interview**: Interview with Sophia Kreitzberg, "Voices from Ellis Island."

402 **Then there is the joke**: Joseph Epstein, "Death Benefits," *Weekly Standard*, May 21, 2007. Although Epstein tells the Moishe Pipik story as a joke, he still believes that the "impatience of officials at Ellis Island altered lots of Eastern European surnames."

402 **Nearly all of these**: On the name change myth, see Alan Berliner's documentary, *The Sweetest Sound*, reviewed in *WSJ*, June 25, 2001.

403 **The inclusive nature**: This issue came up in the development of the plans for the museum in the 1980s. See Holland, *Idealists, Scoundrels, and the Lady*, 184–185.

403 **Historians are supposed to**: *NYT*, September 7, 1990; Mike Wallace, *Mickey Mouse History and Other Essays on American Memory* (Philadelphia: Temple University Press, 1996), 70–71. Wallace is wrong to claim that "for all Reagan's celebration of the Statue as the 'mother of exiles' he was then doing his best to slam the open door shut." Anti-immigration measures were never part of Reagan's politics or rhetoric. The major piece of immigration legislation during the Reagan years, the Immigration Reform and Control Act of 1986, did not call for immigration restriction, but instead created an amnesty program for illegal immigrants already in the country, as well as measures designed to punish employers who employed illegal immigrants. The number of immigrants remained remarkably steady during the Reagan years, going from 530,639 in 1980 to 643,025 in 1988, before jumping to over 1 million in each of the next three years. Peter Schuck has written that the 1980s produced immigration policies that were "remarkably liberal and expansive by historical standards." Wallace, 58; Peter H. Schuck, *Citizens, Strangers, and In-Between: Essays on Immigration and Citizenship* (Boulder, CO: Westview Press, 1998), 92.

404 **In addition**: Wallace, *Mickey Mouse History*, 57. For other academic critics of Ellis Island, see Barbara Kirshenblatt-Gimblett, *Destination Culture: Tourism, Museums, and Heritage* (Berkeley: University of California Press, 1998), 177–187, and John Bodnar, "Symbols and Servants: Immigrant America and the Limits of Public History," *Journal of American History* 73, no. 1 (June 1986). For a more positive academic appraisal of Ellis Island, see Judith Smith, "Celebrating Immigration History at Ellis Island," *American Quarterly*, March 1992.

404 **Art professor**: Erica Rand, *The Ellis Island Snow Globe* (Durham, NC: Duke University Press, 2005), 177.

404 **It is hard for**: Ira De A. Reid, *The Negro Movement: His Background, Characteristics, and Social Adjustment, 1899–1937* (New York: Columbia University Press, 1939), 42; *NYT*, May 30, 1986.

404 **David Roediger's**: David R. Roediger, *Working Toward Whiteness: How America's Immigrants Became White: The Strange Journey from Ellis Island to the Suburbs* (New York: Basic Books, 2005).

405 **For historian**: Jacboson, *Roots Too*, 204–205.

405 **Another group was**: Samuel Huntington, *Who Are We? The Challenges to America's National Identity* (New York: Simon & Schuster, 2004), 37–39, 46; Seelye, *Memory's Nation*, 628–629.

405 **In the years since**: *NYT*, September 7, 1990.

406 **Ellis Island's iconic status**: For a print version of the TD Ameritrade advertisement, see *NYT*, April 26, 2006, A11.

407 **In the 1990s**: *New Jersey v. New York* 523 U.S. 767 (1998); *NYT*, April 3, 1997, January 13, 1998, May 26, 1998, August 13, 2001; *WP*, May 27, 1998.

408 **To help with fundraising**: On the Arrow advertising and fundraising campaign, see http://www.weareellisisland.org. A collection of recent photographs of the abandoned southern section of the island can be found in Stephen Wilkes, *Ellis Island: Ghosts of Freedom* (New York: W.W. Norton, 2006).

408 **In a different context**: *NYT*, August 31, 2001.

408 **Whether Ellis Island**: *NYT*, April 3, 1997; Mayor Rudolph Giuliani, "Remarks at Naturalization Ceremony on Ellis Island with President Bush," July 10, 2001, http://www.nyc.gov/html/rwg/html/2001b/ellis_island.html.

EPILOGUE

410 **"We should not let"**: *Time*, December 15, 1980.

410 **Wolf may have believed**: *NYTM*, March 22, 1998.

411 **In this most recent**: Matt Towery, "Immigration: The Ellis Island Solution," *Townhall.com*, May 31, 2007.

412 **Unhappy with**: Samuel Huntington, *Who Are We? The Challenges to America's National Identity* (New York: Simon & Schuster, 2004), 189, 225.

413 **As Barbara Jordan**: "Testimony of Barbara Jordan, Chair, U.S. Commission on Immigration Reform, Before a Joint U.S. House of Representatives Committee on the Judiciary Subcommittee on Immigration and Claims and U.S. Senate Committee on the Judiciary Subcommittee on Immigration," June 28, 1995. See also, Mark Krikorian, "Immigration and Civil Rights in the Wake of September 11th," Testimony prepared for the U.S. Commission on Civil Rights, October 12, 2001, http://www.cis.org/articles/2001/msktestimony1001.html.

414 **Much of the discussion**: Seyla Benhabib, *The Rights of Others: Aliens, Residents, and Citizens* (Cambridge, UK: Cambridge University Press, 2004), 2, 11.

415 **The plenary power**: On recent trends in immigration rights and citizenship, see Peter H. Schuck, *Citizens, Strangers, and In-Between: Essays on Immigration and Citizenship* (Boulder, CO: Westview Press, 1998), 19–87; Linda S. Bosniak, *The Citizen and the Alien: Dilemmas of Contemporary Membership* (Princeton, NJ: Princeton University Press, 2006), 37–76; and Hiroshi Motomura, "Immigration Law After a Century of Plenary Power: Phantom Constitutional Norms and Statutory Interpretation," *Yale Law School*, December 1990. Michael Walzer makes a strong case for the retention of some boundaries for national member-

ship in *Spheres of Justice: A Defense of Pluralism and Equality* (New York: Basic Books, 1983), 31–63.

415 **In a 2001 Supreme Court**: *Zadvydas v. Davis*, 533 U.S. 678 (2001). See Trevor Morrison, "The Supreme Court and Immigration Law: A New Commitment to Avoiding Hard Constitutional Questions?" July 31, 2001, http://writ.news. findlaw.com/commentary/20010731_morrison.html.

418 **Before the rise**: Schuck, *Citizens, Strangers*, 80; Krikorian, "Immigration and Civil Rights in the Wake of September 11th."

Index